STUDIES IN BENFORD'S I

Arithmetical Tugs of War, Quantitative Partition Models, Prime Numbers, Exponential Growth Series, and Data Forensics

Alex Ely Kossovsky

About the Author:

Alex Ely Kossovsky is the author of the books **"BENFORD'S LAW:** Theory, the General Law of Relative Quantities, and Forensic Fraud Detection Applications", **"THE BIRTH OF SCIENCE:** Kepler's Celestial Data Analysis and Galileo's Terrestrial Experiments", and **"SMALL IS BEAUTIFUL:** Why the Small is Numerous but the Big is Rare in the World". Kossovsky is the inventor of a patented statistical algorithm in data fraud detection analysis, registered at the US Patent Office. The author specialized in Applied Mathematics and Statistics at the City University of New York, and in Physics and Pure Mathematics at the State University of New York at Stony Brook.

Email Address: akossovsky@gmail.com

TABLE OF CONTENTS

SECTION 1

BENFORD'S LAW

SECTION 2

ARITHMETICAL TUGS OF WAR

SECTION 3

QUANTITATIVE PARTITION MODELS

SECTION 4

RANDOM CONSOLIDATIONS AND FRAGMENTATIONS PROCESSES

SECTION 5

PRIME NUMBERS AND DIRICHLET DENSITY

SECTION 6

EXPONENTIAL GROWTH SERIES

SECTION 7

INNOVATIVE FORENSIC FRAUD DETECTION APPLICATIONS IN BENFORD'S LAW

SECTION 8

ORDER OF MAGNITUDE AND BENFORD'S LAW

SECTION 9

ROBUST ORDER OF MAGNITUDE OF DATA

SECTION 1

BENFORD'S LAW

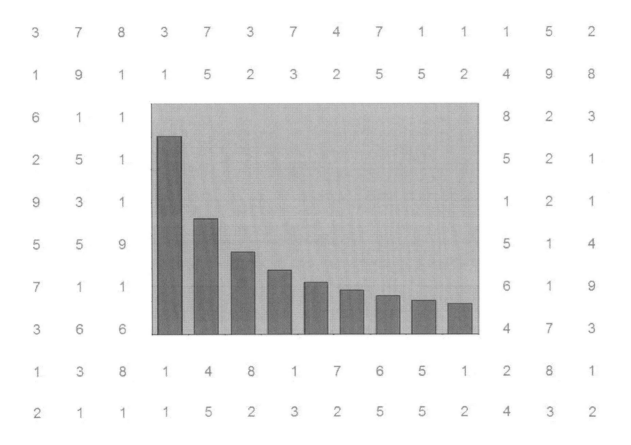

[1]   The First Digit on the Left Side of Numbers

It has been discovered that the first digit on the left-most side of numbers in real-life data sets is most commonly of low value such as {1, 2, 3} and rarely of high value such as {7, 8, 9}.
As an example serving as a brief and informal empirical test, a small sample of 40 values relating to geological data on time between earthquakes is randomly chosen from the data set on all global earthquake occurrences in 2012 – in units of seconds. Figure 1.1 depicts this small sample of 40 numbers. Figure 1.2 emphasizes in bold and black color the 1st digits of these 40 numbers.

285.29	185.35	2579.80	27.11
5330.22	1504.49	1764.41	574.46
1722.16	815.06	3686.84	1501.61
494.17	362.48	1388.13	1817.27
3516.80	5049.66	2414.06	387.78
4385.23	2443.98	2204.12	1224.42
1965.46	3.61	1347.30	271.23
3247.99	753.80	1781.45	593.59
1482.64	1165.04	4647.39	1219.19
251.12	7345.52	1368.79	4112.13

Figure 1.1: Sample of 40 Time Intervals between Earthquakes

285.29	185.35	2579.80	27.11
5330.22	1504.49	1764.41	574.46
1722.16	815.06	3686.84	1501.61
494.17	362.48	1388.13	1817.27
3516.80	5049.66	2414.06	387.78
4385.23	2443.98	2204.12	1224.42
1965.46	3.61	1347.30	271.23
3247.99	753.80	1781.45	593.59
1482.64	1165.04	4647.39	1219.19
251.12	7345.52	1368.79	4112.13

Figure 1.2: The First Digits of the Earthquake Sample

Clearly, for this very small sample, low digits occur by far more frequently on the first position than do high digits. A summary of the digital configuration of the sample is given as follows:

Digit Index: { 1,   2,   3,   4,   5, 6, 7, 8, 9 }
Digits Count totaling 40 values: { 15, 8,   6,   4,   4, 0, 2, 1, 0 }
Proportions of Digits with '%' sign omitted: {38, 20, 15, 10, 10, 0, 5, 3, 0 }

Assuming (correctly) that these 40 values were collected in a truly random fashion from the large data set of all 19,452 earthquakes occurrences in 2012; without any bias or attempt to influence first digits occurrences; and that this pattern is generally found in many other data sets, one then may conclude with the phrase "not all digits are created equal", or rather "not all first digits are created equal", even though this seems to be contrary to intuition and against all common sense.

The focus here is actually on the first meaningful digit – counting from the left side of numbers, excluding any possible encounters of zero digits which only signify ignored exponents in the relevant set of powers of ten of our number system. Therefore, the complete definition of the **First Leading Digit** is the first non-zero digit of any given number on its left-most side. This digit is the first significant one in the number as focus moves from the left-most position towards the right, encountering the first non-zero digit signifying some quantity; hence it is also called the **First Significant Digit**. For 2365 the first leading digit is 2. For 0.00913 the first leading digit is 9 and the zeros are discarded; hence even though strictly-speaking the first digit on the left-most side of 0.00913 is 0, yet, the first significant digit is 9. For the lone integer 8 the leading digit is simply 8. For negative numbers the negative sign is discarded, hence for -715.9 the leading digit is 7. Here are some more illustrative examples:

6,719,525   →   digit 6
0.0000**7**61   →   digit 7
-0.**2**81264   →   digit 2
875   →   digit 8
3   →   digit 3
-5   →   digit 5

For a data set where all the values are greater than or equal to 1, such as in the sample of the earthquaqe data, the first digit on the left-most side of numbers is also the First Leading Digit and the First Significant Digit, and necesarily one of the nine digits {1, 2, 3, 4, 5, 6, 7, 8, 9}; while digit 0 never occurs first on the left-most side.

[2] Benford's Law and the Predominance of Low Digits

Benford's Law states that:

Probability[First Leading Digit is d] = $LOG_{10}(1 + 1/d)$

$LOG_{10}(1 + 1/1) = LOG(2.00)$ = 0.301
$LOG_{10}(1 + 1/2) = LOG(1.50)$ = 0.176
$LOG_{10}(1 + 1/3) = LOG(1.33)$ = 0.125
$LOG_{10}(1 + 1/4) = LOG(1.25)$ = 0.097
$LOG_{10}(1 + 1/5) = LOG(1.20)$ = 0.079
$LOG_{10}(1 + 1/6) = LOG(1.17)$ = 0.067
$LOG_{10}(1 + 1/7) = LOG(1.14)$ = 0.058
$LOG_{10}(1 + 1/8) = LOG(1.13)$ = 0.051
$LOG_{10}(1 + 1/9) = LOG(1.11)$ = 0.046

1.000

Figure 1.3 depicts the distribution. Figure 1.4 visually depicts Benford's Law as a bar chart. This set of nine proportions of Benford's Law is sometimes referred to in the literature as **'The Logarithmic Distribution'**. Remarkably, Benford's Law is confirmed in almost all real-life data sets with high order of magnitude, such as in data relating to physics, chemistry, astronomy, economics, finance, accounting, geology, biology, engineering, governmental census data, and many others.

Digit	Probability
1	30.1%
2	17.6%
3	12.5%
4	9.7%
5	7.9%
6	6.7%
7	5.8%
8	5.1%
9	4.6%

Figure 1.3: Benford's Law for 1st Digit

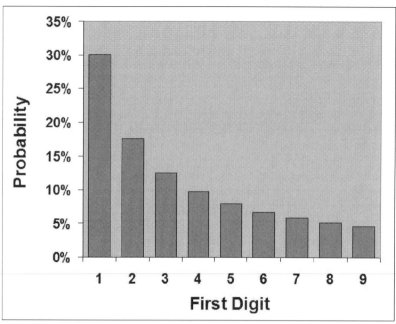

Figure 1.4: Benford's Law – Probability of First Digit Occurrences

Benford's Law was discovered by an indirect inference about patterns in occurrences of digits within numbers of actual data. The two discoverers were Simon Newcomb in 1881, and (independently) Frank Benford in 1938. What caught their attention was the observation about the differentiated physical wear and tear of the pages in old books of tables of logarithms - commonly used by engineers and scientists before the advent of calculators and computers. These old books seemed to be more strained by use and quite worn in the first pages relating to first digits 1, 2, and 3, and progressively less so throughout the book for higher digits, culminating in the last pages relating to first digits 7, 8, and 9 which seemed to be in relatively excellent condition, as if they haven't been much in use.

Naturally Newcomb and Benford inferred that the engineers and scientists reading those logarithmic books were overall more in need of using the first pages of low [first] digits than the last pages of high [first] digits, reflecting the spread of the [first] digits in the real world; namely that such differentiated use in the pages for real-life numbers in actual data indirectly indicates that numbers beginning with low digits occur quite frequently, while numbers beginning with high digits occur less frequently.

NOTE: log(x) or LOG(X) notation in this book would always refer to our decimal base 10 number system, hence the more detailed notation of $LOG_{10}(X)$ is often - but not always - avoided. The natural logarithm base e would be referred to as $LOG_e(X)$ or simply ln(x).

A useful expression yielding the first leading digit of any positive number X is given by:

$$\text{First Digit of X} = \text{INT}(X / 10^{\text{INT}(\text{LOG } X)})$$

The INT function refers to the integer just below X, or to X itself if X is exactly an integer. For example, INT(7.2) is 7, INT(9) is 9, INT(0.85) is 0, and INT(-5.2) is -6. The LOG function refers to the decimal logarithm of X with base 10. This expression should prove quite useful in computer implementations of Benford's Law, and especially in MS-Excel which uses the identical names of LOG and INT for these two functions.

Clearly, all numbers from **1** up to **2** such as 1.00, 1.15, 1.49, 1.76, 1.93, and 1.99 are with first digit 1; all numbers from **10** up to **20** such as 10.0, 13.8, 16.8, 18.2, and 19.6, are with first digit 1; and all numbers from **100** up to **200** such as 100, 141, 176, 195, and 198 are with first digit 1.

In general, the count of numbers within a given data set with first digit 1 is equivalent to the count of data points falling within … [0.01, 0.02), [0.1, 0.2), [1, 2), [10, 20), [100, 200) … and so forth. The count of numbers within a given data set with first digit 7 is equivalent to the count of data points falling within … [0.07, 0.08), [0.7, 0.8), [7, 8), [70, 80), [700, 800) … and so forth. Surely infinitely many other sub-intervals on the left and on the right of the above-mentioned five sub-intervals should also be included, such as [0.001, 0.002) for say the number 0.00176, or [7000, 8000) for say the number 7231, and so forth.

[3]   The Base Invariance Principle in Benford's Law

Benford's Law is valid for all positional number systems of whatever base B; including of course our base 10 system. This important generalization is accomplished by using the appropriate logarithm base B in the expression for the probabilities for the first leading digit d (in the range of 1 to B - 1). The generalized Benford's Law for any base B is:

$$\text{Probability[First Leading Digit is d]} = \text{LOG}_B(1 + 1/d)$$

As an example, for a positional number system with base 4, the three possible first digits are {1, 2, 3}. Here $\text{LOG}_4(1 + 1/1) = 0.50$; $\text{LOG}_4(1 + 1/2) = 0.29$; and $\text{LOG}_4(1 + 1/3) = 0.21$.

For a positional number system base 6 with {1, 2, 3, 4, 5} as the set of all possible first digits, Benford's Law predicts the digital probabilities of {38.7%, 22.6%, 16.1%, 12.5%, 10.2%}. This set is calculated as $\{\text{LOG}_6(1+1/1), \text{LOG}_6(1+1/2), \text{LOG}_6(1+1/3), \text{LOG}_6(1+1/4), \text{LOG}_6(1+1/5)\}$.

For base 2 number system, all numbers necessarily starts with digit 1, hence the probability of digit 1 leading is 100%. This is also confirmed via the expression $\text{LOG}_2(1 + 1/1) = \text{LOG}_2(2) = 1$, namely 100%, and here Benford's Law is reduced to a tautology.

[4] The Scale Invariance Principle in Benford's Law

Remarkably, Benford's Law is valid assuming the use of any scale in measuring the physical phenomenon generating the data set under consideration. Surely, there is nothing special about kilograms, meters, seconds, hours, inches, or miles; they are all arbitrary; and philosophically, for the law to be considered universal and consistent, the law should hold true for any units and scales. The fact that Benford's Law is indeed independent of societal scale system renders it universality. This property of the law is called 'The Scale Invariance Principle'.

As an example, the data on the time in the units of seconds between all successive earthquakes worldwide for the year of 2012 shall be converted into units of minutes and into units of hours.

1st Digits of Earthquake Data in Seconds - {29.9, 18.8, 13.5, 9.3, 7.5, 6.2, 5.8, 4.8, 4.2}
1st Digits of Earthquake Data in Minutes - {28.6, 17.7, 12.8, 10.5, 8.6, 6.8, 5.8, 5.0, 4.2}
1st Digits of Earthquake Data in Hours   - {29.2, 16.9, 12.3, 9.8, 8.1, 7.0, 6.2, 5.7, 4.9}

The small deviations seen here for these three scales are due to the fact that the earthquake data set is not perfectly Benford and that it has finite number of data points. Had it been perfectly Benford and with infinite number of data points (or at least with a truly huge number of points) then there would be no (almost) deviations at all. It should be noted that a scale change from seconds to minutes say, entails multiplying each time interval quoted in seconds by the factor of 1/60. Hence, the more general view of the Scale Invariance Principle is that digit distribution of Benford-type data remains nearly unchanged under a multiplicative transformation of all the data points by the same multiplicand (i.e. the factor).

[5] Integral Powers of Ten (IPOT)

Integral Powers of Ten (IPOT) play a crucial role in the understanding of Benford's Law.

An integral power of ten is simply 10^{INTEGER} with either negative or positive integer, as well as zero power. For example, the IPOT numbers 0.001, 0.01, 0.1, 1, 10, 100, are directly derived from 10^{-3}, 10^{-2}, 10^{-1}, 10^{0}, 10^{1}, 10^{2}.

Adjacent integral powers of ten are a pair of two neighboring and consecutive IPOT numbers 10^{INTEGER} and $10^{\text{INTEGER}+1}$ such as 1 & 10, or 100 & 1000, and so forth.

[6] Second Digits, Third Digits, and Higher Order Digits in Benford's Law

Benford's Law also gives considerations to higher order digit distributions, in addition to the 1st digit order. Assuming the standard base 10 decimal number system for example, then the 2nd leading digit (2nd from the left) of 7834 is digit 8, of 0.03591 it's digit 5, and of 4093 it's digit 0. The 3rd leading digit (3rd from the left) of 3271 is digit 7. The 4th leading digit (4th from the left) of 981054 is digit 0. For all higher orders, digit 0 is also included in the distributions, but digit 0 is not part of the 1st order.

There exist some slight probabilistic dependencies between the digits; in essence a slight positive correlation between them. For example, the conditional 2nd digit order is a bit skewer (in favor of low digits) whenever the 1st digit is low, and it is a bit more equal (in favor of high digits) whenever the 1st digit is high.

The unconditional probabilities according to Benford's Law for the 2nd, 3rd, and 4th orders are:

Benford's Law 2nd - {11.97, 11.39, 10.88, 10.43, 10.03, 9.67, 9.34, 9.04, 8.76, 8.50}
Benford's Law 3rd - {10.18, 10.14, 10.10, 10.06, 10.02, 9.98, 9.94, 9.90, 9.86, 9.83}
Benford's Law 4th - {10.02, 10.01, 10.01, 10.01, 10.00, 10.00, 9.99, 9.99, 9.99, 9.98}

Digit distribution for the 2nd digit order is only slightly skewed in favor of low digits, as opposed to the much more dramatic skewness of the 1st digit order where digit 1 is about 6 times more likely than digit 9. Digit distribution for the 3rd digit order shows only tiny deviations from the 10% possible equality. As even higher orders are considered, digits rapidly approach digital equality for all practical purposes, attaining the uniform and balanced 10% proportions for all the 10 possible digits 0 to 9.

Intuitive explanations of higher order digit behavior, including chart-based visualizations of the causes that lead to digital equality and order dependencies, are provided in Kossovsky (2014) chapter 100. A more general vista via The General Law of Relative Quantities (GLORQ), modeling each digital order on a particular bin scheme, provides a more profound understanding of the tendency towards equality for higher orders; as outlined in Kossovsky (2014) chapter 139.

[7] Physical Order of Magnitude of Data (POM)

Rules regarding expectations of compliance with Benford's Law rely heavily on measures of order of magnitude and variability of data, therefore this chapter as well as the next two chapters shall be devoted to these prerequisite and essential topics.

Physical order of magnitude of a given data set is a measure that expresses the extent of its variability. It is defined as the ratio of the maximum value to the minimum value. The data set is assumed to contain only positive numbers greater than zero.

Physical Order of Magnitude (POM) = Maximum/Minimum

The classic definition of order of magnitude involves also the application of the logarithm to the ratio maximum/minimum, transforming it into a smaller and more manageable number.

Order of Magnitude (OOM) = LOG_{10}(Maximum/Minimum)

Order of Magnitude (OOM) = LOG_{10}(Maximum) - LOG_{10}(Minimum)

Since such logarithmic transformation has a monotonic one-to-one relationship with max/min, it does not provide for any new insight or information, but could rather be looked upon in a sense as simply the use of an alternative scale, still measuring the same thing. For this reason the complexity of logarithm can be avoided altogether by referring only to the simple POM measure.

The more profound reason for using POM instead of OOM is its feature as a universal measure of variability, totally independent of societal number system in use, as well as being independent on the arbitrary choice of base 10, derived from the chanced or random occurrence of us having 10 fingers. This is the motivation behind the use of the term 'physical', expressing real and physical measure of variability, divorced from any numerical inventions, and especially so when data relates to the natural world such as in scientific figures and physical information.

OOM is perhaps more appropriate for a single isolated number, where it is re-defined as simply LOG_{10}(Number) without any reference to maximum, minimum, or any ratio. If we can assume that that number is an integral whole number without any [trailing] fractional part, then an alternative meaning of this OOM definition is simply expressing how many digits approximately are necessary to write the number. Surely in general, the bigger the [integral] number the more digits it takes to write it! For example, LOG_{10}(8,200,135) = 6.9, which is about 7, and that's exactly how many digits the number involves. As another example, LOG_{10}(10,000,000) = 7.0 which is exactly one digit less than the number of digits involved in writing the number, namely 8 digits.

Let us return the focus to data sets, containing many numbers, a minimum, and a maximum.

Suppose that an extensive database on the size (height) of a certain **giraffe species** somewhere on the Africa Savanna is with a minimum of 5.0 meter; a maximum of 6.0 meter; and an average of 5.5 meter. Here the range of variability for the giraffe is (6.0 – 5.0) = 1.0 meters.

Suppose that an extensive database on the size (length) of a particular **ant species** somewhere in the Amazon Basin is with a minimum of 0.003 meter; a maximum of 0.009 meter; and an average of 0.006 meter. Here the range of variability for the ant is (0.009 – 0.003) = 0.006 meters.

Which species should be considered as more varied in size, the ant or the giraffe? The giraffes vary over 1 meter, while the ants vary over only 0.006 meter, and Standard Deviation of the giraffes is much bigger than Standard Deviation of the ants, therefore on the face of it, the giraffe has more variability than does the ant. Yet, such an absolute approach, imposing the same universal benchmark of the scale of the meter on the two very different species is quite arbitrary. What is necessary here is to measure auto-variability, namely the variability of the data relative to itself, not relative to any arbitrary measure of the meter or other artificial scales, nor a comparison relative to other data sets regarding other species.

That apparent 'huge' variability of 1.0 meter for the giraffe is actually quite small in comparison with its own average of 5.5 meter, namely merely (1.0)/(5.5) or 18%.

That apparent 'tiny' variability of only 0.006 meter for the ant is actually quite huge in comparison with its own average of 0.006 meter, namely a whopping (0.006)/(0.006) or 100%.

The POM expression as the ratio of maximum to minimum also represents an auto-variability measure regarding the data set itself, and it is independent of any other data sets regarding different organisms and entities, as well as independent of any societal number system or base in use. POM is certainly also unit-less, independent of any societal and arbitrary units and scales, such as the meter, centimeter, kilometer, inch, or mile.

For the giraffes, POM measure is 6/5 or merely 1.2. For the ants, POM measure is a 0.009/0.003 or 3.0. Therefore, POM measure of the ants is approximately three times larger than POM measure of the giraffes!

In the extreme case where all the numbers in the data set are identical, having the value R say, variability is then nonexistent, and POM = maximum/minimum = R/R = 1.

[8] A Robust Measure of Physical Order of Magnitude (CPOM)

It is perhaps unfortunate that the literature in statistics does not seem to contain any robust definition of order of magnitude. Such a measure should prove steady and consistent for all types of data sets, strongly resisting outliers, preventing them from overly influencing the numerical measure of data variability.

In order to accomplish exactly that, and also to preserve the advantage of avoiding dependencies on arbitrary societal number systems and particular bases, the basic (independent) structure of POM shall be used, but with the added modification of simply eliminating any possible outliers on the left for small values and on the right for big values. This is accomplished by narrowing the focus exclusively onto the core 80% part of the data. This brutal purge eliminates without mercy any malicious and misleading outliers as well as any innocent and proper data points which happened to stray just a little bit away from the core part of the data. The measure shall be called Core Physical Order of Magnitude and it is defined as follows:

Core Physical Order of Magnitude (CPOM) = $P_{90\%}/P_{10\%}$

The definition simply reformulates POM by substituting the 10th percentile (in symbols $P_{10\%}$) for the minimum, and by substituting the 90th percentile (in symbols $P_{90\%}$) for the maximum.

The 10th percentile is the value below which about 10% of the data points may be found. The 50th percentile is the median, below which about half of all the data points may be found. The 90th percentile is the value below which about 90% of the data points may be found.

As an example, 50 data points, sorted low to high, shall be examined, as shown follow:

2	23	24	25	26	27	28	29	32	33
33	33	34	36	37	38	38	39	40	41
42	47	48	50	51	52	53	55	56	57
59	60	63	67	68	75	76	77	78	79
80	84	86	91	94	103	107	114	**213**	**567**

There are three obvious outliers within this data set, namely {2, 213, 567} shown above in bold font for emphasis. Calculating POM mindlessly without any worries whatsoever about possible distortions from outliers, leads to (567)/(2) = 283.5. But this value greatly exaggerates the variability of the data set which spans mostly the much narrower range of 23 to 114 in the approximate. The 10th percentile here is 26.9 and the 90th percentile is 94.9. It follows that CPOM = $P_{90\%}$ / $P_{10\%}$ = (94.9)/(26.9) = 3.5. This is by far a more realistic value for the true variability of the data set, focusing on the core 80% of the data.

Figure 1.5 depicts the histogram of the above data set, showing the 3 outliers as short, thick, and black lines, as well as the 10th and 90th percentile points utilized in the definition of CPOM.

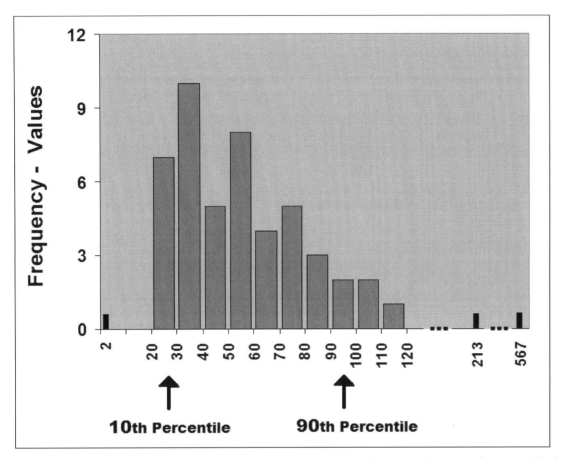

Figure 1.5: A Measure of Core Variability Focusing on the Central 80% of Data – CPOM

An alternative measure for the core variability of any given data set could be defined without involving any percentiles at all. This measure would simply eliminate first all outliers before any calculation of the originally-defined POM measure is performed, and then applying the newly observed minimum and maximum values after the elimination of outliers took place.

Hence for the above data set, once {2, 213, 567} are eliminated, POM for the rest of the data is simply $(114)/(23) = 4.9$. For this data set it is perhaps straightforward to classify {2, 213, 567} as outliers, and almost all data analysts should strongly agree with this classification. Yet, decisions and considerations regarding which data points constitute outliers and which belong to the core data for this or that particular data set could be considered at times as subjective and personal, depending on the data cases. Hence the motivation for having a universal and objective guideline here; a benchmark which would encompass all data sets equally, and which is provided by CPOM.

By strictly applying the 10th and 90th percentiles for all data sets, the procedure avoids vague definitions and arguments about outliers. Surely, the price paid for such universality in the procedure is the occasional eliminations of innocent data points near the edges which are actually very much part of the data, but are being swept away by the crude cleansing method of CPOM.

This is akin to a malignant cancer surgery where the surgeon is keen on making sure that no cancer cells whatsoever are left in area, so that remission is to be avoided at all cost, and therefore the surgeon is cutting some more all around the tumor, even in the healthy tissue and cells immediately surrounding it. For the data set above, the crude outlier-surgery of CPOM removes also the innocent points {23, 24, 25, 26, 103, 107, 114} which are actually authentic part of the data, but this is the price we are willing to pay in order to standardize the procedure.

For a more liberal procedure, with stronger emphasis on avoiding losing authentic data points, and less emphasis on eliminating all possible outliers, the wider range from the 5th percentile to the 95th percentile might be considered, applying the ratio $P_{95\%}$ / $P_{5\%}$ as the measure for the core 90% of the data.

In general, the rejection of outliers appearing in a given data set may be justified, or it may actually be misguided. For example, if over 50,000 students at a large university are surveyed with regards to height, and the top value is say 7.25 meter, then this outlier is certainly some kind of an error in recording and should be excluded from further analysis. If the top value is say 2.37 meter, then this 2.37 outlier is actually an integral part of the data set, and especially so if the well-known tall student is ordered to appear at the administration office, rudely interrupting his exciting basketball game at the court, and another measurement is taken, confirming his 2.37 meter height as well as his existence. This is not simply a matter of mere semantics, and there is a compelling argument not to classify this 2.37 value as an outlier, although in reality it depends on the context.

[9] Measures of Variability: Order of Magnitude & Standard Deviation

Two numerical measures in the field of Descriptive Statistics have been defined in order to express variability of data sets, namely Standard Deviation and Order of Magnitude. Yet, in spite of their supposedly common aim and shared purpose, they differ profoundly in what they actually measure. Standard Deviation is defined as the square root of the average of the squared differences between the data points and their average. This implies that the Standard Deviation is a measure that is used to quantify the amount of variation or dispersion of the data values from their average. A low standard deviation indicates that the data points tend to be close to the average, while a high standard deviation indicates that the data points are spread out over a wider range of values and thus are quite far from the average.

$$Standard\ Deviation = \sqrt{\frac{1}{N} * \sum_{i=1}^{N} (X_i - \text{Average})^2}$$

Index i takes on the values of 1, 2, 3, ... to N, and N stands for the number of points in the data set. The notation X_i stands for one data point indexed i.

For example, for the data set {2, 3, 7, 8}, the average is (2 + 3 + 7 + 8)/(4) = 5.

$$Standard\ Deviation = \sqrt{\frac{1}{4} * [\,(2 - 5)^2 + (3 - 5)^2 + (7 - 5)^2 + (8 - 5)^2\,]}$$

$$Standard\ Deviation = \sqrt{\frac{1}{4} * [\,(-3)^2 + (-2)^2 + (2)^2 + (3)^2\,]} = \sqrt{\frac{1}{4} * [26]} = 2.55$$

The simultaneous applications of the diametrically opposing operations of the square root and the squaring of the differences between the data points and their average, yield approximately the average distance between the data point and their average; namely measuring how far on average the data points are from the central point of the entire data. For example, distances between the four data points {2, 3, 7, 8} and their average of 5 are: {|2 − 5|, |3 − 5|, |7 − 5|, |8 − 5|} or simply {3, 2, 2, 3}. The average of these distances is (3+2+2+3)/(4) = (10)/(4) = 2.50, and which is very close to the 2.55 value of the Standard Deviation.

The two Normal Distributions depicted in Figure 1.6 are both with the same Standard Deviation value of 0.33, but one is with an average of 2 and the other is with an average of 5. Clearly, Physical Orders of Magnitude are approximately (3)/(1) = 3.0 and (6)/(4) = 1.5 respectively, with twice as much POM for the left curve as compared to POM of the right curve, and in spite of the fact that both are with the same value of the Standard Variation. This scenario is similar to the scenario of the data on the ant and giraffe species of chapter 7, where POM for the ants is much larger than POM of the giraffes, and in spite of fact that Standard Deviation for the ants is much lower than the Standard Deviation of the giraffes.

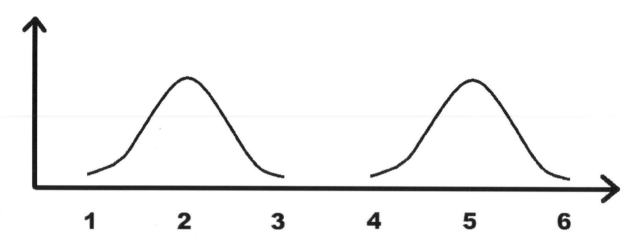

Figure 1.6: Two Normal Distributions with Equality in Standard Deviation but Distinct POMs

The two Normal Distributions depicted in Figure 1.7 are centered around the same value of 3 which is the average for both, but they differ with regards to the value of the Standard Deviation. Clearly, Physical Orders of Magnitude and Standard Deviations here correlate perfectly. The lower and long curve is of higher Standard Deviation as well as of higher Physical Order of Magnitude. The upper and short curve is of lower Standard Deviation as well as of lower Physical Order of Magnitude. Here both measures are truly aiming at the same feature of the data, and might be thought of in a sense as differing only with regards to the distinct scales or units of measurements in use (i.e. the structure of the mathematical definition of the measure.)

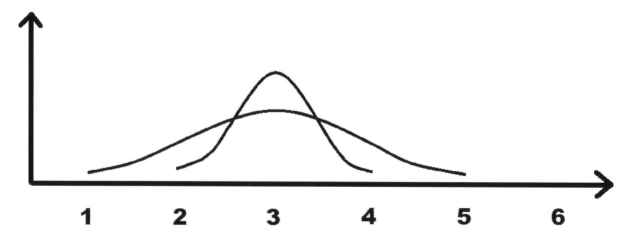

Figure 1.7: Perfect Correlation between Standard Deviation and POM if Normals are Aligned

The two Normal Distributions depicted in Figure 1.8 are centered around different values; the left curve is with an average of 1.5; and the right curve is with an average of 5. In order to obtain equality in the values of Physical Order of Magnitude, the values of the Standard Deviation must adjust accordingly. The edges of the left curve are approximately 1.2 and 1.8 having lower Standard Deviation of approximately 0.11, while the edges of the right curve are approximately 4 and 6 having higher Standard Deviation of approximately 0.33. POM for the left curve is approximately (1.8)/(1.2) = 1.5, while POM for the right curve it is approximately (6)/(4) = 1.5.

In order to obtain here equality in POM, the left curve is being narrowed, and the right curve is widened. For the data on the ant and giraffe species of chapter 7, in a sense, the curve of the ants hasn't been narrowed enough to yield equality in Order of Magnitude. Or perhaps, the curve of the giraffes hasn't been widened enough to yield equality in Order of Magnitude.

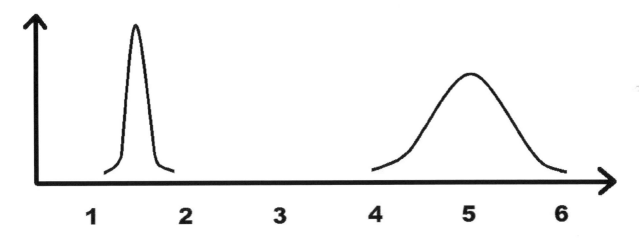

Figure 1.8: Adjusting Standard Deviations of Non-Aligned Normals to Obtain POM Equality

With regards to compliance with Benford's Law, the only variability measure that matters is Order of Magnitude (OOM), or rather Core Physical Order of Magnitude (CPOM), while Standard Deviation is totally irrelevant (when not correlated with OOM that is).

Order of Magnitude is scale-invariant, while Standard Deviation depends on the arbitrary societal definition of scales such as the meter, mile, inch, and kilogram. This feature renders Order of Magnitude universality, a feature totally lacking in Standard Deviation. Order of Magnitude measure is unit-less, while Standard Deviation must be expressed in terms of the unit of the underlying data set itself.

For example, changing the scale of a given data set from kilometers to meters entails multiplying each data value by the factor of 1000, and in general by the scale conversion factor of **k**. Such conversion does not affect Order of Magnitude at all since **k** cancels out in the numerator and in the denominator as in:

(maximum)/(minimum) $\rightarrow$ (**k***maximum)/(**k***minimum) = (maximum)/(minimum).

Standard Deviation on the other hand is directly affected by scale conversion. Obviously, the original Average is transformed into **k***Average due to the scale conversion, hence:

$$Rescaled\ S.D. = \sqrt{\frac{1}{N} * \sum_{i=1}^{N} \big((k * X_i) - (k * \text{Average})\big)^2} = \sqrt{\frac{1}{N} * \sum_{i=1}^{N} k^2 (X_i - \text{Average})^2} =$$

$$\sqrt{k^2 * \frac{1}{N} * \sum_{i=1}^{N} (X_i - \text{Average})^2} = \sqrt{k^2} \sqrt{\frac{1}{N} * \sum_{i=1}^{N} (X_i - \text{Average})^2} = k * Original\ S.D.$$

Therefore, Standard Deviation of the re-scaled data is simply **k** times the original value of the Standard Deviation (as is the case for the Average). In conclusion, the value of the Standard Deviation depends on the scale in use. These results are in perfect harmony with the fact that Benford's Law itself is scale-invariant, which implies that the statement of the prerequisites or conditions for compliance with the law must also be scale-invariant; and this is nicely accomplished by stating the conditions properly with the scale-invariant measure of the Order of Magnitude.

[10] Sum of Squares Deviation Measure (SSD)

It is necessary to establish a standard measure of 'distance' from the Benford digital configuration for any given data set. Such a numerical measure could perhaps tell us about the conformance or divergence from the Benford digital configuration of the data set under consideration. This is accomplished with what is called **Sum Squares Deviations (SSD)** defined as the sum of the squares of the 'errors' between the Benford expectations and the actual/observed values (in percent format – as opposed to fractional/proportional format):

$$SSD = \sum (\text{observed \% of digit d} - 100 * LOG(1 + 1/d))^2$$

with d running from 1 to 9. For example, for the observed 1st digits proportions as in {31.1, 18.2, 13.3, 9.4, 7.2, 6.3, 5.9, 4.5, 4.1}, SSD measure of distance from the logarithmic is calculated as:

$$\begin{aligned} \textbf{SSD} = & (31.1 - \mathbf{30.1})^2 + (18.2 - \mathbf{17.6})^2 + (13.3 - \mathbf{12.5})^2 + (9.4 - \mathbf{9.7})^2 + \\ & + (7.2 - \mathbf{7.9})^2 + (6.3 - \mathbf{6.7})^2 + (5.9 - \mathbf{5.8})^2 + (4.5 - \mathbf{5.1})^2 + (4.1 - \mathbf{4.6})^2 = \mathbf{3.4} \end{aligned}$$

SSD generally should be below 25; a data set with SSD over 100 is considered to deviate too much from Benford; and a reading below 2 is considered to be ideally Benford.

This SSD measure can be easily generalized for higher order digit distributions, as well as for other bases in positional number systems, as follows:

$$SSD = \sum (\text{Observed \%} - \text{Theoretical \%})^2$$

with summation index running from 1 to 10 for higher digital orders in our decimal number system. For the first digits in number systems with bases other than 10, summation index is running from 1 to (Base – 1).

[11] Two Essential Requirements for Benford Behavior

One of the two essential prerequisites or conditions for data configuration with regards to compliance with Benford's Law is that the value of the order of magnitude of the data set should be approximately over 3; in other words, that $LOG_{10}(Maximum/Minimum) > 3$, and that therefore $(Maximum/Minimum) > 10^3$. This in turn implies that the threshold POM value (separating compliance from non-compliance) is about 1000, namely that POM > 1000 constitutes the condition for compliance.

The above prerequisite for compliance totally ignores the thorny issue of outliers and edges, and in that sense it is too simplistic and even completely erroneous for some data sets. Hence, using the CPOM qualification is essential in judging whether or not a given data set is expected or not expected to comply with Benford's Law. The proper qualification for expectance of compliance with the law in the approximate - obtained via extensive empirical studies - is then as follows:

Core Physical Order of Magnitude $= P_{90\%} / P_{10\%} > 100$

Actually, even lower CPOM values such as 50 and 30 are expected to yield Benford, but falling below 30 does not bode well for getting anywhere near the logarithmic distribution.

Skewness of data where the histogram comes with a prominent tail falling to the right is the second essential criterion necessary for Benford behavior. Indeed, most real-life physical data sets are generally skewed in the aggregate, so that overall their histograms have tails falling on the right, and consequently the quantitative configuration is such that the small is numerous and the big is rare, while low first digits decisively outnumber high first digits.

The asymmetrical, Exponential, Lognormal, k/x [and many other distributions] are typical examples of such quantitatively skewed configuration, and therefore they are approximately, nearly, or exactly Benford - respectively. The symmetrical Uniform, Normal, Triangular, Circular-like, and other such distributions are inherently non-Benford, or rather anti-Benford, as they lack skewness and do not exhibit any bias or preference towards the small and the low.

Symmetrical distributions are always non-Benford, no matter what values are assigned to their parameters. By definition they lack that asymmetrical tail falling to the right, and such lack of skewness precludes Benford behavior regardless of the value of their order of magnitude. Order of magnitude simply does not play any role whatsoever in Benford behavior for symmetrical distributions. For example, first digits of the Normal(10^{35}, 10^8) or the Uniform(1, 10^{27}) are not Benford at all, and this is so in spite of their extremely large orders of magnitude. In summary: Benford behavior in extreme generality can be found with the confluence of sufficiently large order of magnitude together with skewness of data - having a histogram falling to the right. The combination of skewness and large order of magnitude is not a guarantee of Benford behavior, but it is a strong indication of likely Benford behavior under the right conditions. Moderate [overall] quantitative skewness with a tail falling too gently to the right implies that digits are not as skewed as in the Benford configuration. Extreme [overall] quantitative skewness with a tail falling sharply to the right implies that digits are severely skewed, even more so than they are in the Benford configuration. The only one exception to the generic rule above requiring skewness as well as high order of magnitude for logarithmic behavior is the perfectly Benford k/x distribution defined over adjacent integral powers of ten such as (1, 10) having the very low order of magnitude value of 1, as shall be discussed in the next chapters.

Bowley Skewness for example, defined as $[(Q3 - Q2) - (Q2 - Q1)] / [Q3 - Q1]$ is an intuitive measure of skewness but its numerical value fluctuates greatly across data sets. Calculated Bowley Skewness values for numerous logarithmic data sets and distributions do not yield any consistent result, except that all values come out above 0.3 and below 1.0, and which is consistent with the fact that all logarithmic data sets are positively skewed in the aggregate. In sharp contrast, non-logarithmic data sets generally come with decisively lower Bowley Skewness values below 0.25 and above 0. In contrast to Bowley's unstable value for logarithmic data sets, Benford's Law is a very consistent and almost exact measure of skewness, with very little fluctuations across logarithmic data sets.

[12]   Data Skewness is More Prevalent than Benford's Law

All data sets obeying Benford's Law (i.e. logarithmic data) are structured in such a way that in the aggregate there are more small quantities than big quantities. In other words, that in the aggregate the histogram is falling to the right, except perhaps in the beginning on the very left for low values where it temporarily rises for a very small portion of overall data, as well as in few and minors reversals along the way where it rises briefly. This quantitative configuration is called 'positive skewness' in mathematical statistics.

The expression or motto for this quantitative phenomenon regarding size configuration is coined as '**small is beautiful**'. The term 'beautiful' in this context is not meant literally, but rather metaphorically, as it signifies the connotation associated with the adjectives numerous, plentiful, frequent, and most common. The terms 'small' and 'big' refer to relative quantities within the framework of any given data set under consideration, and never to any absolute quantities or some imaginary fixed and universal benchmark values applicable to all existing data sets.

The small is beautiful phenomenon has by far much wider scope and it is much more prevalent in the physical world and in the realm of abstract mathematics than the more particular Benford quantitative configuration. This statement does not imply that Benford's Law is not prevalent in scientific, physical, and numerous other data types, on the contrary, it is highly prevalent. The statement only implies that in almost all the counter examples and exceptions to Benford's Law, the small is beautiful phenomenon is still valid, albeit with different quantitative configurations than that of the Benford one (and which are typically milder, but at times even skewer).

The assertion is derived from concrete experience with real-life numerical examples and from general research in Benford's Law. While this discussion may sound vague, in fact it is rather a very essential overview in the entire quantitative phenomenon of Benford's Law and of real-life data analysis in general. For those statisticians and data analysts who have worked on data sets and the Benford phenomenon for many years, including doing theoretical research, this generic statement seems natural, fundamental, and quite necessary.

All this can be stated more succinctly in three ways:

I) The Benford's Law configuration is a subset of the small is beautiful phenomena.
II) The small is beautiful phenomenon is even more prevalent than Benford's Law.
III) A significant portion of non-Benford real-life data is quantitatively structured in the spirit of the small is beautiful phenomenon.

[13] Data Aggregation Leads to Skewness and Often Even to Benford

It may not be obvious, but surprisingly quite often, real-life data sets consist of numerous, smaller, and 'more elemental' mini sub-sets. Therefore, a given data set which appears to exist independently as a whole, may actually be made of several data components which are aggregated in order to arrive at that larger set of data. The implication of such aggregations to quantitative skewness and digital configuration is profound, since results are almost always skewed in favor of the small, with resultant histogram having a tail falling to the right. Indeed, it can be demonstrated in general that appending various data sets into a singular and much larger data set leads to quantitative skewness in favor of the small whenever these data sets commonly start from a very low value, ideally such as 0 or 1, and terminate at highly differentiated endpoints, so that some span short intervals while others span longer intervals. Let us demonstrate this quantitative tendency in data aggregation by combining the following six imaginary data sets:

Data Set A: {2, 3, 5, 7}
Data Set B: {1, 4, 6, 9, 13, 14}
Data Set C: {2, 6, 7, 9, 11, 15, 16, 21}
Data Set D: {1, 2, 6, 8, 13, 14, 19, 23, 25}
Data Set E: {3, 4, 8, 12, 15, 19, 22, 24, 29, 35, 41}
Data Set F: {1, 5, 8, 11, 12, 17, 19, 24, 27, 32, 38, 43, 47}

Individually, each data set does not show any quantitative preference for the small, yet when these six data sets are merged together to become the aggregated and large data set, the resultant quantitative configuration is decisively in favor of the small and biased against the big.

The combined data set A, B, C, D, E, F:

{2, 3, 5, 7, 1, 4, 6, 9, 13, 14, 2, 6, 7, 9, 11, 15, 16, 21, 1, 2, 6, 8, 13, 14, 19, 23, 25, 3, 4, 8, 12, 15, 19, 22, 24, 29, 35, 41, 1, 5, 8, 11, 12, 17, 19, 24, 27, 32, 38, 43, 47}

The combined data set A, B, C, D, E, F, ordered from low to high:

{1, 1, 1, 2, 2, 2, 3, 3, 4, 4, 5, 5, 6, 6, 6, 7, 7, 8, 8, 8, 9, 9, 11, 11, 12, 12, 13, 13, 14, 14, 15, 15, 16, 17, 19, 19, 19, 21, 22, 23, 24, 24, 25, 27, 29, 32, 35, 38, 41, 43, 47}

It should be noted that we are aggregating the numbers residing within the data sets, but we are not aggregating the data sets in and of themselves. The focus here is on numbers, not on data sets. This means that each number is assigned equal importance (chance), yet not each data set earns equal importance (chance). For example, post-aggregation, set A has only 4 values within the aggregated data, while set F has 13 values, hence set F is more 'probable' and 'important'.

Counting occurrences of these 51 values, we note that:

There are **22** values from   1 to 10.
There are **15** values from 10 to 20.
There are **8** values from 20 to 30.
There are **3** values from 30 to 40.
There are **3** values from 40 to 50.

Figure 1.9 depicts the histogram of these 51 values. Remarkably, quantitative skewness seems to appear out of nowhere! In spite of the fact that each component data set is quantitatively structured in a uniform and balanced manner approximately, yet for the aggregated data set the small decisively outnumbers the big! The dynamics behind such tendency to produce numerous small quantities but only few big ones is the differentiated overlapping of ranges for the aggregated data set. Overlapping here occurs more on the left for small values, and less so on the right for big values. Figure 1.10 depicts these six data sets superimposed, thus allowing us to visualize the concentrated piling up of numerous small values that occurs on the left, in sharp contrast to the diluted piling up of only few big numbers that occurs on the right.

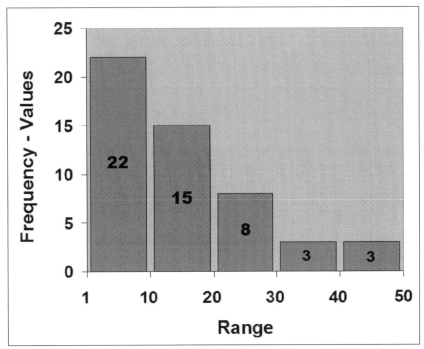

Figure 1.9: Skewed Histogram of the Aggregation of the 6 Data Sets

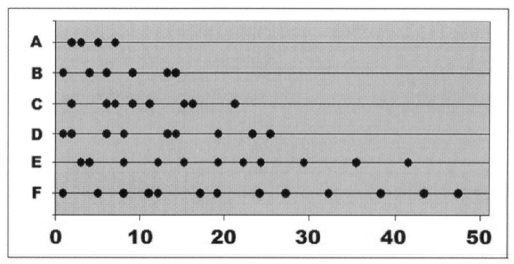

Figure 1.10: Visualizing the Piling up of More Values on the Left for the 6 Data Sets

It should be noted that some types of data aggregations do not lead to any such differentiated piling up of values in favor of the small on the left side of the range. The existence of these exceptions here to the expectation of resultant quantitative skewness is quite significant and should always be acknowledged so as to avoid overgeneralization and rushing to judgment in the considerations of data aggregations. There are two types of (extreme) counter examples. The first one is data aggregation without any overlapping whatsoever, where all the ranges are totally disconnected and distinct. The second one is data aggregation where overlapping is so complete, consistent, repetitive, and perfect, that nothing happens quantitatively; that nothing changes.

Aggregating {2, 6, 8}, {9, 11, 23}, {27, 31, 38, 41, 53}, {57, 61, 64, 72}, {75, 82, 88, 96} into a singular data set provides an example for the first type of exception where component data sets do not overlap at all, and instead simply continue to expand forward on the overall range.

Aggregating {1, 4, 11}, {1, 3, 7, 9}, {2, 4, 5, 8, 10}, {1, 4, 6, 7, 11}, {3, 5, 7, 10} into a singular data set provides an example for the second type of exception where component data sets overlap each other over approximately the same range of values; and where all component data sets begin and terminate at approximately the same points.

Another possible (although very rare) exception here is an inverted differentiation in overlapping of ranges, where more values pile up on the right for big values, and less values pile up on the left for small values. As an example for such rare cases, aggregating {1, 6, 10, 13, 14, 18}, {5, 9, 11, 17, 20, 21}, {3, 8, 13, 15, 19}, {7, 13, 16, 17, 20} into a singular data set results in an inverted quantitative configuration, with many big values but only few small values, and with a histogram having a tail falling to the left, as opposed to falling to the right [i.e. negative skewness].

36

Yet, resultant positive skewness is still the norm in most cases of data aggregations, because the original setup of A to F data sets of Figure 1.10 is the most natural and typical in the aggregations of real-life physical and abstract data sets. The three other alternatives are artificial and imaginary for the most part, and even though they might hold true in some very particular circumstances, they are simply quite rare. For the vast majority of real-life data creation and data aggregation, the natural process is of measurements that start out from 0 or 1 if not from another very low value, while terminating at a wide variety of much bigger values. This fact of life is stated without any concrete proof or argument, and without any experiment or empirical analysis of data. Instead, the author is referring to common sense and general intuition. For those with a lot of experience in real-life data, the above assertion should appear quite natural and compelling as it probably correlates with their knowledge of how data is typically formed and compiled.

Surely, not all data sets are derived from aggregation of smaller data components. For example, data on star sizes (mass) is not derived from any supposed aggregation. Instead, individual stars are observed and examined one by one, and the value of each star is added to the already collected large set of star mass values. For data sets derived from data aggregation, this could occur either in an implicit way or in an explicit way. A good example of implicit data aggregation is address data where the focus of the data analyst is solely on the house number. Here address data for an entire city can be implicitly or indirectly modeled as the aggregation of all the mini data sets of all its existing streets. Let us illustrate this more vividly by listing house numbers for several streets pertaining to a data set of a post office branch in one particular small town, and which typically may be listed as follows:

{1, 2, 3, 4, 5, 6} - Floral Drive
{1, 2, 3, 4, 5, 6, 7, 8 ,9 ,10 ,11 ,12, 13, 14, 15}- Pine Avenue
{1, 2, 3, 4, 5, 6, 7, 8, 9, 10, 11, 12, 13, 14, 15, 16, 17, 18, 19}- Main Street
{1, 2, 3, 4, 5, 6, 7, 8, 9, 10, 11, 12}- South Street
{1, 2, 3, 4, 5, 6, 7, 8, 9}- Lodge Street

All streets necessarily start at house number 1, namely the first house on the left or right corner, but each street terminates at different house number depending on the length of the street, hence many numbers pile up on the left range of small values, leaving the right range of big values diluted and rarer. The aggregation of all house numbers in all the streets in a given city constitutes an implicit data aggregation where the small outnumbers the big, and quantitative skewness can be clearly deciphered in the resultant histogram having a prominent tail falling on the right.

Not all data aggregations lead to Benford exactly, but most are fairly close, and at a minimum their first digits distribution resembles the logarithmic. As one concrete example, address data pertaining to Prince Edward Island in Eastern Canada containing 23,633 addresses is downloaded at the link in the website: http://www.gov.pe.ca/civicaddress/download/. Clearly, the focus is not on the street number or the zip code, but exclusively on the house number. First digits distribution is {30.9, 18.5, 15.1, 10.0, 6.1, 6.0, 5.2, 4.3, 4.0}, and the very low 13.1 SSD value here indicates that this house number data set is in strong compliance with Benford's Law.

[14] Chains of Statistical Distributions and Benford's Law

Two features guarantee quantitative skewness in favor of the small in aggregated data. The first feature is the common beginning of the approximate minimum for all the component data sets around 0 or 1. The second feature is the random or highly varied termination at the end for the maximum, namely that each component data set has its own particular maximum. Such quantitative tendency or mechanism in aggregations of real-life data sets is one of the main causes and explanations of the Benford's Law phenomenon in the real world. Formalism in mathematical statistics draws inspiration from such types of data compilation and points to a slightly different abstract process coined as "chain of two Uniform Distributions", namely the statistical chain **Uniform(min, Uniform(maxA, maxB))**, where min < maxA < maxB.

The Continuous Uniform Distribution **Uniform(a, b)** is the set of all possible values from a to b, all having the same chance of occurring - all with equal probability. It is useful to think of **a** as the parameter signifying the minimum, and of **b** as the parameter signifying the maximum. The parameters **a** and **b** are fixed, constant, and well-known numbers, and there is nothing fuzzy or random about them.

A chain of two Uniform Distributions model in the general spirit of data aggregations is such where parameter **a** is fixed at say 0 or 1, while parameter **b** is considered to be a fuzzy, uncertain, and random number, drawn from another Uniform Distribution with specific parameters, and expressed as Uniform(lowest maximum, highest maximum). Here the term 'lowest maximum' refers to the maximum value in the narrowest component data set, and the term 'highest maximum' refers to the maximum value in the widest component data set. The term 'common minimum' refers to the minimum value within the component data sets. The chain is then:

Uniform(common minimum, Uniform(lowest maximum, highest maximum))

Yet, a chain of two Uniform Distributions differs from data aggregation in one fundamental way, as it assigns equal importance and probability to all component data sets, as opposed to assigning equal importance and probabilities to all the numbers. The focus of the chain of distribution is on the component data sets (i.e. the Uniform Distributions), giving each data set (i.e. Uniform) equal probability within the entire scheme, while the focus of data aggregation is on the numbers, giving each number equal probability. Hence, in this sense at least, the chain of Uniform Distributions is not exactly the proper or the ideal model for data aggregations.

As an example of how data aggregations could (approximately) be modeled by chains of distributions, let us consider the combined data set A, B, C, D, E, F, as shown in Figure 1.10, and let us attempt to crudely represent its resultant distribution by the model of the 2-sequence chain **Uniform(1, Uniform(7, 47))**. Unfortunately, this modeling attempt fails to give a good fit, and its numerical result is somewhat different from the aggregated data set. There are three distinct reasons why the above aggregated data set cannot be represented well via such a chain.

The first reason is that the chain pertains to all values, integral as well as fractional and irrational numbers, while the data aggregation set contains only integral values.

The second reason for the failure to fit the chain model to the aggregated data is that there are gaps and jumps in the values of the integers of the various data sets. They do not increase nicely and smoothly by one integer at a time. In addition, the maximum boundaries are not increasing smoothly by one integer at a time either, but rather they occur with gaps and jumps as well. In addition, not all data sets begin with the minimum value of 1, only data sets B, D, and F do, the rest of the data sets start with either 2 or 3.

The third reason for the failure to fit the chain model to the aggregated data is that such a chain assigns equal probability for Data Set A of {2, 3, 5, 7} with only 4 values, as well as for Data Set F of {1, 5, 8, 11, 12, 17, 19, 24, 27, 32, 38, 43, 47} with 13 values. The chain repeats {2, 3, 5, 7} approximately three times in order to give Data Set A equal importance overall, thus injecting for example the set {2, 3, 5, 7, 2, 3, 5, 7, 2, 3, 5, 7} into the final aggregated data set, in order to accomplish the goal of equivalency in importance for all data sets.

An obvious application of the chains of distributions to real-life data sets is the house number in address data. Assuming that the shortest street in the town is with 6 houses, and that the longest street is with 73 houses, plus presumed smoothness and approximate uniformity in how lengths of streets are distributed, then the model for house number data in this small town could be given by **Uniform(1, Uniform(6, 73))**. Here all streets necessarily start at house number 1, hence parameter **a** is fixed at 1, but each street terminates at a different house number depending on the length of the street, hence parameter **b** should be variable and derived from yet another Uniform Distribution. Yet, the application of the chains of distributions to house number data suffers from the same issue that challenged data aggregation, namely that short streets as well as long streets in the chain model are given equal importance, and this feature of the chain model needs to be adjusted to fit the true configuration of the house number data which assigns longer streets more importance than shorter ones.

Many other types of chains of distributions, applying a variety of distribution forms (not merely the Uniform); with any number of dependent sequences (not merely two); tying up location and/or scale parameter(s), yield quantitative skewness and are often nearly perfectly Benford. This lends the chain of distributions a truly colossal scope of manifestations, occurrences and applications in the context of quantitative skewness in general and Benford's Law in particular.

As another example, 4 Uniform Distributions are chained with regards to parameter **b** only, and with the ultimate **b** value at 55, while parameter **a** is fixed at 0 for all. Formally this chain scheme is written as: **Uniform(0, Uniform(0, Uniform(0, Uniform(0, 55))))**.

This 4-sequence chain process can also be described as a step-by-step simulations scheme as follow: simulate a single value from the Uniform(0, 55) and call it R; then simulate a single value from the Uniform(0, R) and call it G; then simulate a single value from the Uniform(0, G) and call it P; then simulate a single value from the Uniform(0, P) and call it Q. This last Q value is the final value of the entire chain for this single simulation.

Figure 1.11 depicts the histogram of 10,000 simulated values of the chain of distributions Uniform(0, Uniform(0, Uniform(0, Uniform(0, 55)))). The small is clearly more numerous than the big, as quantitative skewness is achieved via the chaining of the non-skewed Uniforms. First digits distribution for this chain is {30.6, 17.5, 11.8, 9.3, 8.4, 7.0, 5.9, 5.1, 4.5}, and the extremely low 1.3 SSD value here indicates that this chain is nearly perfectly Benford. For this particular simulating run, the minimum value came out as 0.0000154, and the maximum value came out as 41.4. Hence POM = (41.4)/(0.0000154) = 2689814, and OOM = LOG(2689814) = 6.4. This might seem to be quite high, yet outliers here are at fault for exaggerating the value of order of magnitude. The 10% percentile is 0.066. The 90% percentile is 9.552. Hence CPOM = (9.552)/(0.066) = 144.5.

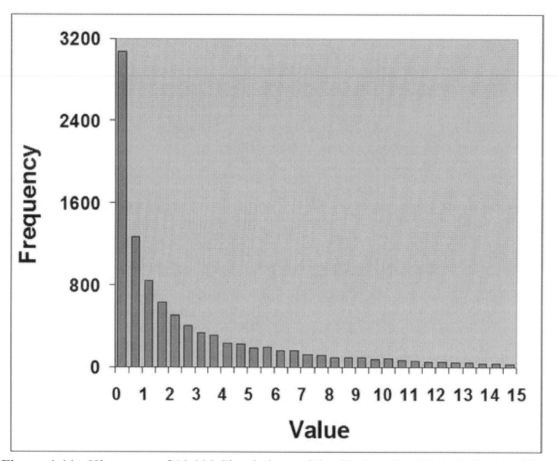

Figure 1.11: Histogram of 10,000 Simulations of the Chain U(0, U(0, U(0, U(0, 55))))

As another chain example, **Exponential(Exponential(Exponential(7)))** is a chain of three Exponential distributions, and which converges to Benford nearly perfectly. In 10,000 simulation runs, first digits distribution were {30.3, 18.1, 12.3, 10.2, 8.0, 6.5, 5.1, 5.0, 4.4}, and the extremely low 1.2 SSD value here indicates that this chain is nearly perfectly Benford.

As an example of a hybrid chain, **Exponential(Normal(Uniform(15, 21), 3))** is a chain composed of three distinct distributions, and which converges to Benford as expected. In 10,000 simulation runs, first digits came out as {29.6, 16.0, 11.9, 10.0, 8.5, 7.4, 6.1, 5.6, 4.8}, and the very low 4.5 SSD value here indicates that this chain is very close to Benford.

The relevance of the chain of distributions model to real-life data sets becomes more apparent considering the causality in life and nature; the interconnectedness in the world; and the dependencies of some entities upon other entities. All this leads to the conclusion that often some physical measurements serve as parameters for other physical measurements. For example, lengths and widths of rivers depend on average rainfall (being the parameter) and rainfall in turns depends on sunspots, prevailing winds, and geographical location, all serving as parameters of rainfall. Weights of people may depend on overall childhood nutrition, while nutrition in turns may depend on overall economic activity, which in turns depends on economic policy, war and peace, weather-related events such as droughts and flooding, and so forth. Exact cause and effect relationships in the deterministic realms such as in physics, chemistry, astronomy, and so forth, often lead to skewed results and scenarios which can be modeled on chains of distributions, and especially so when phenomena is aggregated and examined on a large scale. This is another chief cause and explanation of why so often the Benford digital configuration is found in real-life data.

Order of magnitude of chains of distributions normally does not factor directly or obviously in convergence to Benford, although it does play a hidden role indirectly in facilitating convergence. Hence, when digital results from a wide variety of chains of distributions are compared (of similar format and style but with distinct ultimate parameters), it often shows a strong correlation between the chain's order of magnitude and its convergence to Benford.

Although order of magnitude of chains of distributions normally does not factor in an obvious way regarding convergence to Benford, one exception (among others) to this general statement is the short chain of the form **Uniform(A, Uniform(A, B))**, B > A, where convergence to Benford cannot be achieved at all without sufficient order of magnitude of the set {A, B}, namely having sufficiently large ratio of B/A of at least 10 approximately. Mysteriously, the exact ratio of 13 for B/A for some reason yields the best result here, having the strongest compliance with Benford. Surpassing the value of 13 yields worsening results with wider deviation from Benford, even though order of magnitude increases when the ratio of B/A is made larger.

For example, Uniform(3, Uniform(3, 6)) is of extremely low order of magnitude as seen in the ratio B/A = 6/3 = 2, and clearly the only values this chain generates are between 3 and 6 exclusively, hence digits {3, 4, 5} are the only possible first digits, while digits {1, 2, 6, 7, 8, 9} never even occur on the first digital place.

The opposite pole of the above unsuccessful chain example can be given perhaps by the superior chain Uniform(3, Uniform(3, 9000)) which is of extremely high order of magnitude as seen in the ratio B/A = 9000/3 = 3000. Clearly, the values this chain generates are widely spread between 3 and 9000. In one specific Monte Carlo computer simulations with 20,000 runs, digit distribution came out as {25.6, 18.8, 14.6, 11.9, 9.3, 7.2, 5.5, 4.1, 3.0}, with a moderate SSD value of 36.9. The lowest value obtained was 3.03, and the highest value obtained was 8929.89, resulting in POM value of 8929.89/3.03 = 2944.3. The 10th percentile obtained was 193.9, and the 90th percentile obtained was 5310.8, resulting in CPOM value of 5310.8/193.9 = 27.4. Given sufficient order of magnitude with ratio B/A of over 10, then degree of convergence to Benford still depends on the exact combination of A and B values, although in an apparently quite peculiar and inexplicable manner. The best B/A ratio yielding the closest configuration to

Benford is always ≈13, and overshooting this ratio yields worsening digital configurations! The plot of SSD versus B for any fixed A value fluctuates cyclically and smoothly up & down.

Specific Monte Carlo computer simulations with 20,000 runs each, based on the chain model U(1, U(1, B)) yielded the following results:

U(1, U(1, 155))	{26.8, 15.3, 12.3, 10.5, 9.0, 8.1, 6.8, 6.0, 5.1}	SSD = 22.4	B/A = 155.0
U(1, U(1, 67))	{32.0, 20.7, 15.6, 10.7, 6.9, 4.3, 3.5, 3.3, 3.1}	SSD = 41.6	B/A = 67.0
U(1, U(1, 45))	{37.0, 22.1, 12.5, 6.5, 5.1, 4.8, 4.4, 3.9, 3.7}	SSD = 93.2	B/A = 45.0
U(1, U(1, 36))	{39.5, 20.4, 9.0, 6.9, 5.7, 5.0, 4.9, 4.5, 4.1}	SSD = 125.6	B/A = 36.0
U(1, U(1, 29))	{41.0, 16.6, 8.6, 7.3, 6.3, 5.7, 5.3, 4.7, 4.5}	SSD = 144.5	B/A = 29.0
U(1, U(1, 22))	{39.5, 13.2, 10.2, 8.6, 7.3, 6.7, 5.5, 4.7, 4.4}	SSD = 113.9	B/A = 22.0
U(1, U(1, 18))	{35.9, 14.0, 12.0, 9.1, 7.7, 6.7, 5.8, 4.7, 4.2}	SSD = 47.8	B/A = 18.0
U(1, U(1, 15))	{33.6, 15.9, 12.3, 10.0, 8.1, 6.7, 5.6, 4.3, 3.5}	SSD = 16.8	B/A = 15.0
U(1, U(1, 13))	**{32.6, 17.4, 12.8, 10.4, 8.1, 6.6, 5.2, 4.0, 2.8}**	**SSD = 11.8**	**B/A = 13.0**
U(1, U(1, 11))	{33.2, 19.6, 14.3, 10.2, 7.8, 5.9, 4.4, 3.0, 1.8}	SSD = 32.4	B/A = 11.0
U(1, U(1, 10))	{35.9, 19.5, 14.5, 10.5, 7.9, 5.5, 3.6, 2.0, 0.6}	SSD = 73.9	B/A = 10.0
U(1, U(1, 8))	{42.0, 22.1, 15.0, 10.3, 6.2, 3.5, 1.0, 0.0, 0.0}	SSD = 251.6	B/A = 8.0
U(1, U(1, 4))	{69.9, 23.6, 6.5, 0.0, 0.0, 0.0, 0.0, 0.0, 0.0}	SSD = 1939.5	B/A = 4.0

As can be deciphered from the above results, the approximate ratio of B/A = 13 yields the best digital configuration – judged by SSD score which is the lowest for 13.

Specific Monte Carlo computer simulations with 20,000 runs each, based on the chain model U(24, U(24, B)) yielded the following results:

U(24, U(24, 800))	{23.6, 20.3, 16.9, 12.5, 9.2, 6.8, 4.3, 3.4, 3.1}	SSD = 84.9	B/A = 33.3
U(24, U(24, 500))	{28.8, 22.6, 16.0, 8.6, 5.8, 5.2, 4.8, 4.1, 4.0}	SSD = 49.1	B/A = 20.8
U(24, U(24, 400))	{29.9, 21.9, 13.6, 7.6, 6.6, 5.9, 5.3, 4.7, 4.5}	SSD = 26.9	B/A = 16.7
U(24, U(24, 312))	**{29.6, 18.8, 11.4, 9.2, 7.8, 6.7, 6.1, 5.3, 4.9}**	**SSD = 3.4**	**B/A = 13.0**
U(24, U(24, 230))	{25.2, 14.3, 14.2, 10.8, 9.3, 7.8, 7.4, 5.8, 5.3}	SSD = 45.8	B/A = 9.6
U(24, U(24, 180))	{16.1, 16.6, 17.0, 13.2, 10.5, 8.7, 7.1, 5.8, 5.0}	SSD = 240.9	B/A = 7.5
U(24, U(24, 55))	{ 0.0, 51.1, 35.1, 12.5, 1.3, 0.0, 0.0, 0.0, 0.0}	SSD = 2716.6	B/A = 2.3

As can be deciphered from the above results, the approximate ratio of B/A = 13 yields the best digital configuration – judged by SSD score which is the lowest for 13.

Another remarkable convergence to Benford for this very short chain of only two sequences can be found in the following case where the ratio B/A = 13, and which yields an extremely good fit: **U(17, U (17, 221))**, {30.2, 17.1, 11.8, 9.7, 8.0, 7.0, 6.0, 5.4, 4.9}, SSD = 1.1, B/A = 13.0.

The chain **Uniform(0, Uniform(0, B))** on the other hand theoretically possesses an infinite order of magnitude for any value of B. This is so since B/0 = ∞, and in fact its partial convergence to Benford is accomplished with any B value. Nonetheless, the degree of convergence to Benford depends on the value of B, with SSD values [cyclically] fluctuating between 25 and 55. Surely when the thorny issue of outliers and edges is taken into consideration, and the CPOM measure is applied, the true variability of the chain is understood to be finite, and not even very large, and which explains why the chain does not fully converge to Benford, but only partially so.

The author's 1st conjecture is that an infinitely long chain of distributions should obey Benford's Law exactly. For more discussions and other related conjectures regarding chains of distributions see Kossovsky (2014) chapters 54, 102, and 103. The following is a very brief summary in extreme generality of these three chapters:

Scale parameters such as λX or X/λ (divisions & multiplications) as well as Location parameters such as $X - \mu$ (subtractions), usually respond vigorously to chaining, prefer the small over the big, and obey Benford's Law (i.e. chain-able). Shape parameters such as X^k (powers) usually do not respond to chaining at all, show no preference for the big, for the small, or for any size, and disobey Benford's Law (i.e. not chain-able).

More precisely, a parameter that does not continuously involve itself in the expression of centrality [such as the mean, median, or midpoint] is not chain-able at all; and a parameter that does continuously involve itself in the expression of centrality is indeed chain-able.

The meaning of the phrase 'not continuously involved' means that the partial derivative ∂(center)/∂(parameter) goes to 0 in the limit for high values of the parameter, namely that centrality such as the average, the median, and resultant range in general, are not affected as parameter is further increased; that beyond a certain limit the parameter does not sway centrality.

$$\lim_{\text{parameter} \to \infty} \partial(\text{center})/\partial(\text{parameter}) = 0$$

Hence, if a parameter does not play any role in the determination of centrality and span of range beyond the initial few low values then it's not chain-able at all. Since most scale and location parameters [continuously] play significant role in centrality, they are generally chain-able. Since most shape parameters do not play any role in centrality, they are generally not chain-able.

In addition, a 2nd conjecture in made, predicting the manifestation of Benford's Law exactly for the very short chain of distributions with even just 2 sequences, assuming the chain uses a distribution which obeys Benford's Law for the inner-most parameter in its ultimate sequence. For example, the chain Uniform(0, Exponential Growth Series) is predicted to closely obey Benford's Law, even though it only has two sequences. This is so since Exponential Growth Series is in and of itself quantitative skewed and it obeys Benford's Law almost perfectly.

In extreme generality and concisely in symbols, the 2nd conjecture states that:

Any Distribution(Any Benford) = Benford

A related extrapolation of the 2nd conjecture states that with each new added sequence (elongating the chain), the chain evolves and becomes even skewer, as well as becoming a notch closer to the digital configuration of Benford's Law.

[15] The Random and the Deterministic Flavors in Benford's Law

Not all logarithmic data sets are created equal, but rather they come with two distinct flavors, the random flavor and the deterministic flavor. The essential distinction between the two flavors is the way digits behave throughout the entire range of the data locally, on smaller sub-intervals.

If a given data set with a range say between 3 and 2789 is nearly perfectly Benford, could we then conclude that small parts of the data are also Benford, just because they have been cut out from a whole Benford configuration? For example, is the sub-set of the data belonging to the sub-range between 3 and 1652 also Benford? Is the sub-set of the data belonging to the sub-range between 10 and 100 also Benford? Does the whole endow its Benford property to its parts?

Globally, from the minimum on the very left part of the range, all the way to the maximum on the very right part of the range, digits proportions are as in LOG(1 + 1/d) overall, as predicated by Benford's Law. Yet, for random-flavored data, local mini digit distributions on smaller sub-intervals show a remarkably consistent pattern of differentiation, as digits develop from near digital equality on the left for low values, to approximately the Benford digital configuration around the middle, and finally to extreme digital inequality on the far right for big values, where low digits overwhelm high digits, and where digit 1 typically usurps leadership by earning 40% or even more than 50% proportion in some cases. This pattern in random-flavored data is coined as "**Digital Development Pattern**".

In order to be able to observe Digital Development Pattern, it is necessary to partition the relevant section of the x-axis into sub-intervals standing between integral powers of ten, such as 0.01, 0.1, 1, 10, 100, and so forth. This partition is the most natural one in the context of Benford's Law since these points signify the beginnings and the ends of all the first digit cycles.

Figure 1.12 depicts Digital Development Pattern for the 2012 global earthquake data set containing 19,451 data points. Figure 1.13 depicts Digital Development Pattern for the 2009 US population data set containing 19,509 data points (i.e. cities and towns). These two digital development examples are good representatives for all typical real-life random-flavored data sets.

The vast majority of real-life data is of the random flavor. A tiny minority, such as deterministic exponential growth series, come with very consistent local digit distributions throughout the entire range of data, namely that of the Benford digital configuration, which is found equally on the left, in the center, and on the right, without any development or changes whatsoever.

Figure 1.14 depicts local digital configurations for exponential 5% growth applied for 244 growth periods, starting from 10.3, and ending at 986,501 on the 244th period; thus spanning the range of about 10 to 1,000,000. As can be seen in this table, the Benford configuration is consistently found locally everywhere along the entire range of data, as well as globally.

| From: | 1 | 10 | 100 | 1,000 | 10,000 |
To:	10	100	1,000	10,000	100,000
1	8.6%	11.3%	15.7%	44.0%	98.6%
2	12.5%	10.2%	14.7%	23.5%	1.4%
3	18.8%	9.8%	13.4%	14.1%	0.0%
4	8.6%	10.2%	11.4%	7.5%	0.0%
5	13.3%	11.0%	10.1%	4.9%	0.0%
6	10.2%	12.6%	9.6%	2.5%	0.0%
7	9.4%	12.1%	9.5%	1.8%	0.0%
8	7.0%	10.2%	8.5%	1.0%	0.0%
9	11.7%	12.7%	7.1%	0.6%	0.0%
Data Points	128	1250	8234	9741	72
% of Data	0.7%	6.4%	42.3%	50.1%	0.4%

Figure 1.12: Digital Development Pattern in Global Earthquake Random Data

| From: | 10 | 100 | 1,000 | 10,000 | 100,000 |
To:	100	1,000	10,000	100,000	1,000,000
1	5.3%	19.1%	37.3%	46.0%	62.9%
2	8.1%	17.4%	19.7%	20.2%	17.6%
3	7.0%	13.6%	11.6%	10.9%	6.0%
4	9.2%	11.5%	8.6%	6.4%	4.1%
5	11.5%	9.9%	6.3%	5.8%	3.0%
6	13.9%	8.8%	5.3%	3.8%	3.0%
7	13.9%	7.6%	4.3%	2.8%	1.5%
8	17.0%	6.1%	4.0%	2.3%	1.1%
9	14.1%	6.0%	2.9%	1.7%	0.7%
Data Points	1065	8202	7285	2654	267
% of Data	5.5%	42.0%	37.3%	13.6%	1.4%

Figure 1.13: Digital Development Pattern in US Population Random Data

From: To:	10 100	100 1,000	1,000 10,000	10,000 100,000	100,000 1,000,000
1	29.8%	29.8%	29.8%	31.3%	29.8%
2	17.0%	19.1%	19.1%	16.7%	17.0%
3	12.8%	10.6%	12.8%	12.5%	12.8%
4	10.6%	10.6%	8.5%	8.3%	10.6%
5	8.5%	8.5%	8.5%	8.3%	6.4%
6	6.4%	6.4%	6.4%	6.3%	6.4%
7	4.3%	6.4%	6.4%	6.3%	6.4%
8	6.4%	4.3%	4.3%	4.2%	6.4%
9	4.3%	4.3%	4.3%	6.3%	4.3%
Data Points	47	47	47	48	47
% of Data	13.9%	13.9%	13.9%	14.2%	13.9%

Figure 1.14: Steady Configuration without Development in Deterministic Exp. Growth

The coining of the terms '**deterministic**' and '**random**' is usually appropriate in most cases, but these terms should not be taken literally, because the distinction here is actually not about randomness in data versus predictable events and deterministic generation of resultant numbers, but rather about localized digital behavior within the entire range of data. The choice of these two terms is due to the fact that almost all random data come with such differentiated local digital behavior, while the particular case of deterministic and predictable exponential growth series comes with the consistent Benford behavior throughout its entire range. But these two terms would seem awkward when random data has (very rarely) the consistent Benford behavior throughout its entire range [*such as in the case of exponential growth series with variable and random growth rate, as well as in the case of k/x statistical distribution*], or when deterministic data comes with digital development. Perhaps future authors would coin the alternative terms of the 'consistent flavor' and the 'developmental flavor' in Benford's Law.

It should be noted that Digital Development Pattern is extremely prevalent in real-life data, even more so than the Benford phenomenon itself! In other words, practically all random data, Benford as well as non-Benford types [*such as those with low order of magnitude say*] clearly exhibit Digital Development Pattern throughout their ranges! There is almost no exception!

The next several chapters shed light on some essential features and properties of the Benford's Law phenomenon, leading to a thorough and clear explanation of Digital Development Pattern.

[16] Benford's Law as Uniformity of Mantissa

In the most simplistic way, mantissa could be described as '**the fractional part of the log**', although this definition is not true for numbers less than 1 having negative log values.

The formal definition of the mantissa of any positive number X is that unique solution to $X = 10^C * 10^{MANTISSA}$. Alternatively, mantissa of X is the solution to $X = 10^{C + MANTISSA}$. Here C which is called the 'characteristic' is obtained by rounding down LOG(X) to the nearest integer, namely the largest integer less than or equal to LOG(X). Equivalently, C is the first integer to the left on the log-axis, and regardless whether LOG(X) is negative or positive.

Taking log to both sides of the above mantissa-defining equation we obtain the simple result:
$$LOG(X) = LOG(10^{C + MANTISSA})$$
$$LOG(X) = C + MANTISSA$$

For X = 870, log is 2.93952, the characteristic is 2, and mantissa is 0. 93952.
$$LOG(X) = 2.93952 = 2 + 0.93952$$

For X = 0.063, log value is –1.20066, the characteristic is -2, and mantissa is 0.79934.
$$LOG(X) = -1.20066 = -2 + 0.79934$$

For $X \geq 1$, mantissa is the fractional part of LOG(X), and the characteristic C is the integral part of LOG(X). For $0 < X < 1$, mantissa is 1 minus the fractional part of the absolute value of LOG(X). More generality, mantissa of X can be viewed as the distance on the log-axis between LOG (X) and the integer immediately to the left of it.

Let us consider the following four numbers and their corresponding log values:

8.5274   log = 0.9308

85.274   log = 1.9308

852.74   log = 2.9308

8527.4   log = 3.9308

The fractional part of the logarithm [i.e. mantissa] is the same for all four numbers, namely 0.9308. Only the integral part of the logarithm changes by one integer at a time. This is so since each consecutive number is obtained by multiplying the previous number by a factor of 10 (i.e. moving the decimal point once to the right), and which does not change digital configuration in any way. The first digit is always 8 for all of these four numbers. The same is true for the second digit being 5, third digit being 2, fourth digit being 7, and fifth digit being 4.

To summarize, mantissa cycles from 0 to 1 on each adjacent integral powers of ten sub-interval, such as [1, 10), [10, 100), [100, 1000), and so forth. Mantissa repeats itself over and over again between those crucial points on the x-axis. But so do all the digits! They repeat themselves there!

In other words, digital configuration (all orders considered) and mantissa are basically two distinct ways to indicate the same concept. Clearly, mantissa has a one-to-one correspondence with digital configuration (all orders considered). Therefore, instead of stating in great details how all the digital orders are distributed, one might as well consider [more concisely] how mantissa is distributed, and as a consequence all digital orders are determined in one fell swoop!

As discussed in Kossovsky (2014) chapters 61, 62, and 63, the main result here is:

Benford's Law implies uniformity of mantissa.
Uniformity of mantissa implies Benford's Law.

Hence, converting any large Benford data set from normal numbers into a set of pure mantissa values, would yield an approximate uniform (flat and horizontal) distribution on the mantissa space of (0, 1), where probability is roughly equal for all mantissa values.

The entire area under the density curve of the mantissa must be set to 1, as in all statistical distributions, and here for mantissa the range is $(1 - 0) = 1$, thus if the data set is Benford and mantissa is uniformly distributed, then density height is a constant at 1, namely $PDF(m) = 1$, and areas [representing digital probabilities for the whatever digits] are calculated directly and simply via the reading of the widths on the M-axis. It follows that:

If a given data set or distribution obeys Benford's Law then:
Probability(M_1 < Mantissa < M_2) = M_2 − M_1

Uniformity of mantissa is often referred to as the "General Form of Benford's Law". This is so since uniformity of mantissa directly implies 1st, 2nd, 3rd, and all higher orders digit distributions, while $LOG(1 + 1/d)$ only refers to the 1st order digit distribution. In addition, the general form also implies all the probability dependencies between the digital orders.

The expression for the first digits in Benford's Law could be re-written as:

$LOG(1 + 1/d)$
$LOG(d/d + 1/d)$
$LOG((d + 1)/d)$
$LOG(d + 1) – LOG(d)$

Probability[First Digit is 1] = $LOG(2) – LOG(1)$
Probability[First Digit is 2] = $LOG(3) – LOG(2)$
Probability[First Digit is 3] = $LOG(4) – LOG(3)$

Hence, the probability set {30.1%, 17.6%, 12.5%, 9.7%, 7.9%, 6.7%, 5.8%, 5.1%, 4.6%} of the first order in Benford's Law can be thought of as differences in the values of the logarithms of the natural numbers 1, 2, 3, 4, 5, 6, 7, 8, 9, 10, namely differences in the set {log(1), log(2), log(3), log(4), log(5), log(6), log(7), log(8), log(9), log(10)}, which are the differences in the set {0.000, 0.301, 0.477, 0.602, 0.699, 0.778, 0.845, 0.903, 0.954, 1.000}. The sequence of these fractions above is also the set of the cumulative values of $LOG(1 + 1/d)$.

There are nine compartments within the [0, 1) mantissa space, where each compartment points to a unique first digit, and whose probability is directly proportion to its width on the M-axis. These nine compartments are: [0, 0.301), [0.301, 0.477), [0.477, 0.602), [0.602, 0.699), [0.699, 0.778), [0.778, 0.845), [0.845, 0.903), [0.903, 0.954), and [0.954, 1.000).

Figure 1.15 depicts the nine distinct compartments of mantissa corresponding to the nine possible first digits. The mix of the two notations 'LOG' as well as "Mantissa" for the horizontal axis is correct in the case where data happens to fall within [1, 10), so that related log of data falls within [0, 1) and in which case the two terms are interchangeable (since here the fractional part of the log is also the log itself).

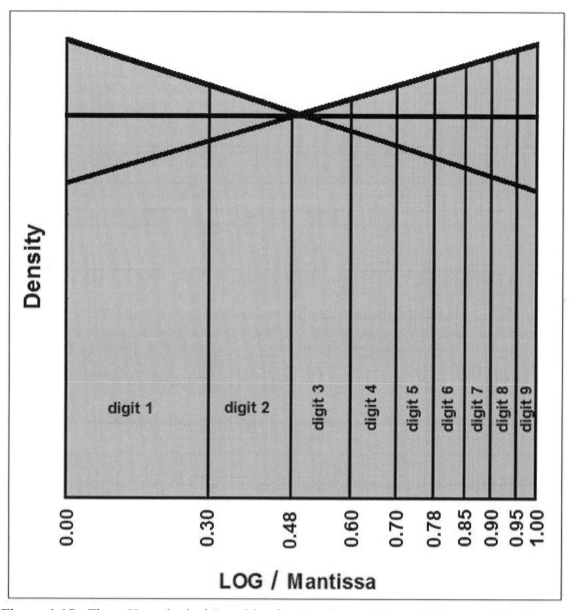

Figure 1.15: Three Hypothetical Densities for Mantissa – Rising, Uniform, and Falling

As an example, digit 2 leads whenever $2 \leq X < 3$, or $20 \leq X < 30$, or $200 \leq X < 300$, and so forth. Taking log of the three parts of the inequality implies that $LOG(2) \leq LOG(X) < LOG(3)$, or $LOG(20) \leq LOG(X) < LOG(30)$, or $LOG(200) \leq LOG(X) < LOG(300)$, and so forth.

These infinite series of inequalities could be re-written as $LOG(2) \leq LOG(X) < LOG(3)$, or $LOG(2*10) \leq LOG(X) < LOG(3*10)$, or $LOG(2*100) \leq LOG(X) < LOG(3*100)$, and so forth, and therefore they could all be condensed into a singular mantissa inequality of the form $LOG(2) \leq M < LOG(3)$ via a subtraction by the respective characteristic in each series. It follows that digits 2 occurs whenever $0.301 \leq M < 0.477$, so that mantissa related to digit 2 occurrences as the first digit resides within the second compartment of $[0.301, 0.477)$.

As it happened (i.e. Benford's Law), the probability of digit 2 occurring as the first digit is easily calculated as $(0.477 - 0.301) = 0.176$, namely the width of the compartment, it's that simple!

Figure 1.15 also depicts three possible densities: rising density, uniform density, and falling density. The case of a rising density corresponds to a more balanced and even distribution of all the nine digits. A much sharper rise in mantissa density than the one seen in Figure 1.15 could correspond approximately to digital equality. The flat and horizontal density corresponds to the Benford digital configuration. The case of a falling density corresponds to an extreme digital inequality in favor of low digits, where digit 1 typically leads by 40% or even 60% - much more so than its 30% allocation according Benford's Law.

[17] Uniqueness of K/X Distribution and Connection to Exponential Growth

Of all the many known distributions in mathematical statistics, and of all those new distributions that would be 'discovered' in the future, there is but only one unique distribution whose logarithmically-transformed values are uniformly distributed - and that distribution is k/x! Indeed, one can define k/x as such. The implication is that Monte Carlo computer simulations of say 100,000 values from the k/x distribution, all transformed into their 100,000 log equivalences, would yield nearly uniform and even distribution on the log-axis. Since the Benford configuration seeks uniformity of mantissa, and since mantissa is just the fractional part of the logarithm when $X \geq 1$, then perhaps the uniformity of the logarithm of the k/x distribution might constitute a reasonable explanation for its Benford configuration - assuming some particular ranges on the x-axis in its definition.

As discussed in details in Kossovsky (2014) chapters 60, 61, 62, 72, and 80, the k/x distribution is the only density that perfectly obeys Benford's Law for a range standing between two adjacent IPOT points, such as (1, 10), (10, 100), or (100, 1000), and so forth. For such adjacent IPOT ranges, there exists no other distribution that perfectly obeys Benford's Law (with all higher orders considered) except k/x distribution. On such particular intervals k/x is unique!

The feature that makes the k/x distribution so unique in this context is that the density of the logarithms values of k/x distribution is uniformly and evenly distributed; and this is so regardless of the range k/x is defined over, hence mantissa could also be uniformly distributed and Benford's Law perfectly obeyed - assuming properly defined ranges. If the range is restricted to two adjacent IPOT such as (1, 10) or (100, 1000) say, then the prerequisite for Benford behavior in having uniformity of mantissa is uniquely achieved via k/x distribution, since its log distribution is always uniform, and the fractional part of its uniform log (i.e. mantissa) starts at 0.0000 and ends at 0.9999 on these adjacent IPOT intervals.

More generally, the k/x distribution is the only density that perfectly obeys Benford's Law for a range (P, Q) such that log difference LOG(Q) – LOG(P) is unity. Such a range can be written as (P, 10*P), and LOG(10*P) – LOG(P) = LOG(10) + LOG(P) – LOG(P) = LOG(10) = 1.

For example, over the intervals (4.379, 43.79) or (851.23, 8512.3), k/x is that unique distribution that yields perfect Benford behavior; no other distribution on this range is Benford.

For any generic adjacent IPOT interval, namely $(10^{INTEGER}, 10^{INTEGER+1})$, log difference is $LOG(10^{INTEGER+1}) - LOG(10^{INTEGER}) = (INTEGER + 1)*LOG(10) - (INTEGER)*LOG(10) = (INTEGER + 1)*1 - (INTEGER)*1 = 1$.

It should be noted that k/x is also perfectly Benford whenever it is defined between any two points A and B such that log difference LOG(B) – LOG(A) is an integer greater than 1, such as say the interval (1.22835, 12283.5) where log difference is the integral value of 4, but k/x is not unique on such wider interval, and there are in principle infinitely many other distributions that are perfectly Benford as well.

As it happened, log values of an exponential growth series are also uniformly spread along the log-axis, albeit in a discrete fashion, and where distances between log values are constant. Let us prove the above assertion using the following notations for exponential progression: B for the base value/quantity at time 0, and F as the factor of growth, such as F = 1.05 for 5% yearly growth. The logarithm of the growing quantity is taken at the beginning of each year or period of growth as follows:

At the 1st year: LOG(B)
At the 2nd year: LOG(B*F)
At the 3rd year: LOG(B*F*F)
At the 4th year: LOG(B*F*F*F)
At the 5th year: LOG(B*F*F*F*F)

At the 1st year: LOG(B)
At the 2nd year: LOG(B) + LOG(F)
At the 3rd year: LOG(B) + LOG(F) + LOG(F)
At the 4th year: LOG(B) + LOG(F) + LOG(F) + LOG(F)
At the 5th year: LOG(B) + LOG(F) + LOG(F) + LOG(F) + LOG(F)

Clearly distances between log values here are separated by a fixed value, namely by LOG(F), and this implies that the discrete distribution or concentration of log of exponential growth is uniformly spread throughout the log-axis. Hence the (surprising) intimate connection between k/x distribution and exponential growth series! The continuous distribution of the logarithm of k/x, as well as the discrete distribution of the logarithm of exponential growth, are both uniformly and evenly spread throughout the log-axis.

The implication of the uniformity of the logarithm density for k/x and exponential growth series to Digital Development Pattern is straightforward and decisive! Since log is uniform throughout the entire range of k/x and exponential growth, it follows that it is also uniform internally on all adjacent integral powers of ten sub-interval, such as [1, 10), [10, 100), [100, 1000), and so forth. This in turns implies that the Benford configuration is found everywhere along the x-axis without any differentiation or development, namely the total absence of Digital Development Pattern!

This finding clearly explains the steady Benford condition found throughout the entire range of exponential 5% growth for 244 growth periods of Figure 1.14. In addition, the Benford configuration in the case of the k/x distribution is also steady and consistent throughout its entire range. The fact that k/x does not come with a development pattern renders it very rare and totally irrelevant to almost all real-life random data. 'Paradoxically', k/x is the only distribution that can perfectly obey Benford's Law (with all higher digital orders considered) for a range standing between two adjacent IPOT points, such as (10, 100) for example, and that renders the case of k/x distribution quite exceptional in the field of Benford's Law, yet, k/x is totally irrelevant to logarithmic real-life random data sets which are practically always accompanied with Digital Development Pattern!

Such is the seductive power of k/x distribution in the context of Benford's Law that some misguided authors and overly enthusiastic students of Benford's Law start their article by basing it on some assumption or feature regarding the k/x distribution and then proceed to draw far reaching conclusions, mistakenly extrapolating the odd case of k/x to all real-life random data. Such regrettable trend has led to several erroneous conclusions, published in respectable journals, and officially certified by expert mathematicians as true. This author has taken on the dissenting role of an agitator as well as a prophet of doom, preaching the virtue of separating the random from the deterministic and of becoming aware of this crucial distinction in the field, and predicting the encountering of contradictions between the empirical and the theoretical in all such misguided pseudo-mathematical endeavors.

[18] Related Log Conjecture

The most obvious and straightforward way for a distribution to obey Benford's Law is to have its log density itself uniformly distributed on some properly defined range, so that mantissa in turn could also be uniformly distributed. For example, for k/x distribution defined over (1, 1000), its logarithm is uniformly distributed over (0, 3), and as a consequence mantissa is also uniform, being simply the aggregation of 3 distinct uniformly distributed portions over (0, 1), rendering overall mantissa uniformly distributed over (0, 1). It is important to note the cyclical manner that logarithm generates its mantissa. As x marches forward on the x-axis from 1 all the way to 1000, its log values - namely LOG(x) - increases smoothly from 0, to 1, to 2, and then to 3. Its mantissa on the other hand, starts at 0, and then gradually approaches 1, then abruptly returns to 0 again, then gradually to 1 again, then abruptly returns to 0 again, and finally approaches 1 gradually once again on its final lag of this emotional rollercoaster long journey.

But this is not the only way to obtain an approximate or nearly perfect uniformity of mantissa. The alternative to a uniform log density is an upside-down U curve log density where the rising part on the left roughly or exactly offsets the falling part on the right, leaving the middle part of approximately flat density as the best representative of the entire curve.

Figure 1.16 depicts a hypothetical histogram curve of related logarithm of a particular random data set. It starts from the bottom on the log-axis itself and ends all the way down on the log-axis as well. On the very left, either high digits are winning slightly or digital equality prevails in the approximate. Further on towards the center, low digits win slightly over high digits. Around the center itself, the Benford configuration is found locally, where the graph is approximately flat, uniform, and horizontal. Finally, at the extreme right side, low digits strongly dominate, and much more so than in the usual Benford configuration. All this suggests some grand trade-off, cancellation, and offsetting effects, between the left and the right regions, and where the Benfordian center can be regarded as the best representative of the entire histogram. Indeed, Related Log Conjecture of chapters 63, 64, 68, 69, and 70 in Kossovsky (2014) applies exactly such an argument and states that if the distribution come with high order of magnitude so that the range of related log histogram is over 3 or 4 log units, and especially if it's over 5 log units say, then the data itself is very nearly Benford.

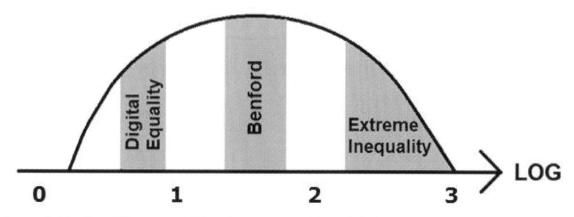

Figure 1.16: Log Histogram of Random Data and its Digital Configuration Throughout

[19] The Mathematical Basis of Digital Development Pattern

A given data set containing only positive numbers can be represented by the logarithms of its values. This is typically done via the decimal base 10 logarithm. This is so since there is a one-to-one correspondence between any positive number and its logarithm. Two distinct numbers point to two distinct logarithms. And two distinct logarithms point to two distinct numbers.

For example, the data set {10, 18, 79, 100, 200, 300, 450, 1004} can be represented by the data set of its related log values, namely by {1.000, 1.255, 1.898, 2.000, 2.301, 2.477, 2.653, 3.002}.

Hence the histogram of a given data set can be substituted by the histogram of its related logarithm values. Often many numerical secrets and essential inner features of the data are revealed by glancing at its log histogram, instead of merely staring at the original numerical histogram. In addition, log histograms often provide for much better and more concise visualization of the data.

With the use of the log histogram, the distinction between the random flavor and the deterministic flavor in Benford's Law can now be stated more precisely and more succinctly, and without confusing the terms with predictability versus randomness.

The **random flavor** is one where related log histogram curves around, starting from [or near] the log-axis itself, rising, then falling, and finally terminating at [or near] the log-axis again further to the right. The rising portion of the log histogram on the left side yields an approximate digital equality locally, and the falling portion of the log histogram on the right side yields extreme digital skewness locally in favor of low digits such as 1 and 2, much more so than in the Benford configuration, hence the empirically observed differentiated digital behavior locally for all random data sets, namely Digital Development Pattern.

The **deterministic flavor** is one where related log histogram is flat and uniform. A data set which consistently obeys Benford's Law throughout all its parts and sub-intervals, namely data with the absence of Digital Development Pattern, is characterized by having a flat and horizontal histogram for its logarithmically-transformed data points. This is the case of exponential growth series and k/x distribution.

The generic shape of the chart in Figure 1.16 is not some invented fantasy concocted in the excited mind of the imaginative and eccentric statistician, but rather an extremely typical log curve found in almost all random data sets, in financial and accounting data, in governmental census data, and in scientific and physical data handmade by Mother Nature herself who strongly favors such aesthetic and round logarithmic curves, and almost always dresses herself up with them. She finds straight and linear dresses (of the log of k/x types) to be totally uninspiring, unattractive, and too boring to wear. The nature of random and statistical data (Benford or non-Benford) in extreme generality is characterized by log-gradualism; that related log very rarely starts abruptly high with an initial value; that it very rarely ends abruptly high with a final value; and that it shows a marked curvature which first rises progressively to a certain plateau and then falls gradually down from there; just as humans are first born small and weak, then gradually grow bigger and stronger, achieving their zenith at the age of about 30 or 40, only to grow old and become weak again, even to shrink a bit, and then die.

Indeed, empirical examinations reveal that related log histograms of the 2012 global earthquake data set and of the 2009 US population census data appear highly curved, and not much different from the generic curve of Figure 1.16 in overall appearance. In sharp contrast, empirical examination of related log histogram of exponential 5% growth reveals indeed that it appears nearly flat and uniform, as theoretically expected.

[20] A Critique on the Mixture of Distribution Model in Benford's Law

The two discoverers of the law also attempted to give explanations for this digital phenomenon, one involving random divisions of two Uniform distributions, and the other involving aggregating particular data structures. Apparently, the mathematical community chose to ignore or reject these two explanations as irrelevant models, not corresponding to the structure of typical real-life data; all the while greatly appreciating the digital discovery itself as relevant and certainly worthy of study. Subsequently, an explanation assuming the existence of some scale invariant digital law was also rejected on the ground that such an assumption cannot be proven. Another attempt at an explanation is the mathematical demonstration that a large mixture of distributions, each defined over the positive x-axis, obeys Benford's Law in the limit as the number of distributions goes to infinity. This is also called 'the distribution of all distributions'.

First, a large collection of random distributions is assembled randomly, such as Uniform(0, 5), Normal(17, 4), Uniform(3, 16), Exponential(0.02), Lognormal(8, 1), Lognormal(5, 6), Exponential(1.9), and so forth. Then a repetitive process generating random numbers in stages starts in earnest. In each stage, one particular distribution is randomly selected [as if uniformly and evenly distributed], and then one random realization from that selected distribution is simulated and collected. This process continues until numerous such random values have been collected.

Translating this abstract model into real-life data collection reveals the irrelevance of the whole idea. Unfortunately the model can only be applied to data blindly and randomly collected from a large variety of sources, and only when very few numbers are picked from each source. The sources should consist of positive values exclusively. Strictly speaking the process should pick only one number from a given source, then another number from another (related or totally unrelated) source, and so forth. The end result of this whole process is a large mixture of unrelated numbers, representing a meaningless data set not conveying any specific message or information, yet digitally structured as Benford. The model cannot be a valid explanation though for the almost perfectly Benford data sets regarding single-issue (*single-source, non-mixed*) physical and scientific phenomena, such as the time between earthquake occurrences, river flow, population count, pulsar rotation frequency, half-life of radioactive material, and so on. The mixture of distributions model does not show any immediate or obvious relation to the logarithmic way Mother Nature generates her physical quantities. It is very hard or rather impossible to argue that river flow, earthquake timing, or pulsar rotation, are the results of some aggregation of numerous invisible, mysterious, and unrelated, mini distributions.

On the other hand, the explanation of the phenomenon provided by the model regarding multiplications processes is not only plausible in general, but it is undoubtedly the proper model in many real-life cases. Isaac Newton's multiplicative expression of $F = M*A$, among many other such expressions for gravitational and electromagnetic forces, including numerous applications in physics, chemistry, other sciences, and engineering involving multiplicands, all remind us of the direct connection between the multiplicative model and real-life physical data. Quantitative partition models as the explanation of the Benford's Law phenomenon in many real-life cases is also certainly correct and plausible. This is especially so in light of the occasional interpretations of 'partition' as 'composition' or 'consolidation', and which endows partition model by far greater scope of applicability. In the same vein, data aggregation and the related chain of distributions model are surely the obvious and plausible models for numerous real-life cases.

In order to demonstrate very clearly the lack of statistical meaning for data derived in the spirit of the mixture of distributions model, a concrete example is given, with only the first 7 numbers of that 'infinite' or sufficiently large collection of values shown in Figure 1.17:

MEASURE	VALUE	UNIT
Time between 2 earthquakes - 1/27/2013 - 7:05 & 7:37	31.9	Second
Depth below the ground - Earthquake -12/30/2013 - 23:87	5.3	Kilometer
Rotation frequency of pulsar IGR J17498-2921	401.8	hertz
Amazon river length	6,437	Kilometer
New York City Population 2013 Census	8,337,342	People
Temperature of the star Polaris	6,015	Kelvin
Mass of exoplanet Tau Ceti f Constellation Cetus	0.783	Solar mass

Figure 1.17: Mixture of Real-Life Data Sets Mimicking the Mixture of Distributions Model

The set of pure values with only the first 7 numbers shown is then:

$$\{31.9, \quad 5.3, \quad 401.8, \quad 6437, \quad 8337342, \quad 6015, \quad 0.783, \quad ... \quad \}$$

But surely this set does not represent seconds, hertz, solar mass, or any other quantity, nor does it convey any specific data-related message. This set does not represent any particular physical entity or physical concept, nor does it stand for any single scale. Mathematically though, this set is demonstrated to be perfectly logarithmic in the limit, yet, it explains nothing but itself.

This author has spent the better part of a week in February 2012 randomly gathering numbers from all sorts of websites, in the first ever serious attempt at empirical verification of this model. In total 34,269 positive numbers were obtained. The results confirmed the theoretical expectation, with first digit distribution coming at: {28.8, 16.4, 12.4, 9.8, 8.3, 7.3, 6.1, 5.7, 5.3}, and with SSD of the low value of 4.6, indicating strong compliance with Benford's Law. Details about this exhausting project can be found in Kossovsky (2014) chapter 110.

A critique on this empirical study might claim that the result thus obtained actually confirms the principle behind data aggregations (such as house number in address data), as opposed to confirming the validity of the mixture of distributions model. The essential distinction between data aggregations and mixture of distributions is the point where each data set or each distribution starts. For data aggregations the start is at 0 or 1, or perhaps at another very low value. For mixture of distributions the starts could be at any value, big or small, and without any restriction. It is possible to argue that most of the relevant websites contain values that typically start at very low values that are not much higher than 0 or 1, and that therefore the empirical study pertains to the model of data aggregations, rather than to the model of mixture of distributions.

SECTION 2

ARITHMETICAL TUGS OF WAR

[1] Multiplication Processes Lead to Skewness and Higher POM

Almost all random and deterministic multiplication processes induce a dramatic increase in skewness where the small becomes relatively numerous and the big becomes relatively rare. Surely, the single product of the multiplication of only two numbers is not to be considered here, since the resultant single product is neither small nor big, but rather it's just itself, and there are no competing sizes here to consider. Instead, a set of N numbers called A is to be multiplied by another set of M numbers called B. The phrase 'multiplication of two sets of numbers' implies that each and every number in set A is to be multiplied by each and every number of set B, producing N*M products. In other words, all possible multiplications between the two sets are attempted, and the entirety of these products constitutes the newly created set of numbers, called A*B. For example:

Set A = {8, 3, 5}
Set B = {11, 47, 26}
A*B = {(8)*(11), (8)*(47), (8)*(26), (3)*(11), (3)*(47), (3)*(26), (5)*(11), (5)*(47), (5)*(26)}
A*B = {88, 376, 208, 33, 141, 78, 55, 235, 130}

Let us examine the quantities within the 10 by 10 multiplication table that we all were forced to memorize in our elementary school years. In this example, the intrinsic characteristics of multiplication processes with regards to resultant sizes shall be explored and compared. Such an analysis is done at the most primitive and basic level, at the arithmetic and quantitative level, before going on to the rigorous mathematical level. The quest is to start out with this very particular example, and then to lend the conclusions derived from this case universality and applicability in almost all multiplication processes.

For the 10 by 10 multiplication table, the entire range from 1 to 100 is to be partitioned into 10 quantitative sections of equal 10-unit width each, namely [1, 10], [11, 20], [21, 30], [31, 40], [41, 50], [51, 60], [61, 70], [71, 80], [81, 90], [91, 100], and then a count is made of the values falling within each section. The goal is to group the 100 products of the multiplication table according to sizes. Figure 2.1 depicts this quantitative partitioning arrangement of the entire multiplicative territory by size. Figure 2.2 depicts the histogram of the values falling within each section. Surprisingly, a decisive trend regarding the occurrences of products within the sections is found here. The section of the smallest quantities (1 to 10) has 27 values falling within it; while the section of the biggest quantities (91 to 100) has only 1 value falling within it.

The sequence of all the values falling within these 10 sections, and presented in order according to sizes is {27, 19, 15, 11, 9, 6, 5, 4, 3, 1}. Clearly, the sections pertaining to bigger quantities have less values falling within them, as the count of values monotonically decreases. The small is definitely more numerous than the big in our standard multiplication table. The crucial lesson learnt from this multiplication process is that surely this tendency is nearly universal, and that it should be present in almost all other multiplication processes, and not only for our particular 10 by 10 multiplication table. There is nothing unique about our standard multiplication table or the set of numbers from 1 to 10, and so this result is extrapolated to almost all other multiplication processes.

X	1	2	3	4	5	6	7	8	9	10
1	1	2	3	4	5	6	7	8	9	10
2	2	4	6	8	10	12	14	16	18	20
3	3	6	9	12	15	18	21	24	27	30
4	4	8	12	16	20	24	28	32	36	40
5	5	10	15	20	25	30	35	40	45	50
6	6	12	18	24	30	36	42	48	54	60
7	7	14	21	28	35	42	49	56	63	70
8	8	16	24	32	40	48	56	64	72	80
9	9	18	27	36	45	54	63	72	81	90
10	10	20	30	40	50	60	70	80	90	100

Figure 2.1: Quantitative Territorial Partitioning of Multiplication Table

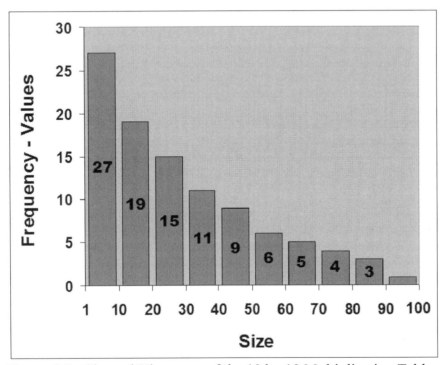

Figure 2.2: Skewed Histogram of the 10 by 10 Multiplication Table

The results of Figures 2.1 and 2.2 can also be interpreted as a particular casino game where two dice having 10 faces each are thrown, and the values of the two faces are multiplied by each other. The value of this product is then declared to be the focus of the gambling game, and an arbitrary benchmark value of say 30 is set determining losses and wins. Players are said to win if the product comes out in the range of 31 – 100 of bigger values, and the casino is said to win if the product comes out in the range of 1 – 30 of smaller values. Such cleaver setup on the part of the casino owner where the small is assigned to the casino and the big is assigned to the gamblers yields steady and enormous profits for the casino. Our 10 by 10 multiplication table can also be interpreted in a more formal sense in the context of mathematical statistics as the random multiplicative process of two **discrete** random Uniform Distributions, namely the product: Uniform{1, 2, 3, 4, 5, 6, 7, 8, 9, 10}*Uniform{1, 2, 3, 4, 5, 6, 7, 8, 9, 10}, instead of regarding it just as a useful tool of the deterministic table of multiplication. More generally, it could also be thought of as the random multiplicative process of two **continuous** random Uniform Distributions, namely the product Uniform[1, 11)*Uniform[1, 11).

It should be noted that for the above result about the 10 by 10 multiplication table, the starting point is the uniform and even distribution of the original numbers about to be multiplied, namely {1, 2, 3, 4, 5, 6, 7, 8, 9, 10}, where neither small nor big is more numerous than the other. Yet, from such even and uniform distribution we arrived (via multiplications) at a decisively skewed and uneven distribution where the small is more numerous than the big.

Exception to this rule is found only in multiplications of rare sets of numbers with almost no variability, namely data sets having extremely low order of magnitude. This case shall be discussed in chapter 5 of this section.

In mathematical statistics, the random product of numerous independent realizations from an identical random variable is known to be the Lognormal Distribution, in the limit, as the number of these multiplied realizations becomes large enough. This seminal result which is derived from the Multiplicative Central Limit Theorem (MCLT) has very lax requirements and only a few restrictions, ensuring broad applicability for almost all types of multiplied variables, namely for almost all multiplication processes. Most significantly, the restriction on having an identical distribution can often be relaxed assuming other easily obtained conditions, thus MCLT can usually be applied to the random product of several distinct random variables. Since the Lognormal Distribution is almost always skewed [that is, for all shape parameters above 0.5 roughly], MCLT guarantees that almost all multiplication processes are skewed.

The scope of applications here is truly enormous, encompassing almost all disciplines. The customary multiplicative form of the vast majority of the equations in physics, chemistry, astronomy, economics, biology, engineering and other disciplines, as well as the numerous expressions relating to their applications and specific results, leads to the manifestation of quantitative skewness and often also to the Benford phenomenon in the natural sciences and in real-life data. These data sets are almost always with high enough variability (i.e. high order of magnitude) and thus they almost never suffer from the above-mentioned exception to the rule.

As discussed in Kossovsky (2014) chapter 90, Isaac Newton gave us F = M*A, not F = M + A. Hence the derivations and results due to the expression Force = Mass*Acceleration are related to multiplication processes, not to addition processes. Newton also gave us the law of universal gravitation $F_G = G*M_1*M_2 / R^2$ which is written in multiplicative and divisional forms, and not in addition and subtraction style such as say $F_G = G + M_1 + M_2 - R^2$.

Resultant quantitative skewness in multiplication processes often leads to digital skewness and Benford as well, hence strengthening the plausible explanation of Benford's Law in terms of multiplication processes and their intrinsic logarithmic properties.

For our 10 by 10 multiplication table from our childhood school days, order of magnitude doubles from 1 to 2 as we multiply; namely, from the range of [1, 10] we arrive at the range of [1, 100], so that variability dramatically increases due to the act of multiplying.
POM of [1, 10] is 10/1, and POM of [1, 100] is 100/1, thus there is a ten-fold increase in POM.
OOM of [1, 10] is LOG(10/1) = 1, and OOM of [1, 100] is LOG(100/1) = 2, thus OOM has been doubled due to the act of multiplying.

In summarizing the generic effects of random multiplication processes on quantities, it can be said that **multiplication processes favor the small over the big - leading to skewed data**. In addition, when the focus is on digital configurations in the first position, it can be said that multiplication processes favor small digits such as {1, 2, 3} over big digits such as {7, 8, 9}.

Random multiplication processes induce two essential results:

(A) A dramatic increase in skewness – an essential criterion for Benford behavior.
(B) An increase in the order of magnitude – another essential criterion for Benford behavior.

[2] Addition Processes Lead to the Symmetrical Normal and Equal POM

Let us demonstrate the sharp contrast between additions and multiplications in terms of resultant quantitative configuration and resultant order of magnitude. This shall be accomplished by converting the 10 by 10 multiplication table into an addition table.

We may view the table as a tool for those too lazy to do additions quickly in their heads, just as the multiplication table is used and memorized as a tool to be used later in life. Or, we may view each discrete set of {1, 2, 3, 4, 5, 6, 7, 8, 9, 10} as a random variable with uniform and equal probability for each integer, and the table as an addition process. Since additions are applied instead of multiplications, such a random vista about this (essentially deterministic) addition table allows us perhaps to extrapolate the results to all random addition processes.

The entire range of [2, 20] is partitioned into six equitable quantitative sections of 3-unit width each: {2, 3, 4}, {5, 6, 7}, {8, 9, 10}, {11, 12, 13}, {14, 15, 16}, {17, 18, 19} *[with the value of 20 conveniently excluded]*. Then, a count is made of the numbers falling within each section - namely grouping them according to quantities.

Figure 2.3 demonstrates such quantitative partitioning of the entire additive territory in details. Figure 2.4 depicts the histogram of the number of values falling within each section.

+	1	2	3	4	5	6	7	8	9	10
1	2	3	4	5	6	7	8	9	10	11
2	3	4	5	6	7	8	9	10	11	12
3	4	5	6	7	8	9	10	11	12	13
4	5	6	7	8	9	10	11	12	13	14
5	6	7	8	9	10	11	12	13	14	15
6	7	8	9	10	11	12	13	14	15	16
7	8	9	10	11	12	13	14	15	16	17
8	9	10	11	12	13	14	15	16	17	18
9	10	11	12	13	14	15	16	17	18	19
10	11	12	13	14	15	16	17	18	19	20

Figure 2.3: Quantitative Territorial Partitioning of the Addition Table

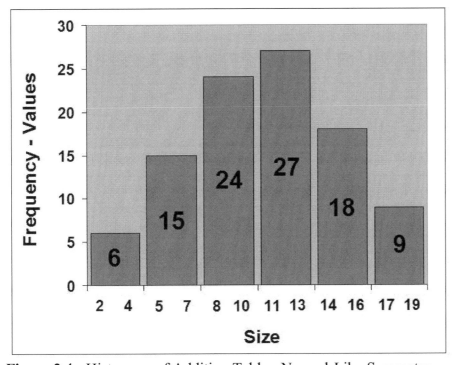

Figure 2.4: Histogram of Addition Table - Normal-Like Symmetry

The histogram is not skewed, and quantitative proportions are concentrated mostly around the middle part of the range, not around the left part (of small values), and not around the right part (of big values). This is not a coincidence, but rather it signifies a very persistent trend found in all addition processes. There is no hope that repeated additions of random variables would eventually attain skewness assuming large number of applied additions, because any such aspiration on the part of addition is frustrated by the Central Limit Theorem (CLT) which points to the symmetric Normal Distribution as the eventual distribution developing after numerous such additions of random variables. A strong hint of this eventuality can actually be seen in Figure 2.4 as it nicely curves around the center and falls off almost evenly on both edges, beginning to resemble the Normal Distribution.

Here for this 10 by 10 addition table, order of magnitude stubbornly refuses to increase. The maximum value is 20, and the minimum value is 2, hence resultant post-addition POM value for the entire table is $(20)/(2) = 10$. This value is exactly the same as the $(10)/(1) = 10$ POM value of the original variable being added, namely of the discrete set of $\{1, 2, 3, 4, 5, 6, 7, 8, 9, 10\}$.

Clearly, when the focus is on quantitative configurations, it can be said in extreme generality that **addition processes favor the medium over the small and over the big**. When the focus is on digital configurations in the first position, nothing in general can be said a priori about addition processes, and digital configuration depends on the specifics of the added variables and especially on defined ranges of the added variables. Addition processes do not favor a priori middle digits such as $\{4, 5, 6\}$ except in some very particular cases.

Random addition processes do not induce any results that are essential in the criterion for Benford behavior:

(A) Lacking any increase in skewness, and even actively increasing the symmetry of resultant distribution, with added concentration forming around the center/medium.

(B) Lacking any increase in order of magnitude beyond the existing maximum order of magnitude within the set of added variables.

Addition processes are associated with mild and slow quantitative increases. For example, if we walk along the diagonal of the 10 by 10 addition table of Figure 2.3 *[from top-left of 1+1 to bottom-right of 10+10]*, the sequence 1+1, 2+2, 3+3, 4+4, 5+5, 6+6, 7+7, 8+8, 9+9, 10+10, increases steadily by 2 units at a time without any acceleration. Figure 2.5 depicts the steady and even march forward of the added ten numbers 2, 4, 6, 8, 10, 12, 14, 16, 18, 20, which are evenly spread along the horizontal axis.

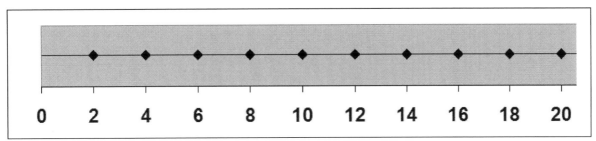

Figure 2.5: Even and Balanced Spread of the Added Numbers **X + X**, for X ϵ {1, 2, 3, ... 10}

In sharp contrast, multiplication processes are associated with strong and rapid quantitative increases. Numbers under 'multiplicative influences' almost always undergo what could be called '**quantitative acceleration**'.

For example, if we walk along the diagonal of the squares in the 10 by 10 multiplication table of Figure 2.1 *[from top-left of 1*1 to bottom-right of 10*10]*, we experience a rapid rise of quantities, from 1*1, to 2*2, to 3*3, to 4*4, to 5*5, to 6*6, to 7*7, to 8*8, to 9*9, and to 10*10, pointing to the sequence of products of 1, 4, 9, 16, 25, 36, 49, 64, 81, 100. Figure 2.6 depicts the accelerated march forward of the square numbers 1 to 100. Walking along the diagonal of the squares from say **5*5** to **6*6**, means that now we are adding **6+6+6+6+6+6** instead of **5+5+5+5+5**, and which represents a quantitative jump in a sense, because not only an extra term is added, but also because each term is higher by one notch. The bigger the number, the more effective multiplication becomes in moving the product forward. In other words, the bigger the multiplicands, the (much) greater are the products.

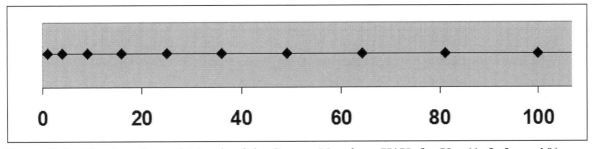

Figure 2.6: The Accelerated March of the Square Numbers **X*X**, for X ϵ {1, 2, 3, ... 10}

[3] Arithmetically-Mixed Real-Life Variables

The 10 by 10 multiplication table involves purely multiplicative processes. The 10 by 10 addition table of the previous chapter involves a purely additive process. These are examples of exclusively multiplication processes or that of exclusively addition processes. Often in real-life data however, multiplication and addition processes mix together within one measurement, and consequently they fiercely compete for dominance, each attempting to exert the greatest influence upon sizes and digits. While addition favors the medium, multiplication prefers the small, and none is willing to compromise.

As an example relevant to real-life accounting and financial data, the single bill for one typical shopper in a large supermarket or at a big retail store may read as follows:

3*($**2.75** *bread*) + **5***($**2.50** *tuna*) + **2***($**7.99** *cheese*) = ($**36.73** *total bill*)

Namely, buying 3 loafs of hearty brown whole wheat Russian bread; 5 cans of white Albacore tuna in olive oil; and 2 packages of low-fat and preservative-free Mozzarella cheese. Here multiplications mix in with additions, and as a consequence these two competing forces pull in opposite directions. One multiplicand within the expression of a typical bill is the catalog of a retail store or the price list of all the items on sale at a large supermarket. The other multiplicand is the quantity purchased of each item (such as the numbers 3, 5, 2 in the above bill of $36.73). The bills of thousands such potential buyers at the supermarket constitute a random variable derived from a mixture of addition and multiplication processes, representing revenue data for the supermarket and expense or cost data for the shopper. Surely, the expectation here is that the quantitative configuration of the catalog or the price list itself could greatly influence relative quantities and sizes of the bills.

As another example relevant to real-life scientific and chemical data, the weight of a complex chemical molecule is derived from the linear combination of its constituent atoms.

For example, Lactose $C_{12}H_{22}O_{11}$ has the molar mass of 342.29648 g/mol. This particular molecular weight is derived from the combinations:

12*(Carbon Mass) + **22***(Hydrogen Mass) + **11***(Oxygen Mass) = Lactose Mass

Here one multiplicand is the atomic weight in the Periodic Table, and the other multiplicand is the number of atoms for each element within the molecule.

In all of these examples, additions and multiplications mix and combine to yield the final value, and each exerts its distinct pull and influence. What should be expected in much arithmetical mixtures in terms of (I) quantitative skewness versus symmetry, and in terms of (II) digital configuration? Who wins and who loses in such tugs of war? Does it simply depend on the relative number of additions versus the relative number of multiplications within the arithmetical expression? The answers to these questions shall become clear in the next several chapters.

[4] Arithmetically-Complex Random Processes

Let us examine the quantitative configuration of 4 arithmetically-complex dice games.

We view each standard 6-sided die as a discrete set of $\{1, 2, 3, 4, 5, 6\}$; consider it as a random variable with uniform and equal probability for each integer; and call it SIX.

Each of the four dice games shown below involves using 10 regular 6-sided dice combinations, as follows:

Game 1: $(SIX_1 * SIX_2 * SIX_3) + (SIX_4 * SIX_5 * SIX_6) + (SIX_7 * SIX_8 * SIX_9) + (SIX_{10})$

Game 2: $(SIX_1 * SIX_2 * SIX_3) + (SIX_4 * SIX_5 * SIX_6) + (SIX_7 * SIX_8) + (SIX_9 * SIX_{10})$

Game 3: $(SIX_1 * SIX_2 * SIX_3) + SIX_4 + SIX_5 + SIX_6 + SIX_7 + SIX_8 + SIX_9 + SIX_{10}$

Game 4: $(SIX_1 * SIX_2) + (SIX_3 * SIX_4) + (SIX_5 * SIX_6) + (SIX_7 * SIX_8) + (SIX_9 * SIX_{10})$

Here each die has an identity and a unique name. They are designated SIX_1, SIX_2, SIX_3, and so forth up to SIX_{10}. The ten dice are thrown simultaneously and independently of each other, and then the arithmetically-complex operations are calculated on the 10 values appearing on the 10 faces of the 10 dice.

Figures 2.7, 2.8, 2.9, and 2.10 depict the histograms for these four games, using 10,000 Monte Carlo computer simulation runs per game.

For games 1, 2, and 3, of Figures 2.7, 2.8, and 2.9, the histogram appears as a hybrid of sorts, standing somewhere between the symmetric Normal histogram of additions, and the generally skewed histogram of multiplications. Perhaps they appear more skewed than symmetrical. This is expected, since these processes are truly mixtures of addition and multiplication operations.

For game 4 of Figure 2.10, the histogram is quite symmetrical, almost mimicking the Normal Distribution. Perhaps this is expected, and almost predicated by the Central Limit Theorem, since the process might be thought of primarily as that of additions. The term (SIX*SIX) is being added over and over again five times. Certainly the term (SIX*SIX) can be thought of as a singular random distribution in its own right, and thus being the addend in this process. The consideration of (SIX*SIX) as a whole entity leads to the simpler vista of having five random variables being added, and this eliminates the appearance of multiplication altogether from the process. In other words, such a vista facilitates the consideration of the final value in this game as derived from a purely addition process. A similar vista though could also be presented for games 1, 2, and 3, but since in those games distinct variables are being added, the Central Limit Theorem is slow to act, and therefore results are not as Normal-like.

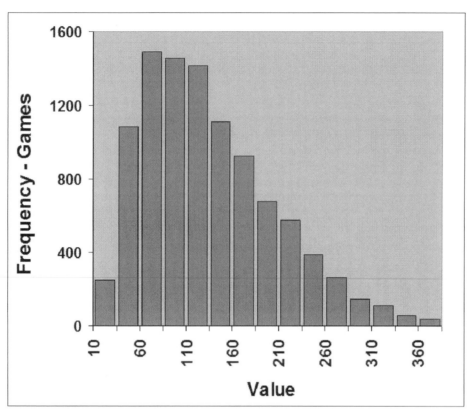

Figure 2.7: $(SIX_1*SIX_2*SIX_3) + (SIX_4*SIX_5*SIX_6) + (SIX_7*SIX_8*SIX_9) + (SIX_{10})$

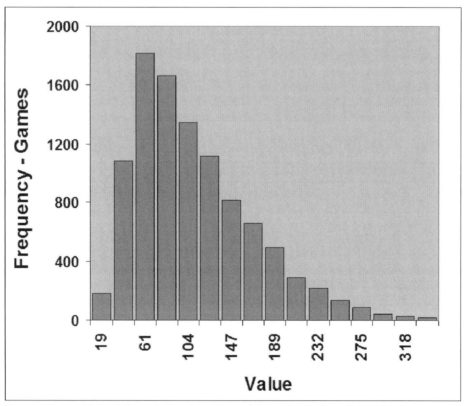

Figure 2.8: $(SIX_1*SIX_2*SIX_3) + (SIX_4*SIX_5*SIX_6) + (SIX_7*SIX_8) + (SIX_9*SIX_{10})$

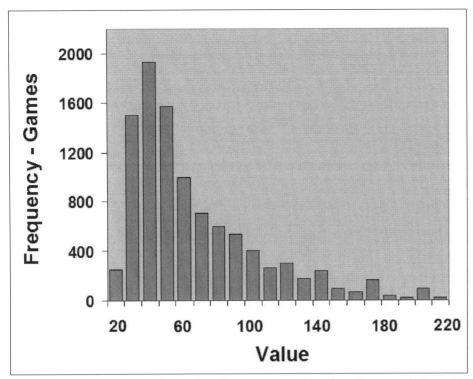

Figure 2.9: $(SIX_1*SIX_2*SIX_3) + SIX_4 + SIX_5 + SIX_6 + SIX_7 + SIX_8 + SIX_9 + SIX_{10}$

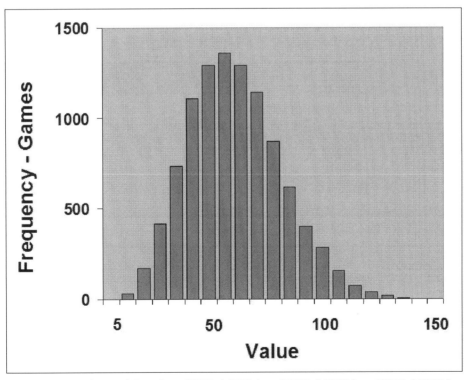

Figure 2.10: $(SIX_1*SIX_2) + (SIX_3*SIX_4) + (SIX_5*SIX_6) + (SIX_7*SIX_8) + (SIX_9*SIX_{10})$

[5] Multiplications of Variables with Low Order of Magnitude

In order to enable a thorough analysis of the entire discourse regarding mixed arithmetical processes and their quantitative and digital properties, it is necessary to digress momentarily and explore random multiplication processes involving variables with low order of magnitude.

Let us examine quantities within the 70 to 79 multiplication table. This table is a variation on the standard 10 by 10 table, using the set of integers {70, 71, 72, 73, 74, 75, 76, 77, 78, 79} instead of {1, 2, 3, 4, 5, 6, 7, 8, 9, 10}. Figure 2.11 depicts the arrangement.

The motivation for choosing this set of integers is to examine skewness for multiplications of variables with very low physical order of magnitude. For this set of integers, POM (70 to 79) = Maximum/Minimum = (79)/(70) = 1.13. In contrast, for the standard 10 by 10 multiplication table, POM (1 to 10) = Maximum/Minimum = (10)/(1) = 10.0. Clearly, POM value for the 70 to 79 set of integers is extremely low, hovering just slightly above the lowest possible 1.0 value of POM for data without any variability whatsoever.

x	70	71	72	73	74	75	76	77	78	79
70	4900	4970	5040	5110	5180	5250	5320	5390	5460	5530
71	4970	5041	5112	5183	5254	5325	5396	5467	5538	5609
72	5040	5112	5184	5256	5328	5400	5472	5544	5616	5688
73	5110	5183	5256	5329	5402	5475	5548	5621	5694	5767
74	5180	5254	5328	5402	5476	5550	5624	5698	5772	5846
75	5250	5325	5400	5475	5550	5625	5700	5775	5850	5925
76	5320	5396	5472	5548	5624	5700	5776	5852	5928	6004
77	5390	5467	5544	5621	5698	5775	5852	5929	6006	6083
78	5460	5538	5616	5694	5772	5850	5928	6006	6084	6162
79	5530	5609	5688	5767	5846	5925	6004	6083	6162	6241

Figure 2.11: Quantitative Territorial Partitioning of 70 to 79 Multiplication

The lowest product in the 70 to 79 table is 70*70 = 4900, and the highest is 79*79 = 6241. The intention is to partition the entire range from **4900** to **6241** into 10 quantitative sections of equal width and compare density within each section (i.e. count of data points falling in).

In order to obtain a smooth and integral partition not involving fractional values as border points, this range shall be widen a tiny bit to span **4900** to **6300**, and therefore the enlarged range is of (6300 – 4900) = 1400 units. Dividing 1400 into 10 equitable sections implies having the width of 140 units for each bin in the histogram.

Figure 2.12 depicts the histogram of the values falling within each section.

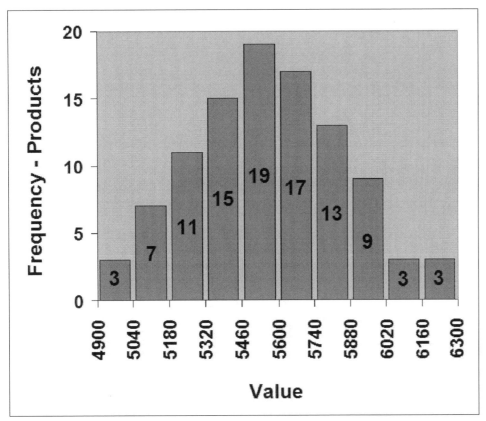

Figure 2.12: Histogram of 70 to 79 Multiplication - Normal-Like Symmetry

The maximum value in the table of Figure 2.11 is 6241, and the minimum value is 4900; hence resultant post-multiplicative POM value for the entire table is $(6241)/(4900) = (79*79)/(70*70) = (79/70)*(79/70) = 1.13*1.13 = 1.27$, and this value represents just a slight increase from the 1.13 POM value of the original 70 to 79 set.

Surprisingly, this multiplication process is not skewed but rather quite symmetrical! Here the medium around the central part of the data dominates, and the histogram in Figure 2.12 is remarkably similar to the Normal Distribution! The only reason for this exception to the usual skewness emerging in almost all multiplication processes is the very low POM value of the 70 to 79 set of numbers. Other sets of numbers with low POM values under multiplicative processes (such as 90 to 99, or 130 to 139 for example) also show the same overall result, where most products fall in the central region of medium values, and where the histogram resembles the Normal Distribution a great deal.

73

[6] Multiplicative CLT and the Lognormal Distribution

Let us provide a brief review of the Central Limit Theorem and Multiplicative Central Limit Theorem:

The **Normal distribution** is obtained in repeated additions of any random variable.
The **Lognormal distribution** is obtained in repeated multiplications of any random variable.

Normal $= X_1 + X_2 + X_3 + \ldots + X_N$, where X_I are independent realizations from an identical random variable X. The Central Limit Theorem guarantees Normality for the sum in the limit as N gets large, and (almost) regardless of the distribution form or parameters of X.

The Central Limit Theorem requires three conditions that ostensibly should be met, namely that realizations from variable X should all be (I) independent from each other, (II) identically distributed as X, and (III) have finite variance. Yet there are quite a few generalizations weakening or even eliminating some or all of the three conditions. These generalizations ensure that CLT, and by extension Multiplicative CLT, have extremely wide applications for real life data. The Lognormal distribution is defined as a random variable whose natural logarithm is the Normal distribution.

$$\text{Lognormal(location, shape)} = e^{\text{Normal(location, shape)}}$$

$$\text{Normal(mean, s.d.)} = \text{LOG}_e(\text{Lognormal(mean, s.d.)})$$

This definition together with the result of the CLT imply that the Lognormal distribution can be represented as a process of repeated multiplications of a random variable, namely as:

$$\text{Lognormal} = e^{[\text{ Normal }]} = e^{[\text{ X1 + X2 + X3 + } \ldots \text{ + XN }]}$$
$$\text{Lognormal} = e^{X1}\, e^{X2}\, e^{X3} \ldots e^{XN}$$

This result, namely that repeated multiplications of a random variable is Lognormal in the limit as N gets large is called the **Multiplicative Central Limit Theorem (MCLT)**.

The shape and location parameters of the Lognormal are the standard deviation and the mean respectively of the 'generating' Normal distribution. Simulations of the Lognormal distribution for the purpose of examining its digital behavior yield an approximate logarithmic behavior whenever the shape parameter is roughly over 0.4, and regardless of the value assigned to the location parameter. Better results are obtained for higher values of the shape parameter. A near perfect logarithmic behavior is observed whenever shape parameter is approximately over 1.0, and this result is totally independent of the value assigned to the location parameter.

One plausible explanation for the prevalence of Benford's Law in the natural sciences is that such physical manifestations of the law are obtained via the cumulative effects of few or numerous multiplicative random factors, all of which leads to the Lognormal as the eventual distribution by way of the Multiplicative Central Limit Theorem. Scientists in each particular discipline have to agree on such a vista of the investigated phenomenon in focus for this to hold true, yet this research method hasn't yet been explored in the field of Benford's Law.

The crux of the matter in applying MCLT as an explanation of Benford's Law in the physical sciences is the pair of answers to the following two questions: Firstly, does Mother Nature multiply her measurements [I] sufficient number of times, or [II] merely two, three, or four times - in which case the use of MCLT may [on the face of it] seem unjustified. The second question is whether Mother Nature typically uses measurements with [I] high order of magnitude or [II] with low order of magnitude.

After some contemplations, and the creations of several hypothetical scenarios of random physical processes, together with some very limited explorations of the typical expressions in physics, engineering, chemistry and so forth, it seems safe to claim that Mother Nature is acting with considerable restraint and that she is typically reluctant to multiply more than say four or five measurements, although she often uses measurements with sufficiently high order of magnitude 'compensating' in a sense for her lack of multiplicands. Does such [partially optimistic and partially pessimistic] conclusion ruin the chance of utilizing multiplication processes as an explanation of the Benford phenomenon in the physical sciences? Certainly not! To see how Frank Benford overcomes Mother Nature's frugality in multiplying her quantities, one needs to differentiate between two very distinct goals, namely: (1) that the density of the logarithm of resultant ['multiplied'] data converges to the Normal, (2) that digit configuration converges to Benford. Mathematical empiricism strongly suggests that often convergence in the digital realm occurs faster than convergence in the stricter sense of MCLT where the goal is the emergence of the Normal shape for the log values. MCLT requires a minimum number of products of distributions in order to obtain something close enough to the Lognormal, but often in nature there are typically only 2, 3, or 4 such products, which is not sufficient for MCLT applications. Nonetheless, given that the individual variables are of sufficiently high order of magnitude, convergence in the digital sense is possible well before convergence to the Lognormal is achieved, enabling students of Benford's Law to let go of MCLT altogether.

For a decisive demonstration of the validity of the multiplicative approach as an explanation of Benford's Law in the physical sciences, let us consider Monte Carlo computer simulation results of the product of merely two Uniform distributions having high order of magnitude. In one simulation run with 35,000 realized values from the multiplicative process of the distributions Uniform(1, 650) and Uniform(1, 1200), first digits distribution came out as:

Uniform(1, 650)*Uniform(1, 1200)   {27.4, 19.7, 15.3, 11.4, 8.7, 6.5, 4.4, 3.3, 3.4}
Benford's Law First Digits Order   {30.1, 17.6, 12.5,   9.7, 7.9, 6.7, 5.8, 5.1, 4.6}

The relatively low 28.3 SSD value signifies that results are quite close to Benford. The Uniform(1, 650) has the high POM value of 650/1 = 650, and the Uniform(1, 1200) has the high POM value of 1200/1 = 1200, and such high orders of magnitude guarantee that their product is very close to the Benford digital configuration.

Yet, multiplications of two Uniform distributions having exceptionally low order of magnitude does not yield any convergence to Benford whatsoever. In one simulation run with 35,000 realized values from the multiplicative process of the distributions Uniform(5, 11) and Uniform(8, 31), first digits distribution came out as:

Uniform(5, 11)*Uniform(8, 31)   {52.7, 23.0, 1.9, 0.8, 2.5, 3.4, 4.5, 5.7, 5.6}
Benford's Law First Digits Order   {30.1, 17.6, 2.5, 9.7, 7.9, 6.7, 5.8, 5.1, 4.6}

The very high 790.3 SSD value signifies that results are decisively non-Benford. The Uniform(5, 11) has the very low POM value of 11/5 = 2.2, and the Uniform(8, 31) has the very low POM value of 31/8 = 3.9, and such exceedingly low orders of magnitude preclude convergence to the Benford digital configuration.

The first result regarding two Uniform distributions with high orders of magnitude is a decisive indication of how closely results can get to Benford in merely a single multiplicative process having only two multiplicands. In other words, that Benford behavior can be achieved quite easily and rapidly, long before resultant data becomes Lognormal via MCLT.

Conceptually, as a general rule, Benford digital configuration is expected to be found in the natural sciences whenever **'the randoms are multiplied'**. More specifically, the multiplicative process needs to be carefully scrutinized to determine whether there are sufficiently many multiplicands, or at least sufficiently high order of magnitude within the constituent distributions. The crucial factor here is the resultant order of magnitude of the entire multiplicative process, which depends on the individual orders of magnitude of the distributions being multiplied, as well as on the number of distributions involved. The higher the number of distributions being multiplied and the higher the orders of magnitude of the various individual distributions – the larger is resultant order of magnitude of the multiplicative process itself and the closer is resultant digital configuration to Benford.

[7] The Effects of Random Multiplication Processes on POM

Let us examine theoretically how multiplication processes of random variables affect POM. Given PDF(x), namely Probability Density Function of variable x with MIN_X and MAX_X endpoints; and given PDF(y), namely Probability Density Function of variable y with MIN_Y and MAX_Y endpoints; then $POM_X = MAX_X / MIN_X$ and $POM_Y = MAX_Y / MIN_Y$.

For the random multiplication process X*Y:
The best scenario is multiplying MAX_X*MAX_Y
The worst scenario is multiplying MIN_X*MIN_Y
These two extreme scenarios represent the maximum and minimum values for the multiplication process itself.

Hence, for the random multiplicative process PDF(x)*PDF(y):
$POM_{X*Y} = [MAX_X*MAX_Y]/[MIN_X*MIN_Y] = [MAX_X / MIN_X]*[MAX_Y / MIN_Y] = POM_X*POM_Y$. Since by definition POM > 1 for each variable that is not a constant, it follows that there is a definite increase in resultant POM due to the multiplicative act.

In reality, for any finite number of computer simulated multiplications of two random variables, such as in say thousands or tens of thousands realizations, it is rare to simultaneously obtain the highest values of the two in one single realization; and it is also rare that the two lowest values are gotten simultaneously. The maximum multiplied value in the entire simulation process is a bit lower than the theoretical; and the minimum multiplied value in the entire simulation process is a bit higher than the theoretical; all of which implies that resultant POM of multiplied variables is a bit lower in actual simulations then the theoretical expression POM_X*POM_Y.

Yet, the basic argument presented here is certainly valid, and actual resultant POM_{X*Y} strongly increases (from the POM_X level and from the POM_Y level) due to the multiplicative process even if it always hovers somewhere below the theoretical POM_X*POM_Y expectation.

Needless to say, this result could easily extend to any number of random variables, not merely two, and where resultant theoretical POM is the product of all the individual POMs, namely $POM_{PRODUCT} = POM_X*POM_Y*POM_Z$ and so forth; or in concise mathematical notations as:

$$POM_{PRODUCT} = \prod POM_J$$

For the 10 by 10 multiplication table from our elementary school days, POM of each multiplicand {1, 2, 3, 4, 5, 6, 7, 8, 9, 10} is 10/1 or simply 10, while POM of the entire table is 100/1 or simply 100. This result nicely complies with the above derived expression of $POM_{X*Y} = POM_X*POM_Y = 10*10 = 100$.

For statisticians who are more accustomed to the logarithm definition of OOM as discussed in chapter 7 of section 1, the net effect of multiplication processes is simply the summing up of the two distinct orders of magnitude as in $OOM_{X*Y} = OOM_X + OOM_Y$, leading to the same conclusion, namely that all multiplication processes yield increased OOM and variability.

77

[8] Scrutinizing Random Multiplication Processes

Let us examine multiplication processes of Uniform distributions via Monte Carlo computer simulations in order to learn how they gradually build up sufficiently large cumulative POM and increase the system's skewness; thereby converging to Benford.

[Note: all simulations are with 35,000 realized values.]

Uniform(3, 40)   POM = 13
1st significant digits: {26.7, 27.4, 29.7, 2.6, 2.9, 2.7, 2.5, 2.8, 2.7}   SSD = 514

Uniform(3, 40)*Uniform(2, 33)   POM = 212
1st significant digits: {23.5, 15.9, 13.1, 11.2, 9.3, 8.5, 7.2, 6.1, 5.1}   SSD = 58

Uniform(3, 40)*Uniform(2, 33)*Uniform(7, 41)   POM = 1076
1st significant digits: {32.2, 18.1, 11.3, 8.4, 7.2, 6.6, 5.9, 5.5, 4.9}   SSD = 9

Uniform(3, 40)*Uniform(2, 33)*Uniform(7, 41)*Uniform(1, 29)   POM = 5386
1st significant digits: {30.3, 18.0, 12.9, 9.5, 7.7, 6.5, 5.6, 4.9, 4.6}   SSD = 1

Random Process	POM Variability/Range	SSD Deviation from Benford
U	13	514
U*U	212	58
U*U*U	1076	9
U*U*U*U	5368	1

Figure 2.13: Increase in POM Variability Measure Induces Closeness to Benford

The table in Figure 2.13 demonstrates how larger POM value goes hand in hand with *[induces rather]* closeness to Benford, with the lowering of SSD at each stage of the multiplicative process. The sequence of <u>empirical</u> POM values in Figure 2.13 can be easily checked against the <u>theoretical</u> expression POM$_{PRODUCT}$ = $\prod POM_J$ since the max and the min of Uniforms can be readily obtained as in Uniform(min, max). Referring to the variables Uniform(3, 40), Uniform(2, 33), Uniform(7, 41), Uniform(1, 29); the four individual POM values in the same order as above are {40/3, 33/2, 41/7, 29/1}, namely {13.3, 16.5, 5.9, 29.0}. Successive multiplications yield {13.3, 13.3*16.5, 13.3*16.5*5.9, 13.3*16.5*5.9*29.0}, namely the POM sequence of **{13, 220, 1289, 37369}**. But this theoretical sequence of sharp rise in POM is not what seems in Figure 2.13. As mentioned above, it is rare to simultaneously obtain the highest/lowest values in any one single realization in actual simulations; hence we empirically encounter the somewhat less dramatic rise in POM of the sequence **{13, 212, 1076, 5368}** as seems in the empirical results of Figure 2.13.

The histogram in Figure 2.14 of the log values of this multiplicative process demonstrates in general what could occur at times in random multiplication processes, namely that digits converge to Benford well before any significant achievement of MCLT in terms of endowing log density the shape of the Normal. Here, log of U(3, 40)*U(2, 33)*U(7, 41)*U(1, 29) is a bit asymmetrical with a longer tail to the left, not resembling sufficiently the shape of the Normal distribution, yet digits here are nearly perfectly logarithmic, with an exceedingly low SSD value of 1. Such low SSD value is rarely found in real-life random data, and it indicates nearly perfect agreement with Benford's Law.

[CLT and MCLT strictly requires 'independent and identically distributed variables', while here we consider the product of four non-identical distributions. Nonetheless, the many extensions and generalizations of CLT and MCLT allow us to consider such non-identical distributions as well, so the general principle holds.]

The histogram in Figure 2.15 of the actual simulation values (not log-transformed) depicts the severe quantitative skewness of the resultant distribution. It is exactly this quantitative skewness which drives digits into their Benford digital skewness as a consequence.

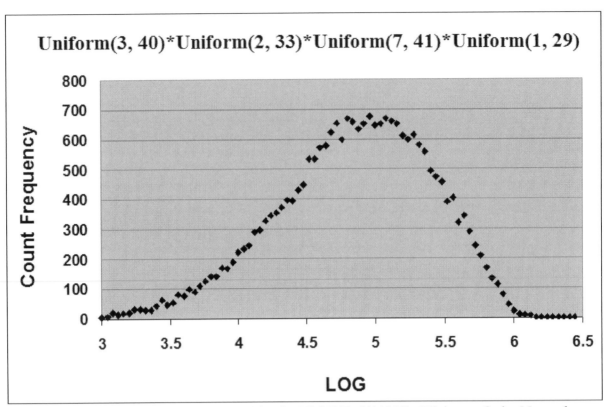

Figure 2.14: Log Histogram of U(3, 40)*U(2, 33)*U(7, 41)*U(1, 29) is not Quite Normal

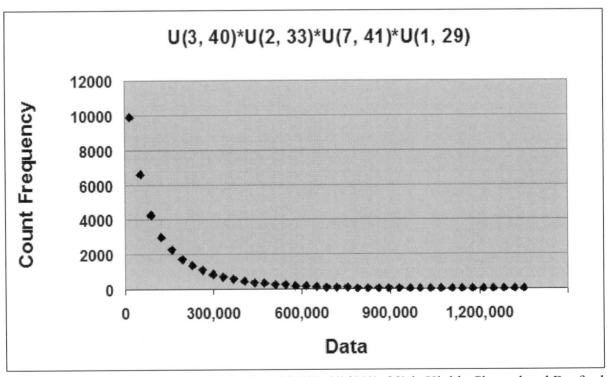

Figure 2.15: Histogram of U(3, 40)*U(2, 33)*U(7, 41)*U(1, 29) is Highly Skewed and Benford

When these Uniform distributions themselves are of **very high POM,** merely a single product of two such high-variability Uniforms yields digital configuration quite close to Benford, all in one fell swoop. As an example, Monte Carlo computer simulations are performed for the product of Uniform(1, 60777333) and Uniform(1, 30222888). Theoretical POM for these two Uniforms are 60777333 and 30222888 respectively. Theoretical POM for their product is 60777333*30222888 = 1836866528197700, although actual simulations of the multiplication process should yield much lower POM value.

Five simulations are performed for Uniform(1, 60777333)*Uniform(1, 30222888); with 35,000 runs each; the results are:

First Digits: {29.7, 13.6, 11.5, 10.3, 8.7, 7.6, 6.8, 6.5, 5.3}   SSD = 22.5   POM =   661343
First Digits: {29.0, 13.5, 11.6, 10.3, 9.0, 8.1, 7.0, 5.9, 5.6}   SSD = 25.9   POM =   455568
First Digits: {29.4, 14.0, 11.5, 10.1, 8.9, 7.6, 7.0, 6.1, 5.4}   SSD = 19.5   POM = 26872700
First Digits: {29.2, 13.3, 11.8, 10.1, 9.1, 7.8, 6.9, 6.2, 5.7}   SSD = 26.0   POM =   114111
First Digits: {29.4, 13.6, 11.5, 10.3, 8.7, 7.7, 6.8, 6.3, 5.5}   SSD = 22.8   POM = 3550739

Yet, MCLT does not even begin to be significantly effective for this very short multiplicative process! The histogram in Figure 2.16 of the log values of this process demonstrates once again that in multiplication processes digits could converge to Benford well before any significant achievement of MCLT is obtained in terms of endowing log density the shape of the Normal distribution (or equivalently that resultant distribution is Lognormal). Here, log histogram of Uniform(1, 60777333)*Uniform(1, 30222888) does not even begin to resemble the Normal, and yet digits here are quite close to the logarithmic, with the relatively low SSD value of about 22!

The histogram in Figure 2.17 of the actual simulation values (not log-transformed) depicts the quantitative skewness of the resultant distribution. It is exactly this quantitative skewness which drives digits into their Benford digital skewness as a consequence.

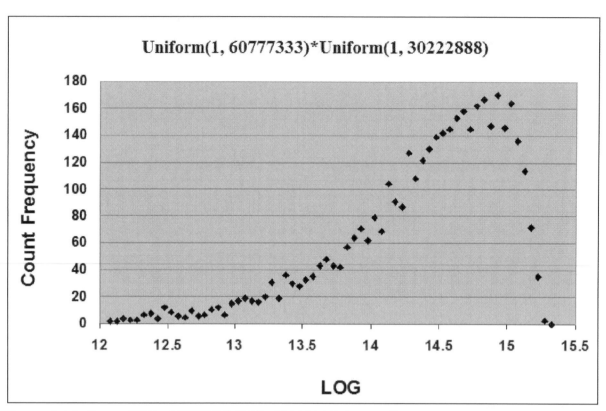

Figure 2.16: Log of U(1, 60777333)*U(1, 30222888) is Totally not Normal but Data is Benford

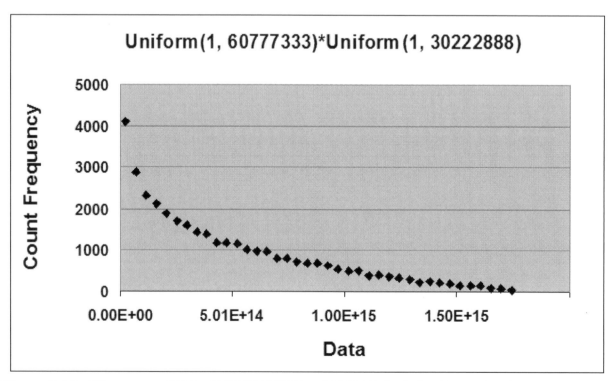

Figure 2.17: Histogram of U(1, 60777333)*U(1, 30222888) is Quite Skewed and Benford

When these Uniform distributions themselves are of **very low POM,** not even the product of several such low-variability Uniforms is capable of yielding digital configuration close enough to Benford, unless a truly great many such Uniforms are multiplied. As an example, Monte Carlo simulations are performed for the product of 6 Uniforms with very low POM as follow: Unif(4, 7)*Unif(8, 11)*Unif(5, 7)*Unif(12, 16)*Unif(237, 549)*Unif(17, 25). Theoretical POMs for these six Uniforms are 7/4, 11/8, 7/5, 16/12, 549/237, 25/17, namely 1.8, 1.4, 1.4, 1.3, 2.3, 1.5, respectively. Theoretical POM for their product is 1.8*1.4*1.4*1.3*2.3*1.5 = 15.3, although actual simulations of the multiplication process should yield much lower POM value.

Five simulations for these six Uniforms are conducted; with 35,000 runs each; the results are:

First Digits: {6.7, 28.5, 30.8, 19.7, 9.5, 3.5, 1.0, 0.2, 0.0}   SSD = 1181.1   POM = 11
First Digits: {6.6, 29.0, 30.9, 19.6, 9.5, 3.3, 0.9, 0.2, 0.0}   SSD = 1197.5   POM = 9
First Digits: {6.7, 28.8, 30.6, 19.8, 9.4, 3.5, 1.0, 0.2, 0.0}   SSD = 1183.3   POM = 10
First Digits: {6.5, 28.9, 30.7, 19.9, 9.3, 3.4, 1.0, 0.2, 0.0}   SSD = 1202.1   POM = 10
First Digits: {6.5, 28.8, 31.2, 19.6, 9.3, 3.3, 1.0, 0.2, 0.0}   SSD = 1213.6   POM = 9

Here, MCLT is properly and nicely applied for this longer process with sufficient number of multiplications (almost), and the histogram of the log is very much Normal-like in appearance. The resultant data set itself is nearly Lognormal, but with a low shape parameter – precluding Benford behavior. Figure 2.18 of the histogram of the log of this process demonstrates what could happen at times in multiplication processes of random variables having very low POM values. Here digits do not converge to Benford in the least, while the significant achievement of MCLT is obvious and visible, endowing log density the shape of the Normal (almost).

The histogram in Figure 2.19 of the actual simulation values (not log-transformed) depicts the approximate quantitative symmetry of resultant distribution. The non-skewed quantitative nature of resultant data precludes any resemblance to the Benford skewed digital configuration.

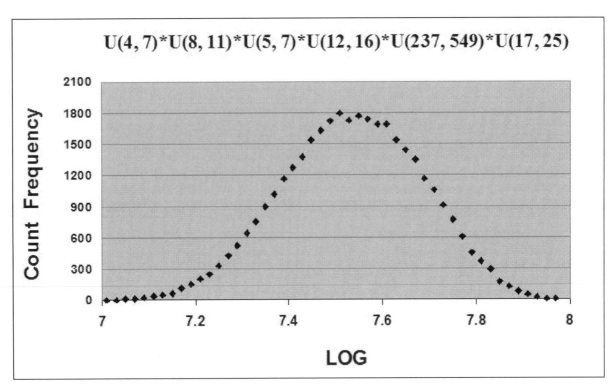

Figure 2.18: Log of Six Multiplied Uniforms with Low POM is Normal but Data is not Benford

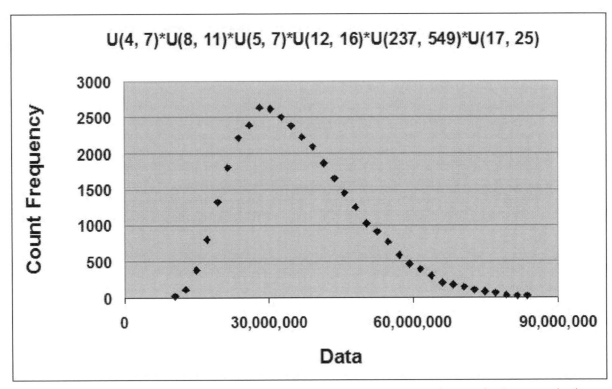

Figure 2.19: Histogram of Six Multiplied Uniforms with Low POM is Nearly Symmetrical

The essential lessons that can be learnt from the three examples above can be stated succinctly:

1) MCLT does not guarantee Benford behavior in multiplication processes!
2) MCLT is not necessary for Benford behavior in multiplication processes!
3) Order of Magnitude is the crucial factor in determining Benford behavior in multiplication processes!
4) The confluence of two factors leads to high resultant POM and thus to Benford behavior and skewness in multiplication processes: (a) POMs of the individual variables, (b) the number of variables that are being multiplied.

POM is a greedy creature when it comes to multiplication processes, it is never satisfied with what it has, hence the more variables we multiply the larger POM gets with each added multiplicand, and indefinitely so. MCLT on the other hand is modest; it has a limited and well-defined goal, namely, Normality of the density of the logarithm. Benford is also very modest; it has a limited and well-defined goal, namely LOG(1 + 1/d). Once log Normality is approximately achieved for MCLT - it is maintained, and throwing in more variables helps only slightly in terms of perfection of convergence. Once LOG(1 + 1/d) is approximately achieved for Benford - it is maintained, and throwing in more variables helps only slightly in terms of perfection of convergence. The grand generic scheme of adding more and more multiplicands to the process (namely multiplying more and more variables one by one in stages) could lead to either MCLT achieving its goal first with log being nearly Normal, while digits are not yet Benford [*the case of low POM values of many multiplicands*], or it could lead to Benford achieving its goal first and digits are nearly LOG(1 + 1/d), while log curve is not yet Normal [*the case of high POM values of few multiplicands*]. But no matter who is first, the grand multiplicative scheme eventually leads to both, to MCLT as well as to Benford, both ultimately achieving their respective goals with certainty. With delicate calibration one could perhaps choose variables/multiplicands that let both MCLT and Benford achieve their respective goals simultaneously at approximately the same stage; namely after approximately the same number of multiplications – but this issue is neither significant nor interesting.

Hence, in light of the above discussion, instead of desperately seeking justifications and validations via the ***[often unrealistic or irrelevant]*** Multiplicative CLT, students of Benford's Law should learn to rely instead on ***[the very realistic, relevant, and easily available]*** Related Log Conjecture [chapter 18 of section 1] and focus on resultant order of magnitude which is the range or width of the histogram of the logarithm of the data. Examinations of several log histograms of multiplied distributions (with sufficient POM) limited to only 2, 3, or 4 such products, show that while they are not really Normal-like, nonetheless the requirements of Related Log Conjecture are almost all nicely met, and thus digits are nearly Benford.

A decisive argument supporting the above discussion can be found in the consideration of log of data in the form of a triangle which yield the Benford configuration as close to perfection as can be measured with any finite set of simulated values for $10^{Triangular}$ whenever log-range is say over 4 approximately [see Kossovsky (2014), chapters 64 and 66 for details]. Moreover, Lawrence Leemis (2000) rigorously proved that a symmetrical triangle positioned onto integral log points is exactly logarithmic. Here the Benford configuration is obtained so perfectly, and without log density being Normal-like in any way. To emphasis this point once again: there is no need whatsoever for log of data to be Normal in order to obtain an exact or near-perfect Benford behavior! Benford's Law and Normality of log are not two sides of the same coin! True, numerous *[Lognormal-like]* physical and scientific data sets which are Benford come with log density that is almost Normal, or at least resembles the Normal a great deal, but this is <u>not</u> what Benford's Law is all about. The law is much more general than this narrow manifestation of it.

Two manifestations of some very short multiplication processes having only 3 multiplicands and resulting in near Benford configuration demonstrate that such situations can occur in scientific and physical data quite frequently. These two scenarios are discussed in Kossovsky (2014) chapters 90 page 393.

Case I: The final speed of numerous particles of random mass M, initially at rest, driven to accelerate linearly via random force F, applied for random length of time T. In other words, each particle comes with (random) unique M, F, and T combination. The equations in Physics describing the motion are: (i) $V_{FINAL}=V_{INITIAL} + A*T$, or since they start at rest simply $V_{FINAL}=A*T$, (ii) $F=M*A$. Thus $V_{FINAL} = (Force*Time)/Mass$. Monte Carlo computer simulations show strong logarithmic behavior here whenever M, F, and T are randomly chosen from Uniform distributions of the form U(0, b) *[intentionally choosing parameter a equals to 0 to ensure high order of magnitude]*. Moreover, even with only 2 random measurements, and with the 3rd measurement being fixed as a constant, we get quite close to the logarithmic [say mass M and force F are random, while time T is fixed - being a constant for all particles.]

Case II: The final position of numerous particles of random mass M, thrown linearly from the same location at random initial speed V_I, under constant decelerating random force F (say frictional) until each one comes to rest. In other words, each particle comes with (random) unique M, F, and V_I combination. The equations in Physics describing the motion are:
(i) $V_{FINAL} = V_I + A*T$, which leads to $T_{REST} = V_I/A$ as the time it takes to achieve rest,
(ii) $F = M*A$, (iii) Displacement $= V_I*T – (1/2)*A*T^2$. Hence
Displacement $= V_I*(V_I/A) – (1/2)*A*(V_I/A)^2$, namely Displacement $= (V_I)^2/(2*(F/M))$. Monte Carlo simulations again show strong logarithmic behavior here as well whenever M, F, and V_I are randomly chosen from Uniform(0, b) *[intentionally choosing parameter a equals to 0 to ensure high order of magnitude]*. Moreover, even with only 2 random measurements, and with the 3rd measurement being fixed as a constant, we get quite close to the logarithmic [say mass M and initial speed V_I are random, while frictional force F is fixed - being a constant for all particles.]

Schematically Case I is of the form: Simulation = (U1*U2)/U3, while Case II is of the form: Simulation = (U1*U1)/(2*(U2/U3)), and so neither can really apply MCLT due to severe lack of multiplicands; nevertheless due to high order of magnitude of multiplied variables digits here are quite close to the logarithmic.

As an additional example of the rapidity and the decisiveness with which multiplication processes could lead to Benford, Monte Carlo computer simulation results of the product of just two Normal(m, sd) distributions would be shown. The Normal is considered to be 'anti-logarithmic' in and of itself, yet as multiplicands it can readily become logarithmic.

Normal(2, 9)*Normal(5, 13) gave {31.7, 17.6, 11.3, 9.7, 7.8, 6.0, 5.8, 5.4, 4.9}
Normal(4, 7)*Normal(2, 3)   gave {28.7, 18.5, 13.1, 10.2, 7.7, 6.6, 5.9, 4.9, 4.3}
Normal(2, 4)*Normal(5, 3)   gave {29.0, 19.3, 13.3, 10.6, 8.1, 6.4, 4.7, 4.9, 3.7}

Benford's Law First Digits   {30.1, 17.6, 12.5, 9.7, 7.9, 6.7, 5.8, 5.1, 4.6}

Admittedly, the choices of parameters above were somewhat intentional, ensuring that the range is of large order of magnitude *[accomplished by crossing the origin, which ensures that the data draws plenty from the interval (0, 1)]*, but the general principle here holds nonetheless, and certainly it could be argued that in nature typical Normals are such, often occurring with high order of magnitude.

Monte Carlo computer simulation results of the product of just two Exponential distributions, a density form already somewhat close to Benford in and of itself, yield much superior results:

Exponential(4)*Exponential(11.0) gave {30.1, 17.6, 12.8, 9.9, 7.3, 7.1, 6.0, 4.7, 4.5}
Exponential(5)*Exponential(0.07) gave {30.2, 17.5, 12.7, 9.3, 9.4, 7.3, 5.8, 3.8, 3.9}
Exponential(13)*Exponential(0.2) gave {30.4, 17.1, 12.9, 9.9, 7.5, 6.9, 5.7, 5.3, 4.3}

Benford's Law First Digits   {30.1, 17.6, 12.5, 9.7, 7.9, 6.7, 5.8, 5.1, 4.6}

[9] The Conjecture on the Benfordized Nature of Multiplication

In Kossovsky (2014) chapter 58, two important results by Hamming (1970) are mentioned relating to products and divisions of several data sets or distributions. Let X be a continuous random variable with an exact logarithmic behavior. Let Y be any other continuous random variable, logarithmic or non-logarithmic. Then the product X*Y, and the ratios X/Y, Y/X, all satisfy Benford's Law as well. The ramification of these two remarkable properties is profound, since it applies to any distribution form Y! These two properties may be thought of as 'propagators' of the logarithmic distribution, guaranteeing to spread it around whenever a logarithmic data set gets 'multiplicatively connected' with any other data type, with the result that in turns it 'infects' other data sets with the logarithmic configuration, and so on. Consequently, in the context of scientific and physical data sets, even 2 or 3 products are sufficient to obtain a perfect Benford behavior given that one measurement is Benford in its own right! Moreover, one can conjecture with total confidence that partial Benfordness in one measurement endows *[at least]* that same degree of Benfordness *[and actually somewhat higher degree]* for the product with any other measurement. This extrapolation for products of random variables is in the same spirit as the extrapolation of the 2nd chain conjecture in Kossovsky (2014) chapter 102, page 460, where it is claimed that any single sequence within a chain of distribution contributes to increase in Benfordness, or that at least there is no backtracking away from it.

If one goes along with SSD as a measure of 'distance' from the logarithmic, then this conjecture can be stated succinctly and formally as:

$$[\text{ SSD of X*Y }] \leq [\text{ SSD of X }]$$
$$\text{and}$$
$$[\text{ SSD of X*Y }] \leq [\text{ SSD of Y }]$$

for any X, Y random variables or data sets of positive numbers.

The mixed inequality/equality sign should be substituted with inequality sign exclusively whenever X or Y are not exactly (perfectly) Benford, signifying an improvement in Benfordness for any product of variables whatsoever. The equality sign is valid for cases where X or Y are exactly Benford to begin with prior to any multiplication, so that SSD is 0 on both sides.

[10] Random Additions of Lognormal Distributions

In sharp contrast to multiplication processes, random additions of random variables affect digital configuration in the opposite way, away from Benford. Random addition processes induce symmetry in resultant histogram, gradually eliminating skewness altogether, and in the limit the emerging shape of resultant histogram is the bell curve of the Normal Distribution.

In order to emphasize that **addition is highly detrimental to Benford behavior**, eight highly logarithmic Lognormal distributions with shape parameter well over 1 are added via Monte Carlo computer simulations, one by one, resulting in significant deviations and retreat from Benfordness - as predicated by the Central Limit Theorem. *[Note: all simulations of Lognormals and their sums are with 35,000 realized valued.]*

Lognormal(shape = 1.5, location = 3.8)
First Digits: {29.7, 17.6, 12.8, 9.6, 8.0, 6.8, 5.8, 5.1, 4.6}   SSD = 0.2

Lognormal(shape = 1.3, location = 4.0)
First Digits: {29.9, 17.8, 12.4, 9.6, 7.8, 7.0, 5.8, 5.1, 4.6}   SSD = 0.2

Lognormal(shape = 1.6, location = 4.3)
First Digits: {30.5, 17.5, 12.6, 9.6, 7.8, 6.4, 5.7, 5.1, 4.8}   SSD = 0.3

Lognormal(shape = 1.2, location = 4.2)
First Digits: {30.3, 17.5, 12.6, 9.9, 7.9, 6.7, 5.7, 4.9, 4.6}   SSD = 0.1

Lognormal(shape = 1.4, location = 4.1)
First Digits: {30.2, 17.4, 12.9, 9.7, 8.0, 6.6, 5.8, 4.9, 4.5}   SSD = 0.3

Lognormal(shape = 1.1, location = 4.4)
First Digits: {30.4, 17.0, 12.1, 9.7, 8.2, 6.9, 5.9, 5.2, 4.6}   SSD = 0.7

Lognormal(shape = 1.3, location = 4.3)
First Digits: {30.3, 17.4, 12.4, 9.9, 7.8, 6.7, 5.6, 5.1, 4.7}   SSD = 0.2

Lognormal(shape = 1.5, location = 4.5)
First Digits: {30.3, 17.4, 12.4, 9.9, 7.8, 6.7, 5.6, 5.1, 4.7}   SSD = 0.2

Benford 1st: {30.1, 17.6, 12.5, 9.7, 7.9, 6.7, 5.8, 5.1, 4.6}   SSD = 0

LogN(1.5, 3.8) + LogN(1.3, 4.0)
{31.6, 17.8, 11.7, 9.1, 7.4, 6.4, 5.9, 5.2, 5.0}
SSD = **3.7**

LogN(1.5, 3.8) + LogN(1.3, 4.0) + LogN(1.6, 4.3)
{30.0, 19.3, 13.2, 9.4, 7.6, 6.3, 5.2, 4.9, 4.2}
 SSD = **4.2**

LogN(1.5, 3.8) + LogN(1.3, 4.0) + LogN(1.6, 4.3) + LogN(1.2, 4.2)
{25.5, 19.1, 14.9, 11.2, 8.4, 6.9, 5.6, 4.5, 3.9}
SSD = **32.6**

LN(1.5, 3.8) + LN(1.3, 4.0) + LN(1.6, 4.3) + LN(1.2, 4.2) + LN(1.4, 4.1)
{23.5, 15.9, 14.5, 12.0, 9.7, 8.0, 6.7, 5.3, 4.5}
SSD = **60.1**

LN(1.5, 3.8) + LN(1.3, 4.0) + LN(1.6, 4.3) + LN(1.2, 4.2) + LN(1.4, 4.1) + LN(1.1, 4.4)
{25.0, 11.7, 11.7, 11.6, 10.7, 9.2, 8.1, 6.5, 5.6}
SSD = **87.2**

LN(1.5, 3.8)+LN(1.3, 4.0)+LN(1.6, 4.3)+LN(1.2, 4.2)+LN(1.4, 4.1)+LN(1.1, 4.4)+LN(1.3, 4.3)
{31.1, 9.7, 8.4, 9.3, 9.9, 9.0, 8.6, 7.4, 6.7}
SSD = **107.7**

L(1.5, 3.8)+L(1.3, 4.0)+L(1.6, 4.3)+L(1.2, 4.2)+L(1.4, 4.1)+L(1.1, 4.4)+L(1.3, 4.3)+L(1.5, 4.5)
{38.2, 11.6, 6.4, 6.5, 7.4, 7.8, 7.8, 7.3, 7.0}
SSD = **166.7**

The additions of these eight Lognormal distributions constitute in essence a tug of war between additions and multiplications, a war decisively won by additions, overcoming multiplications, thus disobeying the law of Benford. This is so since each Lognormal distribution may be represented as a random multiplicative process. For example, assuming that the product of only 6 Uniforms is a sufficient approximation of a Lognormal distribution, then the addition of the 3 Lognormals above LogN(1.5, 3.8) + LogN(1.3, 4.0) + LogN(1.6, 4.3) could be represented by the expression U*U*U*U*U*U + U*U*U*U*U*U + U*U*U*U*U*U where the tension between multiplication and addition is clearly demonstrated.

Figure 2.20 vividly illustrates this tension that exists between additions and multiplications in the context of Benford's Law. Assuming for the sake brevity and lack of space that the product of only 3 Uniforms is a sufficient approximation of a Lognormal distribution (it is not!), namely that Lognormal ≈ U*U*U, then the above arrangement of successive additions of Lognormals, gradually leading to significant deviations from Benford, can be viewed as a gradual victory by additions over multiplications due to the fact that at each successive stage these victorious additions are a notch more overwhelming or numerous than the vanquished multiplications. The damage done by additions here is mostly due to the gradual transformation of the shape of the resultant curve from skewness to symmetry, and not so much from the gradual decrease in POM. Indeed, the final value of 783 POM is still large enough and almost Benford in potential.

The Random Process	POM	SSD
UUU	1133924	0.2
UUU+UUU	37836	3.7
UUU+UUU+UUU	14914	4.2
UUU+UUU+UUU+UUU	4747	32.6
UUU+UUU+UUU+UUU+UUU	2272	60.1
UUU+UUU+UUU+UUU+UUU+UUU	1416	87.2
uuu+uuu+uuu+uuu+uuu+uuu+uuu	979	107.7
UUU+UUU+UUU+UUU+UUU+UUU+UUU+UUU	783	166.7

Figure 2.20: Tug of War between Additions and Multiplications

[11] The Achilles' Heel of the Central Limit Theorem

Yet, the full description of such arithmetical tugs of war is a bit more complex than the apparently simple narrative outlined above in Figure 2.20. First of all, one should bear in mind that CLT works decisively and fast whenever added variables are non-skewed (symmetric). When added variables are skewed, more 'work' (in terms of adding very many of them) is required in order to shape their sum into that symmetric Normal curve from the original non-symmetric configuration. The **CLT's Achilles' heel** - in terms of its rate of convergence to the Normal - is the adverse possibility that added variables are highly skewed and come with very high order of magnitude, constituting a very bad combination for the CLT and a challenge that needs to be overcome. Except for Uniforms, Normals, and other symmetric distributions which converge to the Normal quite fast after very few additions regardless of the value of order of magnitude of added variables, all other asymmetrical (skewed) distributions show a distinct rate of convergence depending on the value of their order of magnitude. For skewed variables, whenever order of magnitude is of very high value, CLT can manifest itself with difficulties, and very slowly, only after a truly large number of additions of the random variables. On the other hand, when skewed variables are of very low order of magnitude, CLT achieves near Normality quite quickly after only very few additions.

As an example of the challenge CLT faces when added variables are highly skewed and of very high order of magnitude; 8 identical Lognormals with very high shape parameter of 2.2 and location parameter 5 are randomly added. The shape parameter of the Lognormal determines its order of magnitude, and for shape value of 2.2 order of magnitude is almost 6 and POM is approximately 1,000,000. These 8 Lognormals are thus with larger POM values than the 8 Lognormals of Figure 2.20 which were with lower shape parameter value of around 1.36 and thus of lower POM. The shape of the resultant data set after the 8 additions of these high POM Lognormals in 35,000 simulation runs shows a decisively non-Normal and highly skewed curve much like the original Lognormal; so that not much has been achieved after 8 additions, and which is in sharp contrast to what is shown in Figure 2.20. Where is CLT? What happened to the mathematicians' solemn pledge of seeing the Normal curve after several random additions? Well, here, when dealing with highly skewed variables with very high POM values, a lot of patience and very strong faith in CLT is necessary. The Average of resultant added data of this simulation is 13325, being much greater than the Median value of 4055, and which is a strong indication of positive skewness. Leading digits of the 8 added Lognormals are still very much Benford-like coming at {29.9, 17.8, 12.5, 9.7, 7.9, 6.7, 5.7, 5.2, 4.7} with an extremely low SSD value of 0.1. It would take many more additions of these highly skewed Lognormals with high POM to arrive approximately at the Normal, and surely our (rational) faith in CLT would eventually be rewarded.

On the other hand, Lognormals with much lower shape parameter (and the implied lower POM value) converge to the Normal quite rapidly in spite of their skewness. As another example, 8 identical Lognormals with the lower shape parameter 0.9 and location parameter 5 are randomly added. Order of magnitude for shape value of 0.9 is only about 2.5, and POM is about 1000. The shape of resultant data set in 35,000 simulation runs appears very much Normal-like, easily confirming the prediction of the CLT. The Average of resultant added data is 1778, and which is near the Median value of 1642, indicating that resultant data is almost not skewed at all. Digits are decisively non-Benford, coming at {62.5, 24.0, 4.5, 0.9, 0.4, 0.6, 1.3, 2.4, 3.5}, with an extremely high SSD value of 1353.

Hence, the table in Figure 2.20 - with its moderate 1.36 average shape parameter for these 8 added Lognormals - should be viewed as a compromise between two extremes, namely between the very high POM (2.2 shape parameter) and the very low POM (0.9 shape parameter). The moderate choice for the shape parameters for Figure 2.20 was made for pedagogical purposes.

If one still doubts the validity of the general tendency or principle demonstrated here with these three distinct schemes of Lognormal addition processes above, then two more Monte Carlo simulations with k/x distributions would convince even the most avowed skeptic of the validity of the principle. The first scheme is of k/x defined over (10, 100) with its low 10 POM value. Here we randomly add 6 of them, namely: k/x + k/x + k/x + k/x + k/x + k/x. First digits are {29.9, 55.1, 13.9, 0.7, 0.0, 0.0, 0.0, 0.1, 0.2} with an extremely high 1675.2 SSD value. The Average is 234.7 while the Median is 231.8, and the curve of resultant additions is extremely similar to the Normal. Benfordness of the original k/x distribution was completely ruined by the process of these random additions. In sharp contrast, the second scheme of k/x defined over (1, 1000000), with its large 1000000 POM value, is highly resistant to CLT for a long while, not showing any resemblance to the Normal (although it would eventually crumble after many more such random additions). Here as well we randomly add 6 of them, namely: k/x + k/x + k/x + k/x + k/x + k/x. First digits are {28.7, 13.8, 10.7, 9.9, 8.5, 7.9, 7.4, 6.8, 6.5} and SSD value is 30.7, which is fairly low. The Average is 432,976, and significantly larger than the Median value of 293,828, while the curve of resultant additions is <u>not</u> similar to the Normal at all.

This resonates (in some very vague sense) as being in harmony with the new definition of logarithmic-ness given in Kossovsky (2014), chapter 136, page 583, where k/x had to be defined over a huge range in order to qualified as being truly logarithmic with respect to all number system bases and with respect to all bin schemes, being resistant to all such base and bin changes, and thus conserving its logarithmic status. One should not get carried away too far with this analogy, because the two issues are fundamentally different; here we deal with temporary and illusionary resistance to CLT, which would eventually crumble no matter how high order of magnitude happened to be, while the definition of logarithmic-ness is about the conservation of k/x logarithmic status under base and bin changes, and which is another matter.

The moral of the story is that '**not all perfectly logarithmic data sets are created equal**'! Those with very large order of magnitude are by far more resistant to detrimental addition processes in the system, while those with just sufficient order of magnitude for a [nearly] perfect Benford behavior are much more vulnerable to random additions. A data set with say 500,000 points generated via simulations from the Lognormal with 3.5 shape parameter and whatever location, is by far superior and much more resistant to random additions and CLT than a comparable set with say 1.2 shape parameter. Both data sets measure equally for all practical purposes in their <u>current</u> logarithmic status and sizes, yet the former is much more resistant and stable as compared with the latter.

[12] Numerical Example of the Achilles' Heel of the Central Limit Theorem

This unique addition game at the casino necessitates having 5 especially-handcrafted 13-sided dice. The faces of these large dice are not marked with usual sequence of the natural integers 1 to 13 as customary with most dice games, but rather each of the 5 dice has the 13 chosen arbitrary numbers of {**3, 7, 12, 19, 24, 35, 42, 76, 92, 176, 331, 564, 978**} written on its sides. The game involves the throwing of these five large dice simultaneously and then the adding of these five values to arrive at the sum upon which a lot of money is bet.

The smallest possible sum here is (3) + (3) + (3) + (3) + (3) = 15.
The biggest possible sum here is (978) + (978) + (978) + (978) + (978) = 4890.

Examples of other possible game occurrences are:

(42) + (42) + (42) + (7) + (92) = 225
(76) + (978) + (92) + (35) + (24) = 1205
(92) + (3) + (564) + (24) + (3) = 686
(3) + (12) + (19) + (35) + (7) = 76

The result of 10,000 Monte Carlo computer simulations of this game is depicted in the histogram of Figure 2.21. The smallest sum in the simulations was 23. The biggest sum in the simulations was 3674. The histogram in Figure 2.21 depicts almost the entire set of resultant sums, from 20 to 3036, in even steps of 232-length bin width, incorporating 9,964 games, and leaving out only 36 games with extremely big sums of over 3036, not shown in the histogram for lack of space and better visualization.

The histogram shows a marked decline with a tail falling to the right, and nearly consistently skewed pattern favoring the small. Even though only 5 additions are performed here, still, on the face of it, this seems very surprising and quite odd, given the expectation of symmetric and Normal-like histograms for almost all types of addition processes, as predicated by the Central Limit Theorem. Yet, since CLT's Achilles' heel predicts that for skewed variables with high order of magnitude, CLT can manifest itself with difficulties, and only after a truly large number of additions of the random variable, it follows that if our set of 13 numbers on the dice is highly skewed and if it is also of high order of magnitude, then these empirical Monte Carlo result obtained here are certainly expected and reasonable. The crux of the matter is then to examine the quantitative structure of the set of the 13 numbers on the dice, and to determine whether or not it is highly skewed and with high order of magnitude.

Indeed, the set of 13 numbers on the dice is clearly biased against the big and in favor of the small, namely skewed. In order to be able to visualize the skewness of the values of the dice, all 13 values are plotted along the horizontal axis as depicted in Figure 2.22. The plot actually distorts the scale by stretching it on the left for small values and compressing it on the right for big values, for brevity and better visualization. In spite of giving the big such advantage over the small, the visual message coming out of the plot is that the small is numerous (condensed) and the big is rare (sparse); i.e. that these 13 numbers are skewed. In terms of the non-modern definition of skewness, namely (average – median)/(standard deviation), here the average of the 13 dice numbers is 181.5, and it is much greater than the median which is only 42.0.

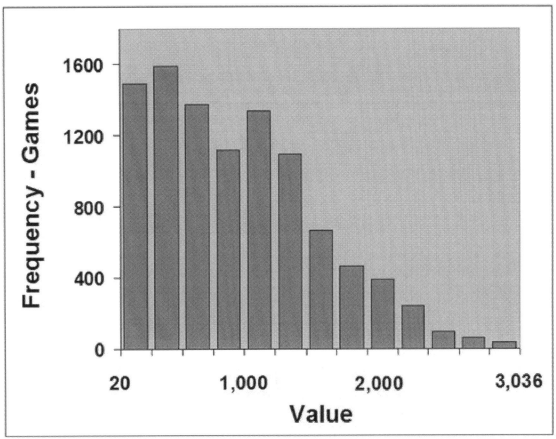

Figure 2.21: Histogram of 10,000 Simulations of the Five 13-Sided Dice Addition Game

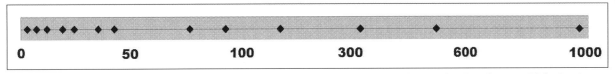

Figure 2.22: Higher Concentration for Small and Diluted Spread for Big for the 13-Sided Dice

In addition, the set of the 13 numbers on the dice is indeed with high order of magnitude. POM is calculated as (978)/(3) = 326, and this value is quite high; especially for example in comparison with a normal 6-sided die of only (6)/(1) = 6 POM value.

Since the set of 13 numbers on the dice is highly skewed and of high order of magnitude, and since only 5 dice are added, the CLT's Achilles' heel clearly explains why CLT could not be manifested here. It would be necessary to build at least about 15 such 13-sided dice for the casino and add 15 dice faces in the game, in order to manifest the CLT well enough. Indeed, computer simulations of 10,000 random additions of 15 such dice (with the same skewed values of high POM above) resulted in nearly symmetrical histogram, resembling the Normal Distribution a great deal, and where the medium came out as the most numerous size by far.

[13] Tugs of War between Multiplication and Addition

In order to further convince ourselves that these two benign-looking arithmetical operations, namely addition and multiplication, do occasionally go to war with each other over digital and quantitative configurations, another more vivid demonstration of their potential conflicts in the context of Benford's Law is given by the following 6 distinct random processes, where U represents the Uniform(5, 33), and where all Us are identical.

U + U
U*U + U*U
U*U*U + U*U*U
U*U*U*U + U*U*U*U
U*U*U*U*U + U*U*U*U*U
U*U*U*U*U*U + U*U*U*U*U*U

As opposed to adding Lognormals as in Figure 2.20 where the number of addends is increasing one by one in the main expression, and where the number of multiplicands (hidden within each Lognormal) is fixed, here the number of multiplicands is increasing while the number of addends is fixed at 2. Six Monte Carlo computer Simulations with 35,000 runs each are performed separately at each stage. One should notice carefully that for one realization of say Uniform(5, 33)*Uniform(5, 33), two separate and independent realizations are needed from Uniform(5, 33), followed by the multiplication of these two realizations; and that we are not referring to the simple squaring of a single realization such as in Uniform(5, 33)^2.

Simulation results are as follows:

U + U	{ 6.3, 19.2, 31.5, 26.8, 13.9, 2.3, 0.0, 0.0, 0.0}	SSD = 1360.0
UU + UU	{21.1, 5.7, 9.4, 11.4, 12.4, 11.8, 10.5, 9.4, 8.2}	SSD = 334.4
UUU + UUU	{42.4, 17.1, 6.7, 4.6, 5.2, 5.6, 6.3, 6.1, 5.9}	SSD = 222.5
UUUU + UUUU	{30.8, 22.4, 14.7, 9.5, 6.7, 5.1, 4.1, 3.5, 3.3}	SSD = 38.9
UUUUU + UUUUU	{26.0, 17.7, 13.9, 11.2 8.9, 7.2, 6.0, 5.1, 4.1}	SSD = 22.3
UUUUUU + UUUUUU	{30.2, 16.2, 11.5, 9.8, 8.3, 7.1, 6.4, 5.6, 5.0}	SSD = 4.3

Finding addition sleeping at the wheel, being fixed at 2, multiplication then strives hard to win by gradually increasing the number of multiplicands. Finally multiplication is able to achieve the Benford condition, where it manifests itself 6 times within the last expression versus only 2 manifestations of addition. Feeling overconfident now and conceited, multiplication then challenges addition to a new tug of war, willing to face 3 addends – given that it is already in possession of 6 multiplicands:

UUUUUU + UUUUUU + UUUUUU
{35.8, 16.4, 9.6, 7.3, 6.5, 6.6, 6.5, 5.8, 5.5}   SSD = 51.9

Upon seeing the severe setback in digital configuration where SSD is now over fifty, multiplication deeply regrets its previous offer, and it then asks addition for permission to use two more multiplicands, achieving a total of 8 multiplicands versus these 3 addends:

UUUUUUUU + UUUUUUUU + UUUUUUUU
{27.2, 18.5, 13.7, 10.4, 8.5, 6.7, 5.6, 4.8, 4.6}   SSD = 12.1

Multiplication is now quite satisfied with this latest improvement in Benfordness, while addition is not much disturbed upon seeing that digits are getting back fairly close to Benford. Addition knows that ultimately CLT is on its side, and that it would eventually win the war between it and multiplication in due time, and especially when POM is low or when added distributions are symmetrical and Normality is achieved fairly quickly. Nonetheless, addition wishes to make one last stand in order to flaunt its clout and demonstrate its ability to control the situation, and it expands addends by just one more, namely a total of 4 addends versus these 8 multiplicands:

UUUUUUUU + UUUUUUUU + UUUUUUUU + UUUUUUUU
{25.3, 15.5, 13.7, 11.3, 9.5, 7.9, 6.3, 5.5, 4.8}   SSD = 35.8

Indeed, by adding one more addend to the process, addition was able to move it decisively away from Benford, managing to increase SSD from 12.1 to 35.8.

Where would all these actions and reactions, attacks and counterattacks; that endless tit-for-tat war of attrition between addition and multiplication lead to? Without a doubt, the CLT is decisively on the side of addition, guaranteeing its eventual triumph. Surely multiplication could win some battles in the short run, but it would lose the war in the long run as resultant data finally becomes Normally distributed and digital configuration turning decisively non-Benford, and regardless of the form and defined range of added distributions.

[14] The Effects of Random Addition Processes on POM

Let us examine theoretically how random addition processes of an <u>identical</u> random variable affect resultant POM. The random process involves repeatedly adding an identical random variable, over and over again, N times.

In theory - POM should not change at all under random additions of N identical random variables X, Y, Z, and so forth. This is so theoretically since the lowest possible value (the worst scenario) for the sum is MINx + MINy + MINz + ... (N times), and since X = Y = Z and so forth, MINx = MINy = MINz = (generic MIN), hence the above sum is simply MIN*N. In the same vein, theoretically, the highest possible value (the best scenario) is MAXx + MAXy + MAXz + ... (N times), or simply MAX*N, hence POM$_{SUM}$ = (MAX*N)/(MIN*N) = (MAX)/(MIN), namely the same value as the POM of the identical variable being repeatedly added. More succinctly:

POM$_{SUM}$ = POM$_{THE\ IDENTICAL\ VARIABLE}$

In reality - when dealing with numerous additions, this almost never occurs in computer simulations, because it is exceedingly rare that the highest added value is gotten by aiming at all the various maximums simultaneously; and it is exceedingly rare that the lowest added value is gotten by aiming at all the various minimums simultaneously. The maximum added value in simulations is lower than the theoretical; and the minimum added value in simulations is higher than the theoretical; all of which implies that POM of added variables is somewhat lower in actual simulations than the theoretical POM value. Surely for an enormous number of simulation runs, numbering in the billions or trillions say, theoretical POM can be achieved on the computer, and certainly in the abstract, in the limit, as the number of runs goes to infinity. This is why in Figure 2.20 successive values of POM are monotonically decreasing, as it becomes increasing more difficult for the process to aim at all the maximums and all the minimums simultaneously while more and more additions are applied. Hence in extreme generality for identical variables: the more variables we randomly add, the lower is the resultant POM obtained for the computer simulated addition process - not numbering in the trillions.

As an example, successive random additions of Uniform(3, 17) distributions are run in 8 stages with 35,000 simulation each. In the first stage only one Uniform is simulated 35,000 times without any addition involved. In the 2nd stage two such Uniforms are added randomly 35,000 times, and so forth. The simulations yield:

# of Vars	1	2	3	4	5	6	7	8
Min Simul.	3.1	6.1	9.6	14.2	19.5	23.1	29.8	35.5
Max Simul.	16.9	33.8	50.7	65.5	79.8	95.0	109.6	123.1
POM Simul.	5.47	5.54	5.28	4.60	4.10	4.11	3.68	3.47
POM Theo.	5.66	5.66	5.66	5.66	5.66	5.66	5.66	5.66

Theoretical POM here is always 17/3 = 5.66, for any number of random additions whatsoever. Clearly, as can be seen in the above table, actual POM gradually and very slowly decreases from

5.47 just below the theoretical level of 5.66, and finally falling to 3.47 on the 8th additive stage. This is the result of running only 35,000 simulations at each stage, instead of billions or trillions simulations which would yield results very close to the theoretical expectations of POM. For only two random additions, very little discrepancy is seen between actual and theoretical, namely 5.66 – 5.54 = 0.12, while for say four random additions, much larger discrepancy is seen between actual and theoretical, namely 5.66 – 4.60 = 1.06. This is so since there is a good chance of obtaining in two additions actual POM value of say (16.98 + 16.95)/(3.01 + 3.02), and such fortunate possibility could occur at times, yet, the chance of obtaining in four additions actual POM value of say (16.98 + 16.95 + 16.98 + 16.99)/(3.01 + 3.02 + 3.03 + 3.01) is much smaller, and such fortunate possibility where all 8 extremes are quite close to the top and bottom values of 17 and 3 is quite rare. Put differently, for 2 additions only 4 rare coincidences are needed; while for 4 additions 8 rare coincidences are needed and this scenario is even less likely. The same reasoning applies to <u>multiplication processes</u> with even more dramatic effects, and where ratio of actual POM to theoretical POM decreases even more sharply and rapidly.

As opposed to the theoretical analysis of POM under multiplication processes where the two variables PDF(x) and PDF(y) were allowed to be considered as distinct, here for the analysis of POM under addition processes, without any loss of generality, the scope is restricted initially to identical variables.

Little reflection is needed to realize that this dichotomy [distinct vs. identical] does not restrict the general principles demonstrated here a great deal, and the purpose of considering initially only identical variables for addition processes is merely to ease the analysis and to be able to arrive at one correct conclusion here by being able to cancel the term N in the numerator and in the denominator. Insisting on adding two variables of distinct types and ranges would lead to:

For the random addition process X + Y:
The best scenario is adding $MAX_X + MAX_Y$
The worst scenario is adding $MIN_X + MIN_Y$
These two extreme scenarios represent the max and min values for the addition process itself.

$POM_{X+Y} = [MAX_X + MAX_Y] / [MIN_X + MIN_Y] \neq [2*MAX] / [2*MIN]$
And this inequality is due to the fact that the variables are distinct, spanning distinct ranges.

Yet, in general, the principle that POM of the sum is about as modest as POM of one typical variable still holds true approximately. Allowing variables to be distinct, yet postulating for convenience that the maximum of the minimums is significantly lower than the minimum of the maximums, and that no single variable is much different than the rest; that no single variable overwhelms results by having a range much different than the rest, we can then obtain the approximate (and quite vague) result:

$POM_{AVERAGE} \equiv (1/N)*[(MAX_X/MIN_X) + (MAX_Y/MIN_Y) + (MAX_Z/MIN_Z) + ...]$

$POM_{SUM} = [MAX_X + MAX_Y + MAX_Z + ...] / [MIN_X + MIN_Y + MIN_Z + ...]$

$POM_{SUM} = (1/N)*[MAX_X + MAX_Y + MAX_Z + ...] / (1/N)*[MIN_X + MIN_Y + MIN_Z + ...]$

$POM_{SUM} = [Average\ Maximum] / [Average\ Minimum] \approx POM_{AVERAGE}$

The main motivation in asserting that $POM_{SUM} \approx POM_{AVERAGE}$, or that at least the two are not much different from each other when no single variable is much different than the rest, is to establish the idea that POM is not being increased due to the act of adding, and that it is still just about at the POM level of any one (average) distribution.

Some concrete numerical examples might assure us that the general line of reasoning above is correct. The following four numerical analyses are of the theoretical highest POM possibilities, or of actual computer simulations with billions or trillions runs. Only one addition process is considered in all of these four examples, namely the random addition process of all the six variables together as in SUM = A + B + C + D + E + F, without stages and without partial additions as was done earlier in the example of Uniform(3, 17).

Variable	A	B	C	D	E	F	**SUM**
Max	43	54	13	77	11	15	**213**
Min	3	4	3	8	3	4	**25**
POM	14.3	13.5	4.3	9.6	3.7	3.8	**8.5**

Here POM of the addition process is 8.5, and it fits rather nicely within the six POM values of the individual distinct variables. Value 8.5 is very close to the average POM calculated as 8.2.

Variable	A	B	C	D	E	F	**SUM**
Max	43	54	957	77	11	15	**1157**
Min	3	4	3	8	3	4	**25**
POM	14.3	13.5	319.0	9.6	3.7	3.8	**46.3**

Here POM of the addition process is 46.3, which is way over POM of variables A, B, D, E, F, but way below POM of variable C. Here the range for variable C is markedly different than the ranges of the others. Value 46.3 is also quite different from the average POM calculated as 60.6.

Variable	A	B	C	D	E	F	**SUM**
Max	43	54	346978	77	11	15	**347178**
Min	3	4	5678	8	3	4	**5700**
POM	14.3	13.5	61.1	9.6	3.7	3.8	**60.9**

Here POM of the addition process is 60.9, which is way over POM of variables A, B, D, E, F, and it closely follows POM of variable C. Value 60.9 is also very different from the average POM calculated as 17.7. Here the huge range for variable C dominates all the others variables.

Variable	A	B	C	D	E	F	**SUM**
Max	5	12	54	310	2144	15335	**17860**
Min	3	3	3	3	3	3	**18**
POM	1.7	4.0	18.0	103.3	714.7	5111.7	**992.2**

Here POM of the addition process is 992.2, which is over POM of variables A, B, C, D, E, but well below POM of variable F. Here the ranges steadily increase as focus shifts from A to F. In addition, the 992.2222 POM value for the Sum of the 6 variables is equal almost exactly to the average POM calculated as 992.2333.

These four numerical examples hint at the possibility that POM of the addition process is always less than the largest POM among all the individual variables, namely that it never exceeds the maximum POM in the set of variables, hence the inequality $POM_{SUM} \leq POM_{MAX}$.

Let us prove this in extreme generality, without postulating anything about the variables whatsoever, allowing them to assume any distribution form, and to be defined on all sorts of distinct ranges. [Note: the general relationship $MAX_X = POM_X*MIN_X$ shall be applied.]

$$POM_{SUM} = \frac{MAX_A + MAX_B + MAX_C + MAX_D + etc.}{MIN_A + MIN_B + MIN_C + MIN_D + etc.}$$

$$POM_{SUM} = \frac{POM_A*MIN_A + POM_B*MIN_B + POM_C*MIN_C + POM_D*MIN_D + etc.}{MIN_A + MIN_B + MIN_C + MIN_D + etc.}$$

Let us artificially create an expression of POM_{MAX} with a similar format:

$$POM_{MAX} = POM_{MAX} * \frac{MIN_A + MIN_B + MIN_C + MIN_D + etc.}{MIN_A + MIN_B + MIN_C + MIN_D + etc.}$$

$$POM_{MAX} = \frac{POM_{MAX}*MIN_A + POM_{MAX}*MIN_B + POM_{MAX}*MIN_C + POM_{MAX}*MIN_D + etc.}{MIN_A + MIN_B + MIN_C + MIN_D + etc.}$$

By definition, $POM_X \leq POM_{MAX}$ for each X variable, hence when comparing the right hand side of the last expression for **POM$_{SUM}$** with the right hand side of the last expression for **POM$_{MAX}$**, it follows that:

$POM_{SUM} \leq POM_{MAX}$

For the 10 by 10 multiplication table from our elementary school days converted into an addition table, POM of each {1, 2, 3, ... ,10} addend is 10/1, namely 10, while POM of the entire addition table is 20/2, or 10. This result nicely complies with the expression for identical variables $POM_{SUM} = POM_{THE\ IDENTICAL\ VARIABLE}$. This result also complies with the general expression $POM_{SUM} \leq POM_{MAX}$.

[15] Summary of the Effects of Arithmetical Processes on Resultant Data

Let us demonstrate in another way the sharp contrast between random multiplication processes and random addition processes. We randomly add four Uniform variables and then contrast all this by randomly multiplying the same set of four Uniform variables. The four variables are: Uniform(6, 75), Uniform(3, 37), Uniform(5, 55), and Uniform(2, 35). We shall give theoretical POM values, as well as actual POM values from computer simulations with 35,000 runs each.

Addition process: Uniform(6, 75) + Uniform(3, 37) + Uniform(5, 55) + Uniform(2, 35)
Multiplication process: Uniform(6, 75)*Uniform(3, 37)*Uniform(5, 55)*Uniform(2, 35)

Addition process: Theoretical POM = 13   Actual POM = 9
Multiplication process: Theoretical POM = 29677   Actual POM = 6626

Addition process: 1st Digits – {61.8, 0.1, 0.3, 1.1, 2.7, 4.7, 7.6, 9.8, 12.0} SSD = 1645
Multiplication process: 1st Digits – {30.3, 18.2, 12.7, 9.5, 8.0, 6.5, 5.4, 4.8, 4.4}   SSD = 0.9

The individual POM values are {75/6, 37/3, 55/5, 35/2}, namely {12.5, 12.333, 11.0, 17.5}.

$$POM_{SUM} = [MAX_X + MAX_Y + MAX_Z + \ldots] / [MIN_X + MIN_Y + MIN_Z + \ldots]$$
$$POM_{PRODUCT} = [MAX_X * MAX_Y * MAX_Z * \ldots] / [MIN_X * MIN_Y * MIN_Z * \ldots]$$

Theoretical POM_{SUM} = (75 + 37 + 55 + 35)/(6 + 3 + 5 + 2) = (202)/(16) = 13
Theoretical $POM_{PRODUCT}$ = (75*37*55*35)/(6*3*5*2) = (5341875)/(180) = 29677

Alternatively, applying the expression $POM_{PRODUCT} = \prod POM_J$ regarding POM values of the four Uniform variables as calculated above, we obtain:
Theoretical $POM_{PRODUCT}$ = (12.5)*(12.333)*(11.0)*(17.5) = 29677

For the process in Figure 2.20 of repeated additions of eight logarithmic Lognormal distributions, POM is steadily decreasing, thereby weakening the intrinsic Benford configuration at each stage. Yet, a worse calamity is awaiting Frank Benford, namely that skewness is also diminishing at each stage of addition, finally attaining that highly symmetrical curve totally contrary to Benford, namely that of the Normal distribution, as predicated by the CLT. Consequently, additive CLT, in one impulsive instant, ruins what MCLT labored on so hard, and for so long, in building up all these individual skewed Lognormals by way of multiplications.

In sharp contrast, Figure 2.13 examines what happens to POM and SSD values in a random multiplicative process of four Uniform distributions, demonstrating how it all converges to Benford – a sort of a reversal and the opposite of the random additive process of the eight Lognormals examined in Figure 2.20 where it all diverges away from Benford.

Let us summarize the effects of random arithmetical processes on resultant data:

Randomly multiplying Uniforms yield Lognormal
Randomly multiplying Normals yield Lognormal
Randomly multiplying Lognormals yield Lognormal
Randomly multiplying Benfords yield Benford
Randomly multiplying non-Benfords yield Benford

Randomly adding Lognormals yield Normal
Randomly adding Uniforms yield Normal
Randomly adding Normals yield Normal
Randomly adding non-Benfords yield non-Benford
Randomly adding Benfords yield non-Benford

Addition Processes: Less POM - More Symmetry - CLT -   Normal - Anti Benford
Multiplication Processes: More POM - More Skewness - MCLT - Lognormal - Pro Benford

The typical multiplicative form of the equations in physics, chemistry, astronomy, geology, and practically all other scientific disciplines, as well as those of their many applications and results, almost always leads to the manifestation of Benford's Law in the physical world. This is so because resultant order of magnitude is sufficiently high in almost all cases.

[16] Multiplication's Ability to Explain Numerous Logarithmic Processes

Let us apply what was learnt in this section regarding tugs of war between additions and multiplications to actual stochastic and physical models with respect to Benford behavior.

In Kossovsky (2014), chapter 97 titled "The Remarkable Versatility of Benford's Law" discusses the seemingly unrealistic quest to somehow unite all the diverse physical processes, causes, and explanations of Benford's Law into a singular concept, and to show that that fundamental concept is logarithmic. Well, a partial success in at least a limited measure can be found in the generic arithmetical structure of multiplications as examined in this section.

Each logarithmic process has already or should have its own rigorous mathematical proof, and one should at least initially treat them separately, yet, having an overall vista, namely the understanding of the common thread running across all these processes helps in seeing the entire forest instead of just individual and unconnected trees. Let us list eight well-known logarithmic processes that could all come under the protective umbrella of the generic multiplicative process:

1) Random Linear Combinations and accounting data - Kossovsky (2014) chapters 16 and 46.
2) Random Rock Breaking - Kossovsky (2014) chapter 92.
3) Random Multiplications of Random Variables - chapter 8 of this section.
4) Final speed of particles randomly driven to accelerate – chapter 8 of this section.
5) Final position of particles under constant decelerating force – chapter 8 of this section.
6) Exponential Growth Series of the standard deterministic (fixed) factor.
7) Exponential Random Growth Series having a random factor (random log walk) –
   Kossovsky (2014) chapters 22 and 78.
8) Super Exponential Growth Series - Kossovsky (2014) chapter 99.

Surely, other articles will be written about Benford's Law in the future, giving rigorous mathematical proofs that this or that process relating - directly or indirectly, explicitly or implicitly - to multiplication processes is logarithmic. They would all come under the same generic protective umbrella of multiplications. Surely the mathematicians ought to provide distinct rigorous proof for each case separately, yet, given that multiplication is repetitive and that it overwhelms the system (e.g. not being mixed with additions to a great extent as well as having or generating plenty of order of magnitude), the expectation is then to find a decisive logarithmic behavior. It must be stressed though that there are several other processes and data structures that are logarithmic for reasons fundamentally different from multiplications, such as quantitative partitioning, mixture of distributions, data aggregations, chain of distributions, random planet and star formation models as in Kossovsky (2014) chapter 94, and so forth. For this reason, the quest to unite all the diverse physical processes, causes, and explanations in Benford's Law may as well be illusionary.

[17] Random Linear Combination is not a Generic Explanation of Benford

Typical **Random Linear Combinations** (RLC) models in Kossovsky (2014) chapters 16 and 46 are of the form: [Price List]*Dice, or of the form: [Price List]*Dice1 + [Price List]*Dice2. The term 'Price List' signifies a very short list of prices with only 6 to 9 items typically, such as {$2.25, $4.75, $7.75, $9.50, $10.25, $35.00}, or {$2.25, $4.75, $35.00}, as well as the list {$2.25, $3.25, $4.75, $7.75, $9.50, $10.25, $25.00, $35.00, $37.00}. The term 'Dice' signifies the standard 6-sided die as the discrete set of {1, 2, 3, 4, 5, 6} with uniform and equal probability for each integer. Dice here represents the number of quantities per item bought by the shopper as shown on the face of the randomly thrown die. RLC is a model for the supermarket's revenue data, or equivalently, a model for the shopper's expense data.

RLC derives its logarithmic tendency exclusively due to the multiplicative nature involved, namely, the product of the Price List by the Dice, and often this is so in spite of the additive terms involved, therefore logarithmic behavior here is in harmony and consistent with all that was discussed in this section. **Indeed, RLC is <u>not</u> any generic or new explanation of the phenomenon of Benford's Law in and of itself in any sense;** rather RLC is just another manifestation of the ability of the multiplicative process to serve as a generic and authentic explanation of the Benford phenomenon in many real-life cases.

Surely Price List could be structured logarithmically in and of itself, but that's an exogenous issue. And clearly, for a presumed shopper determined to purchase numerous distinct items, say 4 as in [Price List]*Dice1 + [Price List]*Dice2 + [Price List]*Dice3 + [Price List]*Dice4, revenue data may not be Benford if Price List is with low order of magnitude or if it's symmetrical. This is so because the final bill here is structured more as additions than as multiplications, and such state of affairs might let CLT ruin any chance toward Benfordness by pointing to the Normal as the emerging distribution. Benford behavior could still be found here for the 4-item purchase, or even for 5 or 6 items, and so forth, given that Price List is highly skewed <u>and</u> that its order of magnitude is very large, because in such cases the Achilles' heel of the Central Limit Theorem would likely prevent the process from achieving any rapid convergence to the Normal distribution. One should keep in mind that even in cases where the Price List is totally symmetrical, and even though Dice is indeed symmetrical, their product [Price List]*Dice is skewed and non-symmetrical as in all multiplication processes, and therefore if such skewness is strong enough, and if Price List and Dice combine to yield high order of magnitude for their product, then it might invoke the Achilles' heel of the CLT, prevents the Normal from emerging too rapidly, and leads to Benford behavior even for a shopper buying 4, 5, or 6 items.

The understanding gained in this section helps to explain one result in Random Linear Combinations involving continuous variables [as opposed to discrete Price Lists] that hitherto appeared quite peculiar. This result appears in Kossovsky (2014) chapter 46, pages 182 & 183, where **Uniform(0, UB)*Dice1 + Uniform(0, UB)*Dice2 + Uniform(0, UB)*Dice3** strongly deviates from the logarithmic regardless of the particular value assigned to UB, while **Uniform(0, UB)*Dice** was found to be approximately logarithmic for any UB value.

Clearly, in such tugs of war between multiplication and addition as in the 3 Uniforms model, addition should have the upper hand here via CLT, assuming that Uniform(0, UB) and Dice do not combine to yield very high order of magnitude for their product, or that the skewness of Uniform(0, UB)*Dice is not strong enough, and that therefore the rate of convergence of CLT here is rapid; that the Achilles' heel of the Central Limit Theorem is not relevant here, and thus results are decisively non-logarithmic – as indeed was confirmed empirically there.

On the other hand, the single term Uniform(0, UB)*Dice is exclusively multiplicative in nature; totally lacking any additions; the CLT is not involved here at all; and therefore results are nearly logarithmic.

On another note, most of the discrete examples about RLC given in Kossovsky (2014) chapter 46 come out quite close to the logarithmic – in spite of that even and balanced split between addition and multiplication such as in the model [Price List]*Dice1 + [Price List]*Dice2. The (partial) reason for this is that the Price Lists themselves in that chapter are often quite skewed quantitatively and their orders of magnitude are relatively large; rendering the model resistant to CLT by way of the Achilles' heel of the CLT.

[18] Distinct Models of Revenue Data Lead to Varied Digital Results

In light of what was discussed in this section, the claim made on page 191 in Kossovsky (2014), that not only [General Store Shopping] but also [Car] is logarithmic under certain conditions, supposedly because both processes generate enough variability in resultant data, cannot be valid, unless Car's components are exceedingly skewed and come with some extremely large orders of magnitude, so that the process is resistant to CLT by way of the Achilles' heel of the CLT. Surely, such scenario for Car is highly unlikely. In addition, Car has to overcome the big hurdle of having so many additive terms within its expression, rendering it highly vulnerable to CLT's tenacious drive towards Normality.

[General Store Shopping] = N_1*Item1 + N_2*Item2 + N_3*Item3

[Car] = 4* Brake + 1*Engine (with its many smaller components) + 2*Bumper + 4*Door + 1*Fuel Tank + 3*Mirror + 1*Transmission System + 4*Bearing + 4*wheel + 1*Air Conditioning + 1*Steering System + 4*Tire + 2*Air Bag + etc.

The model of General Store Shopping expressed as the addition of <u>three</u> addends appears to resemble a bit more addition processes than multiplication processes, hence given that order of magnitude of the price list or the catalog is not high enough, or that it is nearly symmetrical, results might be significantly influenced by the CLT and thus Normal-like, resulting in deviation from Benford.

On the other hand, the distribution of General Store Shopping expressed as the addition of only <u>two</u> addends [*as is usually the case in Kossovsky (2014) chapter 46*] resembles multiplication processes and addition processes in equal measure, and thus results might be close to Benford, especially when Price List is highly skewed and its order of magnitude is high.

In summary: Random Linear Combinations and accounting data are normally full of tensions and dramatic tugs of war between additions and multiplications. Therefore each particular manifestation or model of Random Linear Combinations and accounting data should be judged and evaluated individually according to the relative strength of the competing addition and multiplication forces and their abilities to exert influence upon the system, as well as according to the quantitative configuration of the price list involved.

[19] Revenue Data and the Achilles' Heel of the CLT

The typical bill for a shopper at a large supermarket or at a big retail store purchasing several different items is a mixture of random combinations of multiplications and additions. For example, the bill could be expressed as follows:

3*($2.75 *bread*) + **5*($2.50** *tuna*) + **2*($7.99** *cheese*) = (**$36.73** *total bill*)

Since the typical shopper buys only very <u>few distinct items</u>, it follows that very few additions are involved, and therefore CLT cannot manifest itself well, especially if the price list is skewed and of high order of magnitude. Empirical examinations of typical price lists and catalogs of several well-known and big retail shops, supermarkets, and large companies show that they are indeed highly skewed in favor of the small, and that their orders of magnitude are quite large. Therefore, the typical bill should be highly skewed in favor of the small and Benford, just as the underlying price lists and catalogs are, and this conclusion is indeed in perfect agreement with empirical examinations of real-life bills in revenue data sets.

Clearly, for a particularly presumed shopper determined to purchase <u>numerous distinct items</u>, revenue data may not be skewed in favor of the small and may not be Benford. But even for these shoppers, the Achilles' heel of the CLT could still strongly retard convergence to the Normal distribution if the price list is highly skewed and of high order of magnitude.

For more modest retail stores and smaller supermarkets having price lists and catalogs without much skewness and with relatively modest order of magnitude, revenue data may deviate somewhat from Benford. Decisive deviation from Benford in revenue data can be found in very small business operations, such as a very small coffee shop at the corner of an insignificant and short street, serving only 3 brands of overpriced and very bitter coffee, as well as just 2 types of tasteless cakes, and having a very small clientele.

[20] Expense Data Empirical Compliance with Benford's Law

Almost all accounting, financial, and economics-related data – and with only very few and very rare exceptions - are structured in such a way that the small is more numerous than the big, namely having positive skewness, and with the Benford digit configuration.

Empirical examination of actual revenue data at the raw level, detailed bill by bill (as opposed to summaries or ratios) is nearly impossible to obtain due to confidentiality and secrecy issues. Only Quarterly or Annual Financial Statements are public information that can be readily examined, but Benford and skewness are encountered mostly in the raw detailed values (at the individual bill, receipt, and invoice level), and not really in summary values, aggregations, or ratios of values, such as the numbers typically found in the Financial Statements.

Fortunately for statisticians, some governmental expense data is available at the raw and original level, detailing each and every transaction.

The State Of Oklahoma in the USA provides detailed information at the transaction level for all its vendor payments for the fiscal year 2011. The website for this database is at: https://data.ok.gov/dataset/state-oklahoma-vendor-payments-fiscal-year-2011

These payments reflect disbursements from a state fund for the purchase of goods received, services performed, reimbursements, and payments to other governments. Although this data set is purely of expenses and costs, yet, it also reflects strongly on revenue data in general. This is so since every entry here for a given expense is surely also one revenue item for some provider, company, or agent, billing the state and charging it for the product or service rendered.

Examination of this data set reveals that the distribution of expense amounts is highly skewed in favor of the small. Figure 2.23 depicts the histogram of the vast majority of expenses from $0 to $1 million, containing 986,962 items; leaving out only a small minority of 530 very expensive items of over $1 million. The horizontal x-axis scale is mostly a logarithmic one, except for that jump or gap from $0 to $100. Clearly, except for a brief and very gentle rise on the left of the histogram between $0 and $1000, the cheap (considered as small) outnumbers the expensive (considered as big). The temporary and minor reversal of the histogram in the beginning for very low values is quite typical in revenue and expense accounting data, yet, the overall description of relative quantities is decisively in favor of the small.

Examining digital configuration for this data set (of the expenses from $0 to $1 million, containing 986,962 items) yields the following results:

First digits for Oklahoma: {29.7, 17.7, 12.1, 9.7, 8.6, 6.6, 6.1, 4.9, 4.5}
Benford's Law 1st Digits : {30.1, 17.6, 12.5, 9.7, 7.9, 6.7, 5.8, 5.1, 4.6}

The extremely low 0.9 SSD value here is a strong indication that the Oklahoma expenses data set complies with Benford's Law almost perfectly. Figure 2.24 visually demonstrates the strong compliance with Benford's Law for this extremely large data set containing 987,492 bills.

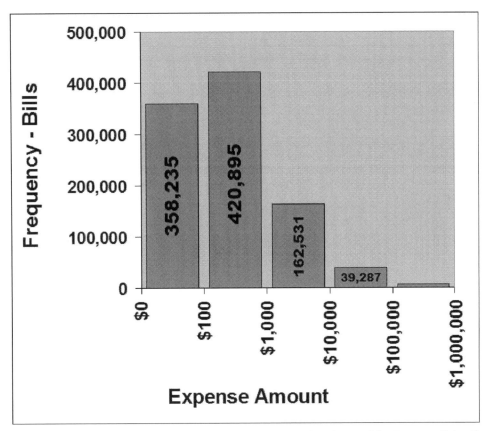

Figure 2.23: Histogram of Oklahoma Expenses is Skewed in Favor of the Small

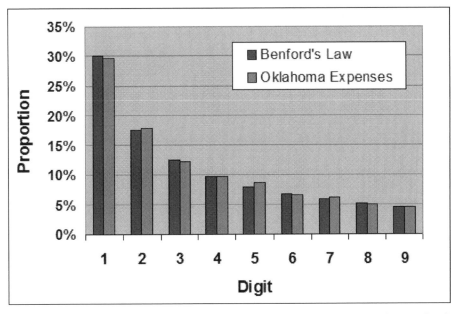

Figure 2.24: Superb Compliance of Oklahoma Expenses with Benford's Law

[21] Molar Mass of Chemical Compounds and the Achilles' Heel of the CLT

Lactose is chosen as a good representative of chemical compounds, demonstrating how molecules are typically formed from elementary atoms in the natural world. The chemical formula for Lactose is $C_{12}H_{22}O_{11}$ and its molar mass is 342.29648 g/mol.

12*(12.0107 *Carbon***) + 22*(1.00794** *Hydrogen***) + 11*(15.9994** *Oxygen***) = 342.29648** g/mol

The approach here differs only slightly from the case of revenue data pertaining to a large collection of purchasing bills. The consideration is of the molar mass, namely the weight of the entire molecule, which is simply the sum of the weights of all the small atoms constituting the large molecule.

Empirical examination of the Periodic Table shows that its histogram is nearly symmetrical and that it's only marginally and very slightly skewed in favor of the small. In addition, order of magnitude of the Periodic Table is found to be fairly large.

The main part of the Periodic Table is quantitatively analyzed, considering the most relevant elements, from Hydrogen with atomic number 1, all the way to Radon with atomic number 86 and atomic weight 222.0; and ignoring the less relevant heavier elements with atomic numbers over 86. Figure 2.25 depicts the histogram of the Periodic Table by the weight variable, from 0 up to 216 (just below Radon) in steps of 27 unified atomic mass units (u) for the bin width.

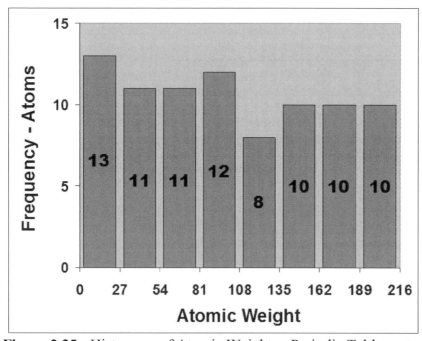

Figure 2.25: Histogram of Atomic Weights - Periodic Table up to Radon

On the face of it, the Central Limit Theorem should guarantee that the weights of molecules are symmetrically distributed; closely mimicking the shape of the Normal Distribution, and that Benford's Law is not obeyed at all. This assumption is based on the observation that the molar mass is structured primarily as an addition process with 3 or 4 typical addends, and even with 5 addends in some cases; while the multiplicative terms in the molar mass expression are with only 2 multiplicands. Yet, this is not the case in this tug of war, and in reality the Achilles' heel of the CLT strongly retards convergence to the Normal distribution. This is so because the multiplicative terms within the expressions are somewhat skewed and of sufficiently high order of magnitude. Indeed, since the Periodic Table itself is of high order of magnitude, these multiplicative terms are of high order of magnitude as well. In addition, these multiplicative terms are also somewhat skewed, even though the Periodic Table is far from being skewed, and this is so since the very act of multiplying - such as in (11)*(Oxygen Mass) – leads to skewness as in all multiplication processes with sufficient order of magnitude (and this act of multiplying also leads to even higher order of magnitude as an extra bonus).

In conclusion: the molar mass of molecules should be skewed in favor of the small and approximately Benford.

[22] Molar Mass Empirical Compliance with Benford's Law

A large list of 2175 commonly used and naturally occurring chemical compounds is provided on the website: http://www.convertunits.com/compounds/. Here we consider a large collection of molecules which are relevant to human use, industrial production, human environment, and general availability. The website criterion for the selection of these 2175 molecules does not follow any particular formal procedure, and instead it simply pulls together information from a variety of chemical and scientific sources, as well as utilizing the informal suggestions of chemists and scientists regarding which molecules should be considered as relevant and important for compilation.

The real-life data on molar mass in the above-mentioned large list of 2175 chemical compounds is indeed quantitatively skewed in favor of the small, as predicted theoretically. Figure 2.26 depicts the histogram up to 900 gram/mole, where small molecules outnumber big molecules in general, except for a brief rise on the very left of the histogram between 0 and 300 where relatively bigger molecules are slightly more numerous than smaller molecules. Such temporary and minor reversal of the histogram in the beginning for very low values is quite typical in many other physical, scientific, financial, and accounting data sets.

Examining compliance with Benford's Law for this set of 2175 chemical compounds we get:

First digits of Molar Mass: {31.9, 25.2, 16.1, 8.4, 5.7, 4.3, 2.9, 3.2, 2.3}
Benford's Law 1st Digits : {30.1, 17.6, 12.5, 9.7, 7.9, 6.7, 5.8, 5.1, 4.6}

Admittedly, the 102.9 SSD value for the Molar Mass data is somewhat high, indicating weak or partial compliance with Benford's Law. Figure 2.27 depicts the partial compliance of this list of molar mass values with Benford's Law.

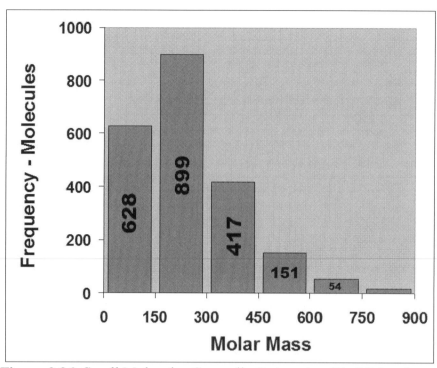

Figure 2.26: Small Molecules Generally Outnumber Big Molecules

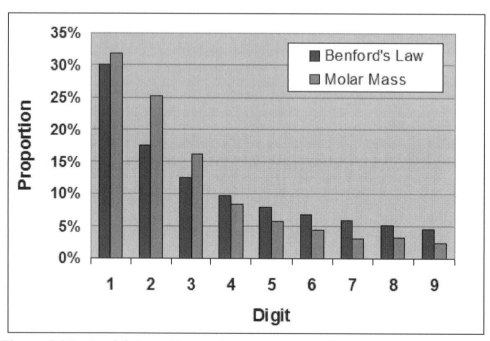

Figure 2.27: Partial Compliance of the Molar Mass Data with Benford's Law

[23] Application of the Distributive Rule in Random Arithmetical Processes

One should be cautious in interpreting the application of the Distributive Rule in algebra to expressions describing random statistical processes. For example, given the probability distributions A, B, C, D, X; the process of adding A, B, C, D, followed by the multiplication of that sum by X is symbolically written as **X*(A + B + C + D)**, namely that the process is principally a product of two multiplicands, and on the face of it may be quite close to Benford assuming large enough orders of magnitude for the individual distributions. Yet, the application of the Distributive Rule here would yield **X*A + X*B + X*C + X*D**, namely that the process is principally an addition of four addends, and on the face of it may be non-Benford due to CLT, especially if distributions are symmetrical or if they are with low orders of magnitude.

The above interpretation is mistaken. These two expressions above signify an identical statistical process which is multiplicative in nature, in spite of their distinct appearances! It is important to keep in mind that in the latter expression of X*A + X*B + X*C + X*D, all four X symbols denote an identical and singular value of the variable, in other words, one particular realization of X in the stimulations. A truly distinct random process could be gotten by having four distinct and independent realizations of X, such as in the expression $X_1*A + X_2*B + X_3*C + X_4*D$, where it truly resembles more additions than multiplications; most likely be non-Benford; and closely resembling the Normal Distribution. Surely, the Distributive Rule cannot be applied here since in general $X_I \neq X_J$.

[24] An Historical Note on Simon Newcomb's Ratio of Uniforms Assertion

In light of the new results and better understanding obtained in this section, Simon Newcomb's two-page short article in 1881 now appears to contain not only the correct digital proportion in real life typical data sets (i.e. Benford's Law), but also a remarkable insight into why it should be so in data relating to the physical sciences - namely 'almost' one correct explanation for the existence of the law!

Newcomb writes: *"As natural numbers occur in nature, they are to be considered as the ratios of quantities. Therefore, instead of selecting a number at random, we must select two numbers, and inquire what is the probability that the first significant digit of their ratio is the digit n."*

While the ratio [or product] of just two independent random variables may not be quite exactly "Benford" or "Newcomb", and digital configuration still depends somewhat on the type of distributions involved and especially on their orders of magnitude, yet it could be very close to the logarithmic, as was seen in this section. Clearly, ratios and products belong to the same class of random arithmetical processes, while additions and subtractions belong to another class.

SECTION 3

QUANTITATIVE PARTITION MODELS

[1] Partitions Almost Always Lead to Quantitative Skewness

The mathematical field of Integer Partition investigates the ways an integral quantity can be expressed as the sum of (smaller) integers. It deals with very simple and straightforward questions such as: "In how many ways and exactly how can the quantity 7 be broken into integral parts?" The answer to this question is the exhaustive list of all possible integral partitions of 7 as shown in detail below:

7
6 + 1
5 + 2
5 + 1 + 1
4 + 3
4 + 2 + 1
4 + 1 + 1 + 1
3 + 3 + 1
3 + 2 + 2
3 + 2 + 1 + 1
3 + 1 + 1 + 1 + 1
2 + 2 + 2 + 1
2 + 2 + 1 + 1 + 1
2 + 1 + 1 + 1 + 1 + 1
1 + 1 + 1 + 1 + 1 + 1 + 1

If the focus is on the quantity 5 instead of the quantity 7, then the question posed is: "In how many ways and exactly how can the quantity 5 be broken into integral parts?" The answer to this question is the exhaustive list of all possible partitions of 5, and this is shown below together with some philosophical comments about sizes:

5   ← few big parts
4 + 1   ← few big parts
3 + 2
3 + 1 + 1
2 + 2 + 1
2 + 1 + 1 + 1   ← many small parts
1 + 1 + 1 + 1 + 1   ← many small parts

These two examples demonstrate a very profound, universal, and yet extremely simple principle regarding how a conserved quantity can be partitioned into parts, namely the observation that: **'One big quantity is composed of numerous small quantities'**, or equivalently: **'Numerous small quantities are needed to merge into one big quantity'**.

Partitioning a fixed quantity into parts can be done roughly-speaking in two extreme styles, either via a breakup into many small parts, or via a breakup into few big parts. For example, two extreme styles of integer partition of 7 are {1, 2, 1, 1, 1, 1} with many small parts, and {3, 4} with fewer but relatively bigger parts. A more moderate style perhaps would be to have a mixture of all kinds of sizes, consisting of many small ones, some medium ones, and few big ones. The above conceptual outline is one of the chief causes why so often real-life data sets are skewed quantitatively, having numerous small values, but only very few big values.

The principle merits additional visual presentations. Figure 3.1 demonstrates all possible integer partitions of 5, where the small clearly outnumbers the big in the entire scheme. Naturally, 3 could be designated as the middle-size quantity here. Consequently, 1 & 2 are designated as the small ones, while 4 & 5 are designated as the big ones. According to such classification of sizes, there are only 2 big quantities, but as many as 16 small quantities. In any case one should not lose sight of the main aspect seen in Figure 3.1 which shows a mixture of all sorts of sizes, but with a strong bias towards the small, discriminating against the big. This is of course true in all Integer Partitions, and not only for integer 5. Figure 3.2 which organizes all the parts of Figure 3.1 nicely according to size, clearly demonstrates the above principle, and if modified as a proper histogram then it can be said to be positively skewed with a tail falling on the right.

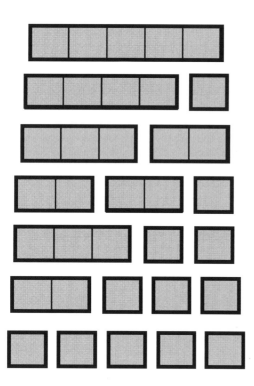

Figure 3.1: The Small Decisively Outnumbers the Big in Integer Partitions of 5

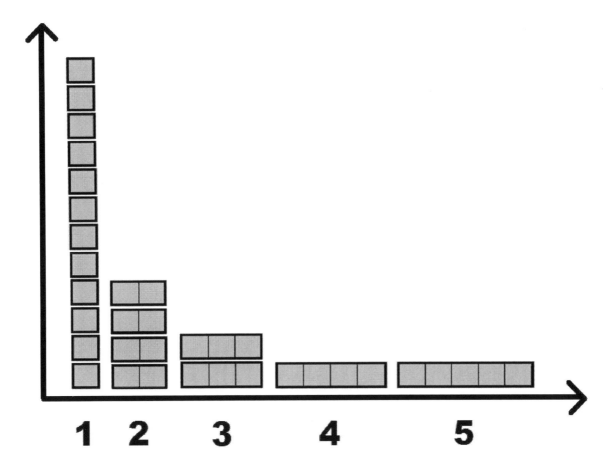

Figure 3.2: The Organization of All Possible Integer Partitions of 5 – A Histogram of Sorts

The abundance of small integers and the rarity of big integers is an intrinsic feature of Integer Partitions, and this is certainly not restricted to the examples of Integer Partition of 7 and Integer Partition of 5. As an additional example, Integer Partition of 13 is similarly analyzed. There are 101 distinct possible partitions for integer 13; as compared with only 15 possible partitions for integer 7, and only 7 possible partitions for integer 5. Only five partition examples out of the complete set of 101 possibilities are shown below as follows:

$13 = 7 + 4 + 1 + 1$
$13 = 4 + 3 + 3 + 2 + 1$
$13 = 10 + 1 + 1 + 1$
$13 = 5 + 5 + 3$
$13 = 7 + 6$

Figure 3.3 depicts the histogram of <u>all</u> the integers from 1 to 13 within <u>all</u> 101 possible partitions. There are 556 integers in total residing within these 101 partitions of 13. The histogram is consistently and monotonically falling to the right, except at the very end for integer 12 with frequency 1 and for integer 13 with frequency 1. Since integer 12 and integer 13 occur exactly once in the entire scheme, namely as $13 = 13$ and as $13 = 12 + 1$, the histogram is actually flat and horizontal there for the tiny part on the right-most part of the histogram.

All histograms in Integer Partitions of whatsoever integer N are flat at the end of the tail on the right-most part for the largest two integers. This is so since N = N and N = (N – 1) + 1.

Yet the histogram never retreats and it never rises, and thus in the case of Integer Partition it can be said in general that the small is consistently more numerous than the big (with the irrelevant and insignificant exception of the largest two integers).

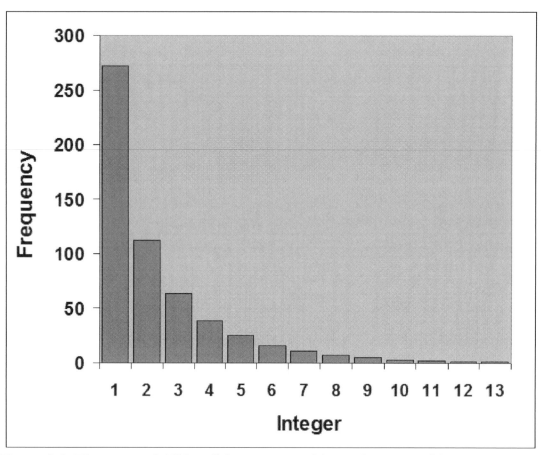

Figure 3.3: Histogram of All Possible Integer Partitions of 13 – Small is Numerous

If we ask a friend to partition a given integral quantity into smaller integral parts only once, in any way he or she sees fit, then no grand conceptual principles can be applied. The friend could favor the big, or could favor the small. If possible, the friend might even partition the quantity into completely even pieces perhaps, where all the parts are of the same quantity, resulting in one size only. Clearly, no quantitative prediction can be made whatsoever for a single partition. Yet, if we ask the friend to randomly repeat partitioning many times over, or if he or she has the time and the patience to perform all possible partitions and to present them as one vast data set, then the skewed configuration where the small is numerous is inevitable! Such resultant blind bias toward the small is observed as long as he or she does not favor any particular sizes, treating all sizes equally and fairly. This generic bias towards the small is almost a universal principle in most other partition models, including those where fractional parts are allowed.

In Figure 3.4 the entire area in the shape of an oval representing the original value is partitioned in two distinct ways according to size. The first partition in the left panel divides the oval-shape area into big parts, and therefore the low number of only 5 parts is obtained. The second partition in the right panel divides the [same] oval-shape area into much smaller parts, and therefore the high number of 13 parts is obtained. Staring at Figure 3.4 reinforces the inevitability of the consequence of skewnes where the small is numerous in all partitions of a conserved quantity into parts having a variety of sizes. This territorial example broadens the quantitative application of Integer Partitions into fractional partitions as well.

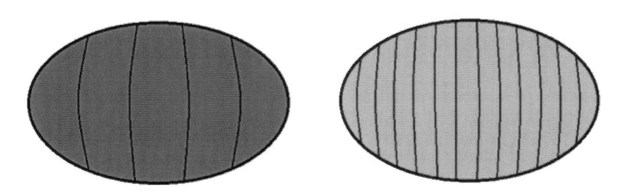

Figure 3.4: A Given Quantity is Partitioned into Either Few Big Parts or Many Small Parts

Figure 3.5 provides another visual and intuitive demonstration that a given random partition containing the big, the small, and the medium; and where all sizes mix together randomly, many more small pieces should be found than big pieces. Figure 3.5 focuses on the area as the quantitative variable, but the lesson learnt from it is generic and applicable to any other types of quantities. Figure 3.5 depicts one possible random partition in the natural world where approximately 1/3 of the entire oval area consists of big parts (around the left side); approximately 1/3 of the entire oval area consists of small parts (around the center); approximately 1/3 of the entire oval area consists of medium parts (around the right side), namely endowing equal portions of overall quantity fairly to each size without any bias. Surely in nature there exists no such order and grouping by size around the left-center-right sections or along any other dimensions. In nature, the big, the small, and the medium are all mixed in chaotically. But the order for the sizes along the left-center-right sections shown in Figure 3.5 is made for pedagogical purposes, to reinforce visually for the reader the profound quantitative consequences affecting sizes in typical partition models. The example given via Figure 3.5 broadens the quantitative application of Integer Partitions into fractional partitions as well.

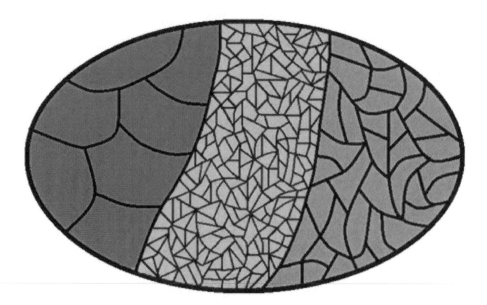

Figure 3.5: An Equitable Mix of Small, Medium, and Big Yielding 'Small is Beautiful'

Hence, the data of Figure 3.5 is structured in such a way that adding all small values yields approximately the same sum as adding all big values (or medium). Yet, in spite of such equitable allocation of quantitative portions to the 3 sizes, the small outnumbers the big. Vaguely counting the number of enclosed areas for each size leads to the decisive conclusion that the small is by far more numerous than the median, and that the median is definitely more numerous than the big. Careful count of the enclosed areas in the oval shape above shows that there are **220** small parts in the middle, **35** medium parts on the right, and only **7** big parts on the left.

It should be emphasized that in fact, real-life data sets frequently deviate from the 33.3% fair allocation of overall quantitative portion for the 3 sizes. Also evidently, a great deal depends on the exact definition of what should constitute small, big, and medium. Since real-life data sets almost never occur nicely with exactly and merely 3 sizes, but rather mostly with numerous distinct values, it is necessary to arbitrarily group all values into 3 camps according to size – assuming one wishes to stick with the small, big, and medium categories.

For example, for the Integer Partition of 5 discussed earlier, the large data set of all possible partitions is {5, 4, 1, 3, 2, 3, 1, 1, 2, 2, 1, 2, 1, 1, 1, 1, 1, 1, 1, 1}. Overall quantity is (5 + 4 + 1 + 3 + 2 + 3 + 1 + 1 + 2 + 2 + 1 + 2 + 1 + 1 + 1 + 1 + 1 + 1 + 1 + 1) = 35. Considering 1 and 2 as small and 4 and 5 as big, the portion of big is $(5 + 4)/(35) = (9)/(35) =$ **25.7%**; the portion of small is $(1 + 2 + 1 + 1 + 2 + 2 + 1 + 2 + 1 + 1 + 1 + 1 + 1 + 1 + 1 + 1)/(35)$ $= (20)/(35) =$ **57.1%**; and the portion of medium is $(3 + 3)/(35) = (6)/(35) =$ **17.1%**. Such state of quantitative affairs is extremely in favor of the small, much more so than the 33.3% equal allocation for all 3 sizes. But surely, the definitions of small, big and medium here are arbitrary.

Considering small as {1}, big as {5}, and medium as {2, 3, 4}, also leads to the small is beautiful conclusion, although of slightly different intensity in beauty. Such re-definition of sizes also yields different allocation of portions from the 35 overall quantity. The portion of big is (5)/(35) = **14.3%**; the portion of small is (1 + 1 + 1 + 1 + 1 + 1 + 1 + 1 + 1 + 1 + 1 + 1)/(35) = (12)/(35) = **34.3%**; and the portion of medium is (4 + 3 + 2 + 3 + 2 + 2 + 2)/(35) = (18)/(35) = **51.4%**.

Figure 3.6 depicts another quantitative configuration that could also occur in the physical world. Here the small constitutes approximately 80% of the entire area (i.e. entire quantity), while the big constitutes only approximately 10% of the area, and the medium only about 10% as well. Portions such as 40%, 50% or even 60% are much more typical for the small in the physical world, while the 80% depicted here is an exaggeration that does not really happen often. Here the small is by far more numerous than the big, over and above the configuration of Figure 3.5.

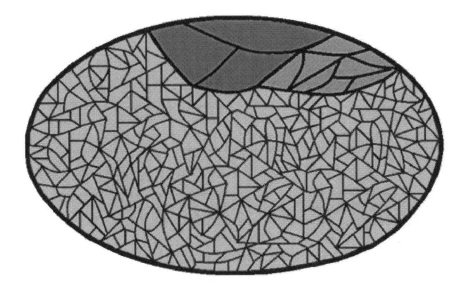

Figure 3.6: Uneven Mix with Too Many Small Parts Yielding 'Small is Exceedingly Beautiful'

Figure 3.7 depicts an unnatural quantitative configuration that rarely occurs in the physical world. Here the big constitutes approximately 85% of the entire area (i.e. entire quantity), while the small constitutes only approximately 5% of the area, and the medium only about 10%. Here the big managed to be more numerous than the small as well as more numerous than the medium by strongly dominating overall area at the expense of the small and the medium.

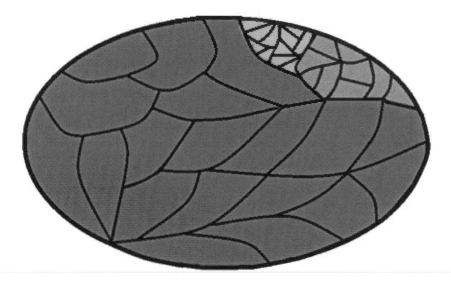

Figure 3.7: Uneven Mix with Too Many Big Parts Yielding Unnatural and Rare Configuration

Apart from any <u>static</u> mathematical arguments, one could actually postulate the equitable mix of all sizes via heuristic argument involving the existence of random consolidation and fragmentation <u>dynamic</u> forces acting upon the quantities. These forces are thought of as physical, chemical, geological, or biological forces, as opposed to any abstract mathematical model of quantitative consolidations and fragmentations. The mechanism driving the system in the direction towards size-equilibrium is the differentiated intensity that these consolidations and fragmentations tendencies occur which depends on the current size configuration of the pieces. The bigger the pieces the stronger is the tendency towards fragmentation. The smaller the pieces the stronger is the tendency towards consolidation. Hence a situation with too many big quantities as in Figure 3.7 is unstable, as nature tends to break the many existing big pieces into smaller ones, much more so than it tends to consolidate the very few existing small pieces into bigger ones, thereby gradually changing the composition of sizes in favor of the small. In a similar fashion, a situation with too many small quantities as in Figure 3.6 is unstable, as nature tends to consolidate the many existing small pieces into bigger ones, much more so than it tends to break up the very few existing big pieces into smaller ones, thereby gradually changing the composition of sizes in favor of the big. The equitable and unbiased mix of Figure 3.5 where all 3 sizes share about 1/3 of the entire territory (quantity) is stable. Perhaps such equitable mix of all sizes is the long term quantitative configuration to which many physical systems tend.

A hypothetical example can be given in the political arena of ancient times. A huge empire is hard to control. Communication and transportation are not well-developed, and so it's difficult for the ruling center in the capital to obtain information about occurrences on the peripheries, and even harder to transport soldiers and material there. Local commanders appointed to rule those far-flung regions are tempted to claim independence and take full control themselves. Hence as a general principle, the larger the empire the greater are the chances and the forces acting upon it towards dissolution and fragmentation. On the other hand, for very small principalities or municipalities ruled by brutal warlords, scheming princes, petty dictators, and such, the opposite forces of conquest, mergers, and consolidations are normally very strong. Here the fear of being absorbed by the well-known proximate neighboring rulers who could easily move armies and material through these short distances is overwhelming and constant. The smaller the political territory the fewer the defenses it can master and the more insecure it feels, and thus the greater the tendency towards conspiracies, attacks, and absorptions. All this leads to plans of preemptive attacks by all sides in order to avoid surprises and conquest, and at times simply due to greed and the desire to rule and exploit larger territory. Such tense state of affairs inevitably ends up with just one successful strongman absorbing his neighboring rivals and ruling all the proximate territory, thus achieving as a consequence much greater political stability. Long term tendencies work against having many big empires or too many small principalities. Thus a happy balance on the political map between the small and the big is achieved, ensuring that this configuration of the territorial sizes is steady and durable.

Surely, not many entities in the world are derived from actual or physical partition processes. Yet, if we substitute the words '**Composition**', '**Constitution**', or '**Consolidation**' for '**Partition**', and think of the entity represented by the data set under consideration as something composed of many parts, then the partitioning vista and its consequences could be appropriate and applicable for some real-life cases. The three oval-shape entities of Figures 3.5, 3.6, and 3.7 and the conclusions from the related analysis are valid whether it is believed that the entities are actively going to be physically partitioned along the lines inside, or that they are passively composed of all these inner parts – and these two points of view are equivalent as far as resultant quantitative configuration is concern. For example, atomic analysis of the typical chocolate consumed by people shows that it is composed mostly of very small hydrogen atoms as well as of medium-size atoms such as carbon, nitrogen, and oxygen, but that it contains only very few and rare big atoms such as zinc, iron, and calcium. A piece of chocolate is certainly not some continuous chunk of primordial matter initially, only to be 'partitioned' later and sorted into its constituents atoms, yet, allowing such fictitious description of the piece of chocolate leads to the same quantitative and size configuration! Clearly, allowing the interpretation of partition as composition enlarges the scope of the analysis.

In other words, applications of partition models in the natural sciences do not have to explicitly assume actual or physical fragmentations of the whole into parts for the results outlined in this chapter to hold. Indeed, a natural entity at times can be thought of as being composed of much smaller parts, or that it exists as the consolidation of numerous separate and smaller parts held together by some physical force or because of any other reason, leading to the same conclusion regarding quantitative and size configuration as for actual partition.

In truth, the exact 3 categories for sizes, namely Small, Medium, and Big shown in Figures 3.5, 3.6, and 3.7 rarely occur as such in nature; they are artificial and arbitrary drawing for the sake of demonstrating the principles of involved regarding partitions. Typically, Mother Nature cannot be as exact, orderly, and regimented when randomly partitioning an overall quantity into parts by sticking only to 3 sizes throughout the entire process. Yet, the neat arrangements of partitions into parts with only 3 possible sizes in the above figures are presented for pedagogical purposes, in order to gain broad insight about the different possibilities of partitions, and in order to inform on the fact regarding which generic configuration is more common and which is relatively rare in real-life data and partitions. There exist two extreme poles above and below such exact and neat partitioning into 3 sizes as in the above figures, as in the following two opposing scenarios:

(I) A partition into identical parts, utilizing only one size for all the parts, as shown approximately in Figure 3.8 (difficulties in artistically drawing the exact same size for all areas led to slight variations). The small is beautiful phenomenon cannot be manifested here a priori since there is only one size involved. Certainly, Mother Nature would almost never partition in such even and controlled manner, except in some exceedingly rare occasions and cases.

(II) A partition into totally distinct parts, utilizing as many sizes as there are parts, as shown in Figure 3.9. This is the most common scenario in refine random partition processes having the flexibility of producing parts of any real, fractional, or integral values whatsoever. True randomness in typical partition processes ensures that Mother Nature would produce her parts in a thoroughly chaotic manner, and without any repetition of any single quantity/size whatsoever.

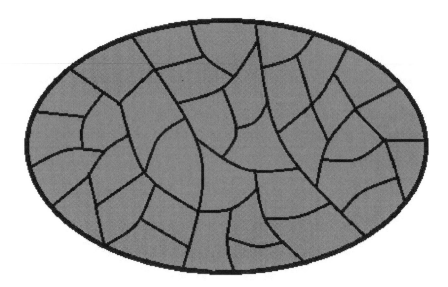

Figure 3.8: Partitioning a Quantity into Equal Parts Utilizing One Size Only

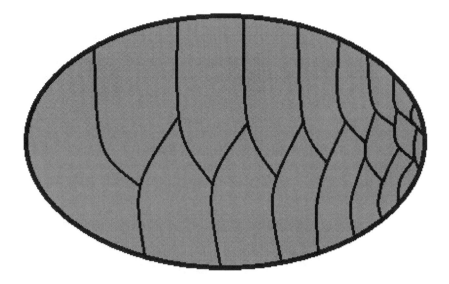

Figure 3.9: Partitioning a Quantity into Totally Distinct Parts – Small May be Beautiful

Yet, guaranteeing that the resultant set of parts contains distinct parts without any repetition of values does not immediately imply the small is beautiful phenomenon (i.e. positive skewness), rather a careful examination of the resultant set of parts must be performed. Surely the fanatical and stubborn statistician can superficially manufacture the phenomenon by generously defining Small on a wide range endowing it many parts, while restricting the definition of Big to a narrower range and thus depriving it of the ability to earn many parts. And surely, another statistician with opposite tendencies, a contrarian, might place the threshold or cutoff boundary points between Small, Medium, and Big in such a way as to greatly benefit Big, letting it earn the majority of parts. Yet, proper definition of sizes requires that we assign equal sub-interval length to all sizes; and that the entire range is divided fairly into equal territorial segments for the various sizes. Each oval-shape type of partition as in Figure 3.9 with totally distinct parts requires careful definition of sizes according to the entire range, as well as subsequent careful count of sizes according to these size definitions – in order to determine whether or not the small is beautiful phenomenon manifests itself there.

The next two chapters regarding refine random partition processes (allowing the parts to be any real numbers without restricting them to integral values) illustrate the main driving force behind the manifestation of the small is beautiful phenomenon for their resultant sets of parts. These two random partition processes will be computer-simulated, and their resultant sets of parts will be shown to exhibit the small is beautiful phenomenon.

[2] Random Dependent Partition is Always in Favor of the Small

The generic small is beautiful principle found in Integer Partitions is applicable to fractional partitions as well, and this fact lends the principle much wider scope and nearly universal applicability in almost all partition schemes. One particular partition scheme in this context is of a well-structured arrangement of repeatedly partitioning a single quantity randomly into many parts, and this is coined as '**Random Dependent Partition**'. This process is best exemplified by randomly breaking a big rock in multiple stages into much smaller pieces, and this example of the generic idea is coined as '**Random Rock Breaking**'. Surely the description of a rock only serves as a vivid example, and the generic idea here is the repeated break up of a single quantity into smaller and smaller values, culminating in the final set of much smaller values.

In spite of the liberal use of the adjective 'random', the process of Random Dependent Partition actually follows a strict partition procedure with exact and carefully executed stages. In the first stage the rock is broken into 2 pieces using a random pair of percentage values, such as 23% and 77% for example. Then the second stage starts with the orderly breaking of each of the 2 pieces in a random fashion using two new random pairs of percentage values, resulting in 4 pieces altogether. In the third stage, each of the 4 pieces is broken into two pieces using four new random pairs of percentage values, resulting in 8 pieces altogether, and so forth.

An essential feature leading <u>rapidly</u> to the small is beautiful phenomenon here is the random manner by which each piece is broken into two smaller pieces, namely that the pair of percentage values are always chosen randomly anew at each stage and for each piece. It is helpful to envision an especially-manufactured roulette in a respectable and honest casino. The roulette wheel contains 99 pockets of numbers and one thrown ball which randomly falls into one of the pockets. These 99 values are marked as {1%, 2%, 3%, … , 97%, 98%, 99%} so that all possible (integral) percentage values can be obtained by chance. For example, if the ball lands inside the 25% pocket, then the 75% - 25% pair of percentages are employed to break the next piece of rock. Figure 3.10 depicts the physical arrangement of such roulette in one imaginary casino.

Figure 3.10: Roulette with 99 Pockets Determining the Percent Breakup in Rock Partition

Let us illustrate Random Dependent Partition with one concrete example of a six-stage process.

A **500** kilogram rock is broken in the first stage via the random 18% - 82% percentage pair, yielding two new pieces [500]**x**[18/100] = **90** and [500]**x**[82/100] = **410**.

In the second stage, firstly, the 90 piece is broken via the random 71% - 29% percentage pair, yielding two new pieces [90.0]**x**[71/100] = **63.9** and [90.0]**x**[29/100] = **26.1**.
Secondly, the 410 piece is broken via the random 86% - 14% percentage pair, yielding two new pieces [410]**x**[86/100] = **352.6** and [410]**x**[14/100] = **57.4**.

In the third stage, the four random pairs of percentage values for the 4 pieces above about to be broken are respectively: 72% - 28%,   29% - 71%,   38% - 62%,   52% - 48%.
This leads to 8 new pieces {**46.0, 17.9, 7.6, 18.5, 134.0, 218.6, 29.8, 27.6**}.

Figure 3.11 depict the process from the original 500-kilogram rock to the end of the third stage.

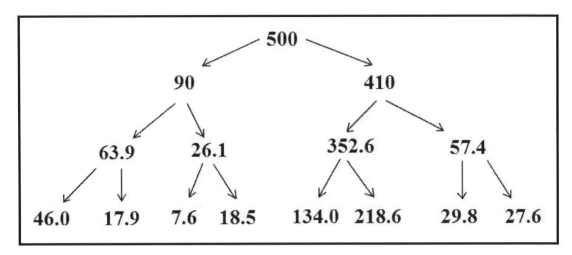

Figure 3.11: Random 500-Kilogram Rock Breaking – From Original Piece to the Third Stage

In the fourth stage, these 8 pieces are broken in a similar fashion, leading to 16 new pieces [*percentage values are not shown for the sake of brevity*]:

8.3	37.7	2.5	15.4	5.1	2.5	15.8	2.8	61.6	72.4	54.7	164.0
17.6	12.2	2.2	25.3								

In the fifth stage, these 16 pieces are broken in a similar fashion, leading to 32 new pieces:

3.8	4.5	17.7	20.0	0.3	2.2	14.9	0.5	0.2	4.9	1.2	1.3
11.2	4.6	2.4	0.3	30.2	31.4	61.5	10.9	45.9	8.7	113.1	50.8
4.2	13.4	5.1	7.1	0.5	1.7	22.6	2.8				

In the sixth stage, these 32 pieces are broken in a similar fashion, leading to 64 new pieces:

1.87	1.94	3.00	1.48	15.25	2.48	4.00	16.00	0.27	0.06	1.00	1.18
2.69	12.24	0.14	0.32	0.07	0.08	1.52	3.39	0.23	0.99	1.07	0.20
10.18	1.01	2.65	1.92	2.10	0.34	0.24	0.10	12.68	17.52	18.55	12.89
19.68	41.82	6.62	4.23	35.35	10.56	3.76	4.98	66.75	46.38	9.15	41.68
2.96	1.27	10.44	2.94	2.36	2.78	6.81	0.28	0.48	0.00	1.63	0.09
13.08	9.48	0.86	1.92								

The final set of the 64 pieces after the sixth stage, sorted low to high, is as follows:

0.005	0.06	0.07	0.08	0.09	0.10	0.14	0.20	0.23	0.24	0.27	0.28
0.32	0.34	0.48	0.86	0.99	1.00	1.01	1.07	1.18	1.27	1.48	1.52
1.63	1.87	1.92	1.92	1.94	2.10	2.36	2.48	2.65	2.69	2.78	2.94
2.96	3.00	3.39	3.76	4.00	4.23	4.98	6.62	6.81	9.15	9.48	10.18
10.44	10.56	12.24	12.68	12.89	13.08	15.25	16.00	17.52	18.55	19.68	35.35
41.68	41.82	46.38	66.75								

A cursory look at the above set of numbers roughly confirms the small is beautiful feature of the pieces, and this observation is decisively confirmed by the detailed histogram of Figure 3.12 which pertains to the 64 pieces after the 6th stage. It should be noted that the first three bins in the histogram are of 1-kilogram width, while the last four bins are of 5-kilogram width. There are 5 pieces weighing over 23 kilogram which are not shown in the histogram as they are very sparsely and thinly spread far out on the horizontal axis, rendering the big there even rarer.

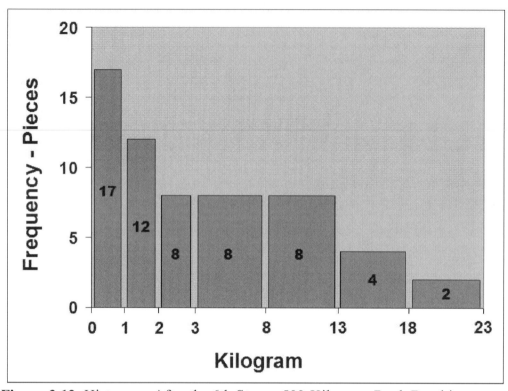

Figure 3.12: Histogram After the 6th Stage –500-Kilogram Rock Breaking

Skeptical or cynical readers might suspect the author for possible manipulations of the roulette simulations so as to arrive at many small pieces but only few big pieces. Yet the above results could be easily re-produced and verified with the aid of a standard personal computer. MS-Excel for example provides random number generator called RAND() which yields random values uniformly and evenly distributed between 0 and 1, such as 0.964125, and 0.176387, and so forth. These random fractions could be interpreted here as the random percentage values. Indeed, this more general usage of fractions from the **real number line on (0, 1)** - without restricting random percentages to integral values - is what constitutes part of the setup and the real definition of Random Dependent Partition.

There are two fundamental differences between the small is beautiful model of Integer Partition as outlined in the previous chapter, and this model of Random Dependent Partition. Firstly, Integer Partition model strictly and exclusively considers only integral values, while Random Dependent Partition liberally includes any fractional and real ones. Secondly, in Integer Partition the model incorporates or aggregates all possible partition scenarios into one vast data set, while in Random Dependent Partition there is no need to aggregate distinct partition scenarios, and instead only a single trajectory or scenario of the breakup of the original rock is considered.

The general <u>dependency</u> of the weights of the pieces at each stage upon the weights of the pieces in the previous stage should be noted carefully. Surely, this dependency on the previous stage also involves dependency on random elements, namely the random percentage values which determine how to break the pieces of the previous stage. Hence the process concocts new values at each stage out of the old values of the previous stage, mixing in random elements as well.

Let us construct a pie chart for the resultant set of 64 rock pieces in order to demonstrate the quantitative breakdown by size as in the oval-like Figures 3.5, 3.6, and 3.7. Defining the 3 sizes fairly, each on a third of the entire range $(66.75 - 0.005)/3 = 22.25$, would endow Small such a huge advantage over the other sizes, that the phenomenon would be very strongly manifested. But since the biggest 66.75 piece is suspected of being an outlier and anomaly, including this anomalous value in the definitions of all the sizes may not be appropriate. In any case, in order to illustrate the persistency and the tenacity of the small is beautiful phenomenon here, it shall be shown to hold true even under a different [yet quite reasonable] set of definitions of sizes having a slight bias against the Small and in favor of the Big, as follows:

Small: Less than 10 kilogram.
Medium: From 10 to 20 kilogram.
Big: Over 20 kilogram.

This size criterion divides the entire set of 64 pieces into 3 classes:

0.005	0.06	0.07	0.08	0.09	0.10	0.14	0.20	0.23	0.24	0.27	0.28
0.32	0.34	0.48	0.86	0.99	1.00	1.01	1.07	1.18	1.27	1.48	1.52
1.63	1.87	1.92	1.92	1.94	2.10	2.36	2.48	2.65	2.69	2.78	2.94
2.96	3.00	3.39	3.76	4.00	4.23	4.98	6.62	6.81	9.15	9.48	

10.18	10.44	10.56	12.24	12.68	12.89	13.08	15.25	16.00	17.52	18.55	19.68

35.35	41.68	41.82	46.38	66.75

Calculations of total weight by size for each of the 3 size classes, as well as the proportion of each size-total within the entire system-weight of 500 kilograms, yields the following results:

Small: 47 pieces weighing a total of 98.9 kilograms, or 98.9/500 = **20%** of overall weight.
Medium: 12 pieces weighing a total of 169.1 kilograms, or 169.1/500 = **34%** of overall weight.
Big: 5 pieces weighing a total of 232.0 kilograms, or 232.0/500 = **46%** of overall weight.

Figure 3.13 depicts the quantitative portion of each piece, sorted low to high in the clockwise direction, and where relative area signifies relative quantity. In addition, the color for Small is light gray; the color for Medium is plain gray; and the color for Big is dark gray. Small is bitter about having only 20% portion of overall weight, and it accuses Big of stealing a portion of its weight, while Medium is not suspected of any theft. This accusation is based on Small's misguided notion that it has a divine right to earn **33.3%** of total quantity in all systems, cases, and situations whatsoever. Big retorts that it didn't do anything of the sort; that this is all the fault of the definer who has created unreasonable criteria for sizes; and that Small should have been assigned the wider range from 0 to 15 kilograms at the expense of Medium, instead of the narrow range of 0 to 10 kilograms. Big's excuse immediately creates animosity between Small and Medium and triggers fierce competition between them in attempts to establish preferential criterion for sizes. Nonetheless, Small is actually quite content, being by far the most numerous size within the set of 64 pieces, having 47 pieces of its own. Big turned out to be the least popular and quite rare, having only 5 pieces out of a total of 64.

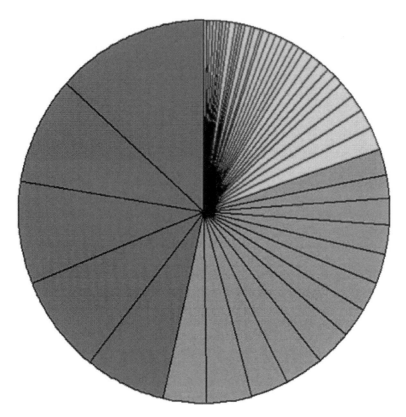

Figure 3.13: Nearly Equitable Mix of Small, Medium, and Big – Random Dependent Partition

[3] Random Real Partition is Always in Favor of the Small

A random and spontaneous partitioning process lacking any dependent stages leads to the small is beautiful phenomenon just as well. Here the phenomenon is encountered without any constraints, free of strict sequential procedures, encompassing integral as well as fractional values. Instead of breaking the original quantity in sequential stages, a comprehensive plan of how and where to cut or break is contemplated beforehand, and then at an opportune and appropriate grand moment the partition is fully executed, cutting or breaking simultaneously in all the planned places and locations. The process is coined as '**Random Real Partition**'. This process is best exemplified by the cutting of one-dimensional long pipe at random locations along its length, and this example of the generic idea is coined as '**Random Pipe Breaking**'.

Let us provide a concrete numerical example. A 15-meter long metal pipe is to be randomly partitioned into 30 parts. This is accomplished by obtaining 29 independent random points along the pipe - prior to the moment of the actual cutting - to serve as marks indicating where the pipe should be partitioned. These marks constitute the grand plan of the entire partition process, and they are generated via independent simulations from the continuous **Uniform(0, 15)**.

Actually, distances from the left edge of the pipe are obtained via 29 computer simulations using the random number generator uniformly distributed between 0 and 1. These 29 simulated numbers then provide random percentage values to determine the positions of the marks along the 15-meter length pipe, as for example (0.28)*(15) = 4.2, or (0.71)*(15) = 10.7, and so forth.

Following the simulation of each such random position, a mark on the physical pipe is made accordingly with a black marker or by gently scratching the location with a saw to indicate where to do the actual cutting later when the grand moment of execution arrives. It is only at the end of the long sequence of 29 simulations and markings that actual cutting and sawing at these marked locations take place, breaking the long pipe into 30 parts of (usually) totally distinct lengths. Figure 3.14 depicts the actual 29 marks along the pipe obtained via computer simulations.

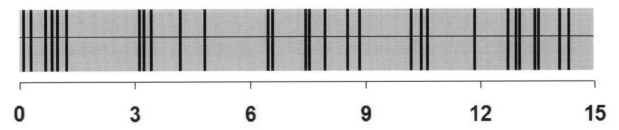

Figure 3.14: The 29 Marks Randomly Scratched along the 15-Meter Pipe Prior to Partition

The 29 random values generating the marks on the pipe, order from low to high; including the left edge of 0.000 as extra, and the right edge of 15.000 as extra, are outlined as follows:

0.000 0.140 0.301 0.710 0.868 1.003 1.243 3.163 3.273 3.453 4.181 4.871
6.500 6.606 7.453 7.550 7.962 8.555 8.847 10.207 10.464 11.652 11.882 12.750
12.947 13.060 13.452 13.564 14.123 14.376 15.000

Calculating distances between the marks (i.e. differences) by subtracting from each value greater than zero its adjacent value on the left, we obtain the set of the lengths of the parts of the pipe:

0.140 0.161 0.409 0.158 0.135 0.239 1.921 0.109 0.180 0.728 0.689 1.629
0.106 0.846 0.097 0.412 0.593 0.292 1.361 0.257 1.188 0.230 0.868 0.197
0.113 0.392 0.112 0.559 0.253 0.624

Ordering these distances low to high, we obtain the final ordered set of the lengths of the parts:

0.097 0.106 0.109 0.112 0.113 0.135 0.140 0.158 0.161 0.180 0.197 0.230
0.239 0.253 0.257 0.292 0.392 0.409 0.412 0.559 0.593 0.624 0.689 0.728
0.846 0.868 1.188 1.361 1.629 1.921

Clearly, the small is beautiful principle is evidently valid for this set of 30 values. Had this pipe been cut deterministically and fairly into 30 equal parts, then each part would have been 15/30 = 0.5 meter long, hence the value of 0.5 serves as the benchmark for the 'truly middle size'. In contrast, for this random partition here, the majority of the parts are shorter than 0.5 meter; fewer are longer than 0.5 meter; and all this is surely in the spirit of the small is beautiful principle.

Figure 3.15 depicts the histogram of these 30 parts after partition, confirming visually the manifestation of the small is beautiful phenomenon for this random partition process.

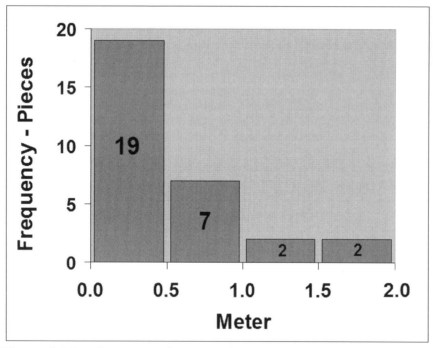

Figure 3.15: Histogram of 30 Parts in Random Real Partition of 15-Meter Pipe

Let us construct a pie chart for the resultant set of 30 pipe parts in order to demonstrate the quantitative breakdown by size as in the oval-like Figures 3.5, 3.6, and 3.7. Defining the 3 sizes fairly, each on a third of the entire range $(1.921 - 0.097)/3 = 0.61$, would endow Small decisive advantage over the other sizes, and the phenomenon would be strongly manifested. But since the biggest 1.921 part may perhaps appear as an outlier, including this value in the definitions of all the sizes might not be prudent. In any case, in order to illustrate the persistency and the tenacity of the small is beautiful phenomenon here, it shall be shown to hold true even under a different [yet quite reasonable] set of definitions of sizes having a slight bias against the Small and in favor of the Big, as follows:

Small: Less than 0.5 meter.
Medium: From 0.5 to 1.0 meter.
Big: Over 1.0 meter.

This size criterion divides the entire set of 30 parts into 3 classes:

0.097 0.106 0.109 0.112 0.113 0.135 0.140 0.158 0.161 0.180 0.197 0.230
0.239 0.253 0.257 0.292 0.392 0.409 0.412

0.559 0.593 0.624 0.689 0.728 0.846 0.868

1.188 1.361 1.629 1.921

Calculations of total length by size for each of the 3 size classes, as well as the proportion of each size-total within the entire system-length of 15 meters, yields the following results:

Small: **19** parts with a total length of 3.993 meters, or $3.993/15 = $ **27%** of overall length.
Medium: **7** parts with a total length of 4.909 meters, or $4.909/15 = $ **33%** of overall length.
Big: **4** parts with a total length of 6.098 meters, or $6.098/15 = $ **41%** of overall length.

Figure 3.16 depicts the quantitative portion of each part, sorted low to high in the clockwise direction, and where relative area signifies relative quantity. In addition, the color for Small is light gray; the color for Medium is plain gray; and the color for Big is dark gray. Here all 3 sizes earn close to the **33.3%** ideal, equitable, and expected proportions of overall length.

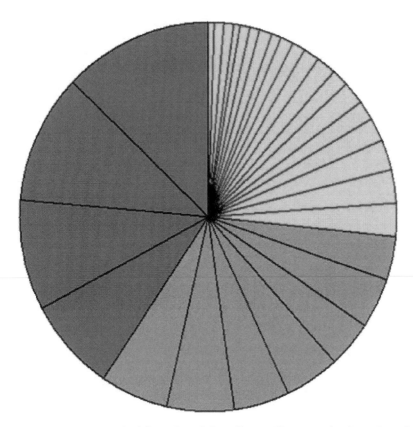

Figure 3.16: Equitable Mix of Small, Medium, and Big – Random Real Partition

The process of Random Real Partition involves random and totally **independent** numerical choices for the partition of a given quantity. Typical examples of the process involve the cutting of one-dimensional long **pipe** at random locations along its length – giving each section and each part of the pipe equal probability in being marked by using the continuous Uniform. Such pipe examples fit naturally with the defined process since it is very easy and straightforward to mark those locations along the one-dimensional pipe beforehand as planned partition. In contrast, **rock** is three-dimensional, and marking surfaces and areas or presenting diagrams and maps for cutting beforehand is extremely cumbersome and complex, and totally impractical.

The term '**real**' in the phrase 'Random Real Partition' refers to the fact that there is no restriction on having only integral values as was the case for Integer Partition, and that the parts or the pieces can attain any fractional, rational, real, and irrational value. In addition, the positions of the marks along the one-dimensional pipe as in Figure 3.14 invoke the concept of the **real line**, or the **real number line** in mathematics, which is the line whose points are the set R of all real numbers (also viewed as a geometric or Euclidean space).

[4] Numerous & Distinct Parts in Partitions are Necessary Conditions

In order to arrive at the resultant small is beautiful quantitative configuration in random partitions, it is necessary at a minimum that the process thoroughly breaks the original quantity into a set of numerous and mostly distinct smaller parts. Some numerical examples follow:

Breaking {15} into {5, 5, 5} or into {1, 2, 3, 3, 3, 3} involve repeated values such as 5 or 3, and thus these partition processes are not conducive to the small is beautiful phenomenon. In addition, these partitions yield very few parts, and this is also not conducive to the phenomenon.

Breaking {15} into {1, 2, 3, 4, 5} is a bit more conducive to the phenomenon since the parts are of distinct quantities without any repetition of values, yet since this process partitions 15 into only 5 smaller parts, it is rendered insufficient to lead to the phenomenon.

Breaking {15} into the set of 30 parts as in the example of pipe breaking of the previous chapter yields the following set of parts (sorted from low to high):

{0.097, 0.106, 0.109, 0.112, 0.113, 0.135, 0.140, 0.158, 0.161, 0.180, 0.197, 0.230, 0.239, 0.253, 0.257, 0.292, 0.392, 0.409, 0.412, 0.559, 0.593, 0.624, 0.689, 0.728, 0.846, 0.868, 1.188, 1.361, 1.629, 1.921}

This pipe partition process is quite conducive to the phenomenon for two reasons. Firstly, the process produces numerous parts, namely thirty. Secondly, the resultant set of parts is of totally distinct quantities, without any repetition of values whatsoever. In other words, when viewed as an ordered set from the low of 0.097 to the high of 1.921, there is constantly some 'growth' in the quantity of the parts; they are steadily increasing; guaranteeing that the entire set of parts are of totally distinct values. As shall be discussed later, the third requirement is that the (percent-wise) growth is steady or varied, but without any overall downward trend.

Certainly, Random Dependent Partition (Rock) with only 1, 2, or 3 stages say, producing only 2, 4, or 8 pieces, does not result in the small is beautiful quantitative configuration, due to scarcity in the number of resultant pieces. In the same vein, Random Real Partition (Pipe) with only 2 to 8 parts say, does not result in the small is beautiful quantitative configuration, due to scarcity in the number of resultant parts. For these two partition processes to arrive at the small is beautiful quantitative configuration it is necessary that enough randomness has been executed throughout the system, and that the system has produced plenty of parts.

Breaking {15} into the set of 30 parts, with 15 parts of value 0.40, 10 parts of value 0.50, and only 5 parts of value 0.80, is a partition process which does indeed produce numerous parts.

{0.40, 0.40, 0.40, 0.40, 0.40, 0.40, 0.40, 0.40, 0.40, 0.40, 0.40, 0.40,
0.40, 0.40, 0.40, 0.50, 0.50, 0.50, 0.50, 0.50, 0.50, 0.50, 0.50, 0.50,
0.50, 0.80, 0.80, 0.80, 0.80, 0.80}

The resultant set of parts is typically not of distinct values, since there are frequent repetitions here, nonetheless, this partition process has been carefully and deliberately calibrated to produce more small parts than big parts, hence it manifests the small is beautiful phenomenon. Surely, it must be acknowledged that this meticulous partition process was not random in any way, but rather deliberately produced and planned.

Breaking {15} into the set of 30 parts, with 10 parts of value 0.20, 10 parts of value 0.50, and 10 parts of value 0.80, is a partition process which does indeed produce numerous parts.

{0.20, 0.20, 0.20, 0.20, 0.20, 0.20, 0.20, 0.20, 0.20, 0.20, 0.50, 0.50,
0.50, 0.50, 0.50, 0.50, 0.50, 0.50, 0.50, 0.50, 0.80, 0.80, 0.80, 0.80,
0.80, 0.80, 0.80, 0.80, 0.80, 0.80}

Yet, the resultant set of parts is typically not of distinct values, since there are frequent repetitions of values. This partition process is definitely not in the spirit of the phenomenon, since all 3 sizes are of equal [10 times] frequency. Surely, it must be acknowledged that also this meticulous partition process was not random in any way, but rather deliberately produced and planned.

THE REST OF THIS CHAPTER FROM HERE IS **OPTIONAL**! READERS CAN (OR SHOULD) SKIP TO THE NEXT CHAPTER #5.

We might view the difficulty of arriving at the small is beautiful phenomenon for this partition process as arising from the fact that there is very little '**growth**' for the parts, and that the '**growth factors**' X_{N+1}/X_N are almost always 1.0; except on the occasion of growth from 0.20 to 0.50 with its growth factor of 2.5; and on the other occasion of growth from 0.50 to 0.80 with its growth factor of 1.6.

It is useful to adopt this particular growth vista for the resultant set of parts (ordered from low to high), artificially viewing the set as exponential growth series. Conceptually of course, there exists no growth in partition processes in any sense whatsoever, and there exists no time dimension here within which quantities 'grow', yet, such a vista is helpful in the description of the conditions necessary for arriving at the small is beautiful phenomenon.

The focus in such growth analysis is on the set of individual factors leading from one value [part] to its next adjacent value [part] on the right, namely on $F = X_{N+1}/X_N$.

The set of 30 parts of the partitioned 15-meter pipe of the previous chapter can be viewed as exponential growth for 29 periods from the smallest 0.097 part to the biggest 1.921 part. The set of 29 growth factors of the ordered set of parts is as follows:

{1.091, 1.028, 1.023, 1.007, 1.200, 1.035, 1.131, 1.017, 1.122, 1.091, 1.168, 1.040, 1.056, 1.018, 1.134, 1.345, 1.043, 1.006, 1.359, 1.061, 1.052, 1.104, 1.057, 1.162, 1.025, 1.369, 1.146, 1.198, 1.179}

Average factor is 1.1126, although for a more precise effective/overall factor we solve for F in the expression First*$(F)^{29}$ = Last, namely 0.097*$(F)^{29}$ = 1.921, leading to F = 1.1084. The crucial feature in the set of these growth factors leading to the small is beautiful phenomenon is that there exists no overall downward trend, and that the growth factors fluctuate randomly up and down around their average value.

As another example, the final set of the 64 pieces of the broken 500-kilogram rock after the sixth stage in chapter 2 is viewed as exponential growth for 63 periods from the smallest 0.005 piece to the biggest 66.75 piece.

{0.005, 0.06, 0.07, 0.08, 0.09, 0.10, 0.14, 0.20, 0.23, 0.24, 0.27, 0.28, 0.32, 0.34, 0.48, 0.86, 0.99, 1.00, 1.01, 1.07, 1.18, 1.27, 1.48, 1.52, 1.63, 1.87, 1.92, 1.92, 1.94, 2.10, 2.36, 2.48, 2.65, 2.69, 2.78, 2.94, 2.96, 3.00, 3.39, 3.76, 4.00, 4.23, 4.98, 6.62, 6.81, 9.15, 9.48, 10.18, 10.44, 10.56, 12.24, 12.68, 12.89, 13.08, 15.25, 16.00, 17.52, 18.55, 19.68, 35.35, 41.68, 41.82, 46.38, 66.75}

The set of 63 growth factors of this ordered set of pieces is as follows:

{ **12**, 1.167, 1.143, 1.125, 1.111, 1.400, 1.429, 1.150, 1.043, 1.125, 1.037, 1.143, 1.063, 1.412, 1.792, 1.151, 1.010, 1.010, 1.059, 1.103, 1.076, 1.165, 1.027, 1.072, 1.147, 1.027, **1.000**, 1.010, 1.082, 1.124, 1.051, 1.069, 1.015, 1.033, 1.058, 1.007, 1.014, 1.130, 1.109, 1.064, 1.058, 1.177, 1.329, 1.029, 1.344, 1.036, 1.074, 1.026, 1.011, 1.159, 1.036, 1.017, 1.015, 1.166, 1.049, 1.095, 1.059, 1.061, 1.796, 1.179, 1.003, 1.109, 1.439}

There is only one repetition in the set of pieces, namely 1.92 and 1.92, leading to only one factor of 1.000. The first extraordinarily large factor of 12 (from piece 0.005 to piece 0.06) is viewed as an outlier and anomaly, thus the smallest 0.005 pieces is excluded from further analysis, and only the remaining 62 'growth periods' are considered beginning from the 2nd-smallest 0.06 piece.

Excluding factor 12, the average factor is 1.1293, although for a more precise effective/overall factor we solve for F in the expression First*$(F)^{62}$ = Last, namely 0.06*$(F)^{62}$ = 66.75, so that F = 1.1198. The crucial feature in the set of these growth factors leading to the small is beautiful phenomenon is that there exists no overall downward trend, and that the growth factors fluctuate randomly up and down around their average value.

Since rock breaking and pipe breaking are performed randomly, and since these partition processes are not limited to integral values but are rather performed on the basis of all possible real numbers, there should normally be no repetition, and the chances of finding many/most/all parts having identical values are exceedingly small, and formally the probability for repeated real values is zero. Given that computer simulations are thorough and refine, then not even a single repetition of parts should be found in these random partition processes. The same can be said about the probability of finding all the parts totally distinct but well-structured and being neatly arranged having the same exact fixed growth factor for all – and which would border on the extraordinary or magical. Randomness ensures that almost nothing is steady and equal in the final resultant quantitative configuration of the set of parts after partition. Even a split concentration of growth factors between two dominant groups, with say one group of a variety of factors between 1.05 to 1.10, and another group of a variety of factors between 1.20 and 1.25, and nothing in between 1.10 and 1.20, would constitute extraordinary and extremely rare occurrence, and this is certainly not expected.

Without arguing mathematically and theoretically a priori against the possibility that random partition processes could result in a clear upward or downward overall trend in the growth factors, what is found empirically and repeatedly is that these growth factors are almost always found to be spread randomly over a range clustered around the value of some average growth factor, showing some gentle downward trend on the left for small pieces, and some gentle upward trend on the right for big pieces, mimicking U-shape histogram of sorts, albeit in a zigzag manner.

In the context of the small is beautiful analysis, the utilization of a single fixed growth factor for the entire partition process shall now be attempted, as it differs little from the U-shape structure of factors. In other words, the assumption is taken that the substitution for all these fluctuating growth factors by a single fixed and deterministic growth factor constitutes a somewhat similar model for random partition processes. Let us then attempt to represent Random Real Partition as in the example of Random 15-meter Pipe Breaking of the previous chapter in terms of a particular deterministic growth model with a constant F_{FIXED} growth factor standing as the average factor for the entire process, and where 30 parts shall be created. In addition, the value of the smallest part is arbitrarily set to 0.045. The value of F_{FIXED} shall be determined by calibrating it in such a way as to yield the total quantity of 15 for the entire deterministic growth process of 29 expansions. Such calibration yields $F_{FIXED} = 1.1365$.

By deriving each consecutive part from its adjacent smaller part on the left as 13.65% growth, we ensure that all the parts are of distinct values, without a single repetition. By fixing the value of all the growth rates at 13.65%, we eliminate the mild upward and downward trends in the set of growth factors, and create a model where they are neither rising nor falling. By choosing many parts, namely 30 values in total, we ensure that there are 'numerous' parts in the system. It is conjectured that these three achievements are enough to guarantee the manifestation of the small is beautiful phenomenon. Indeed, this conjecture can be easily proven mathematically from the expressions of the growing quantity themselves, namely B, BF, BFF, BFFF, BFFFF, and so forth.

Let us examine the resultant parts for this abstract 13.65% growth model from 0.045 base value:

{0.045 0.051 0.058 0.066 0.075 0.085 0.097 0.110 0.125 0.142 0.162 0.184
0.209 0.237 0.270 0.307 0.349 0.396 0.450 0.512 0.582 0.661 0.751 0.854
0.970 1.103 1.253 1.424 1.619 1.840}

Each value here is derived from its adjacent smaller value on the left via multiplication by the factor 1.1365, namely by growing at 13.65%.

The entire range of {0.045, 1.840} is divided fairly into 3 equal sections for the classification of Small, Medium, and Big. Each section is of the width (1.840 - 0.045)/(3) = (1.795)/(3) = 0.598. It follows that Small is from (0.045) to (0.045 + 0.598), and that Medium is from (0.045 + 0.598) to (0.045 + 0.598 + 0.598).

Hence, the 3 classes for size are as follows:

Small: From 0.045 to 0.643.
Medium: From 0.643 to 1.241.
Big: From 1.241 to 1.840.

Utilizing this fair classification to calculate sub-totals per size, we obtain:

Small Total = 0.045 + 0.051 + 0.058 + 0.066 + 0.075 + 0.085 + 0.097 + 0.110 + 0.125 +
0.142 + 0.162 + 0.184 + 0.209 + 0.237 + 0.270 + 0.307 + 0.349 + 0.396 +
0.450 + 0.512 + 0.582 = **4.512**

Medium Total = 0.661 + 0.751 + 0.854 + 0.970 + 1.103 = **4.339**

Big Total = 1.253 + 1.424 + 1.619 + 1.840 = **6.136**

System Total = 4.512 + 4.339 + 6.136 = **15**

This entire set of 30 quantities can be conceptualized as a deterministic partition process of the original quantity of 15 into 30 smaller parts, instead of viewing it as exponential growth series.

According to above fair classification:

Small: 21 parts with a total value of 4.512, or 4.512/15 = **30.1%** of system total.
Medium: 5 parts with a total value of 4.339, or 4.339/15 = **29.0%** of system total.
Big: 4 parts with a total value of 6.136, or 6.136/15 = **40.9%** of system total.

Figure 3.17 depicts the quantitative portion of each part, sorted low to high in the clockwise direction, and where relative area signifies relative quantity. In addition, the color for Small is light gray; the color for Medium is plain gray; and the color for Big is dark gray. Here, all 3 sizes earn in the approximate the **33.3%** ideal, equitable, and expected proportions of overall quantity.

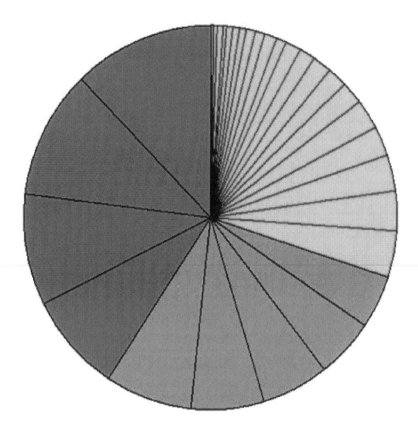

Figure 3.17: Equitable Mix of Small, Medium, and Big in Generic Presentation of Partitions

Hence the attempt to represent a random partition process as simply a set of totally distinct parts with a constant growth factor appears reasonable, while the small is beautiful phenomenon is manifested in both; namely in the partition of the 15-meter pipe with its randomly fluctuating growth factors, as well as in the abstract model of constant quantitative growth.

In order to drive the point that the small is beautiful phenomenon is not much disturbed due to the switch from the original random model to the deterministic model, and that both models yield approximately the same quantitative structure, a comparison chart is added. Figure 3.18 depicts the bar chart of the two sets of parts for comparison, clearly demonstrating that both sets have an almost identical quantitative configuration, and that the small is beautiful phenomenon is manifested in both cases. In a sense, the deterministic model can be thought of simply as applying **'Data Smoothing'** on the random model. Data Smoothing is the use of an algorithm to remove noise from a data set, allowing important patterns to stand out.

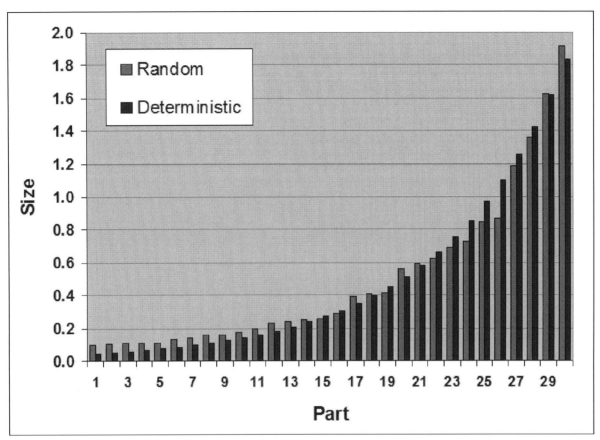

Figure 3.18: Correspondence in Overall Quantitative Configuration - Random & Deterministic

It should be emphasized that the two conditions of this chapter for partitions, namely having numerous and mostly distinct parts, are not entirely sufficient for the manifestation of the small is beautiful phenomenon. The necessary third condition for the manifestation of the phenomenon is that the values of the parts are increasing in a <u>multiplicative manner</u> such as $X_{N+1} = F*X_N$, where F is a constant, or where F varies randomly but without any overall downward trend. The phenomenon is not manifested for a model where the parts are increasing in an <u>additive manner</u> such as $X_{N+1} = X_N + C$, where C is a constant, or where C varies randomly but without any overall downward or downward trend. An additive manner of growth does not yield the small is beautiful phenomenon but rather an equitable and fair size configuration where all sizes are equally numerous. Indeed, an additive manner of growth yields the Uniform Distribution which is size-neutral, and where the small is beautiful phenomenon does not manifest itself.

As a numerical example of the (deterministic) additive partition model, the initial base value (the smallest part) is arbitrarily set to 0.70, and the constantly added value is 0.22. In addition, it is determined that 32 parts shall be created, so that 31 additions of the value 0.22 are executed.

{0.70, 0.92, 1.14, 1.36, 1.58, 1.80, 2.02, 2.24, 2.46, 2.68, 2.90,
3.12, 3.34, 3.56, 3.78, 4.00, 4.22, 4.44, 4.66, 4.88, 5.10, 5.32,
5.54, 5.76, 5.98, 6.20, 6.42, 6.64, 6.86, 7.08, 7.30, 7.52}

143

The entire range of {0.70, 7.52} is divided fairly into 3 equal sections for the classification of Small, Medium, and Big. Each section is of the width $(7.52 - 0.70)/(3) = (6.82)/(3) = 2.27$. Small is from (0.70) to (0.70 + 2.27). Medium is from (0.70 + 2.27) to (0.70 + 2.27 + 2.27). Hence, the 3 classes for size are as follows:

Small: From 0.70 to 2.97.
Medium: From 2.97 to 5.25.
Big: From 5.25 to 7.52.

Utilizing this classification to calculate sub-totals per size, we obtain:

Small Total = 0.70 + 0.92 + 1.14 + 1.36 + 1.58 + 1.80 + 2.02 + 2.24 + 2.46 + 2.68 + 2.90
        = **19.8**

Medium Total = 3.12 + 3.34 + 3.56 + 3.78 + 4.00 + 4.22 + 4.44 +   4.66 + 4.88 + 5.10
        = **41.1**

Big Total = 5.32 + 5.54 + 5.76 + 5.98 + 6.20 + 6.42 + 6.64 + 6.86 + 7.08 + 7.30 + 7.52
        = **70.6**

System Total = 19.8 + 41.1 + 70.6 = **131.5**

This entire set of 32 quantities can be conceptualized as a deterministic partition process of the original quantity of 131.5 into 32 smaller parts, instead of viewing it as an additive growth.

According to above fair classification, all sizes occur with nearly the same frequency:

Small: **11** parts with a total value of 19.8, or 19.8/131.5 = **15.1%** of system total.
Medium: **10** parts with a total value of 41.1, or 41.1/131.5 = **31.3%** of system total.
Big: **11** parts with a total value of 70.6, or 70.6/131.5 = **53.7%** of system total.

Figure 3.19 depicts the quantitative portion of each part, sorted low to high in the clockwise direction, and where relative area signifies relative quantity. In addition, the color for Small is light gray; the color for Medium is plain gray; and the color for Big is dark gray. Here the 3 sizes are not even close to the **33.3%** ideal division of overall quantity, rather, the larger the size, naturally the more proportion it earns.

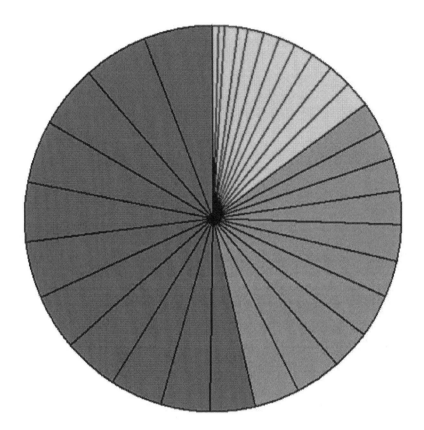

Figure 3.19: Equality in Frequency - Inequality in Quantity - Additive Presentation of Partitions

An alternative perspective for the failure of the (deterministic) additive partition model to manifest the small is beautiful phenomenon is that the set of growth factors (in a multiplicative sense) shows a consistent downward trend. The set of growth factors here are as follows:

{1.314, 1.239, 1.193, 1.162, 1.139, 1.122, 1.109, 1.098, 1.089, 1.082, 1.076, 1.071, 1.066, 1.062, 1.058, 1.055, 1.052, 1.050, 1.047, 1.045, 1.043, 1.041, 1.040, 1.038, 1.037, 1.035, 1.034, 1.033, 1.032 1.031, 1.030}

Clearly, these growth factors of the additive model are monotonically and consistently decreasing in value, showing a distinctive downward trend from the smallest part to the biggest part, thus precluding the small is beautiful phenomenon from manifesting itself.

The opposite pole of this case, namely a partition where the growth factors are with an upward trend, yields the small is beautiful phenomenon even more decisively and forcefully so. Such sets of quantities with constantly increasing growth rates, growing ever faster, are sometimes referred to as 'super exponential growth'.

[5] Conclusion: Random Partitions and the Small is Beautiful Phenomenon

As shall be mentioned later in this section, the significant digits configuration of the resultant set of parts in Random Dependent Partition is derived in a mathematically rigorous way, and this digital result relates to the small is beautiful quantitative configuration found for the set of parts.

In Mathematical Statistics it is shown that distances between random markings along the length of one spacial dimension are distributed as in the Exponential Distribution. Hence, Random Real Partition is intimately connected with the Exponential Distribution. For example, L-meter pipe is cut randomly into N parts via (N – 1) random markings between 0 and L applying the Uniform Distribution. This implies the rate of (N - 1)/(L) markings per unit distance, and pointing to the Exponential Distributions with (N - 1)/(L) Lambda parameter value.

The Exponential Distribution is strictly positively skewed, monotonically falling to the right, and having a strong preference for the small. This result implies that Random Real Partition exhibits the small is beautiful phenomenon as well.

Surely there are in principle many random partition processes other than Random Dependent Partition and Random Real Partition. These two processes are perhaps the most straightforward or natural ways to break a quantity into parts in a random fashion. The concluding general statement regarding the manifestation of the small is beautiful phenomenon in random partition processes is based on very consistent and broad empirical testing and evidences, as well as sound conceptual underpinning. In part it also rests on some rigorous mathematical results.

The general statement is based on three assumptions:

(I) Given that partition is performed on the real number basis and not exclusively on integers.
(II) Given that partition is truly random.
(III) Given that partition thoroughly breaks the original quantity into numerous refined parts.

Then all the (numerous) parts are of distinct values, and the small is beautiful quantitative configuration is found in the resultant set of parts.

[6] Quantitative Partitions Models Lead to Skewness and Often to Benford

Random Real Partition and Random Dependent Partition shall now be revisited and further analyzed, and various other quantitative partition models shall be examined in terms of the quantitative and digital behavior of the resultant set of parts. Almost all the models are of the random type, and only a few are of the deterministic type. The [nearly] universal feature found across [almost] all partition models is having many small parts but only few big parts, namely quantitative skewness, while Benford's Law is valid only in some particular partition cases and under certain constraints. Hence the phenomenon of Benford's Law is viewed as a particular subset of the broader small is beautiful phenomenon in quantitative partitioning.

Significantly, such a vista is true in the context of other causes and explanations of Benford's Law where the small consistently outnumbers the big in [almost all] partial structures of the model under consideration, or well before full convergence to Benford is achieved, while Benford is often found only in full model structures and only after the complete convergence of the model. These results endow the vista much broader scope.

Further in this section it will be shown how either the <u>active</u> process of partitioning or the <u>passive</u> consideration of a large quantity as the composition of smaller parts can be considered as another independent explanation for the widespread empirical observation of Benford's Law, as well as another cause of the more universal observation of the small is beautiful phenomenon.

The following is brief summary of the partition models discussed in this section:

Complete Equipartition: The set of all possible Integer Partitions of N.
Refined Equipartition: The limited set of all Integer Partitions of N restricted to n << N.
Irregular Equipartition: The set of Integer Partitions of N restricted in an arbitrary manner.
Partition as a Set of Marks on the x-axis: The lengths of the intervals between the random marks established on the x-axis constitute the parts in this random partition process.
Singular Balanced Partition: The balanced partitioning of quantity X into many identical parts yielding only one size exclusively.
Random Real Partition as in Random Pipe Breaking: Generalized beyond the standard use of the Uniform Distribution in generating the marks on the x-axis to include also other skewed and symmetrical distributions.
Random Dependent Partition as in Random Rock Breaking: Generalized via the use of a deterministic fixed ratio/percentage of breakup.
Chaotic Rock Breaking: A variation on Random Dependent Partition via the introduction of an extra measure of randomness in the partition process.
Random Minimum Breaking: A variation on Random Dependent Partition where the minimum is repeatedly broken via random ratio/percentage of breakup.
Random Maximum Breaking: A variation on Random Dependent Partition where the maximum is repeatedly broken via random ratio/percentage of breakup.
Balls Distributed inside Boxes: The partitioning of a singular set of numerous discrete balls via their placement into fewer discrete boxes.

[7] Refined Equipartition Parable

A group of 33 well-trained spies are about to be sent to the enemy state for long periods of military sites observations, sabotage activities, political meddling and subversion. The secret service commander thinks that it is best to separate them into independent cells of no more than 5 spies each, in case one spy is uncovered, so that not all would be lost as a result in such adverse instances. In addition, the work is more suitable for small cells, as well as for an individual spy operating totally alone. Not having any specific plans of spying assignments beforehand, the commander first simply divides these 33 future spies in an arbitrary or random manner before their departure. Later, once they are all settled in their new environments, each cell would be assigned specific work according to its size and abilities. Some possible partitions are:

{5, 5, 5, 5, 5, 5, 2, 1}
{2, 4, 5, 5, 1, 3, 1, 1, 1, 5, 5}
{1, 1, 1, 1, 1, 1, 1, 1, 1, 1, 2, 2, 2, 2, 2, 3, 5, 5}
{4, 5, 4, 5, 4, 5, 2, 2, 2}
{3, 3, 3, 3, 3, 3, 3, 3, 3, 3, 3}
etc. etc.

The bored statistician at the secret service headquarters who hadn't gotten any work assignment in several days has decided to entertain himself by fictitiously inventing the abstract large data set of all such possible partitions (arbitrarily aggregated as a singular data set) with the intention of thoroughly analyzing it quantitatively. The aggregated data set that he has in mind is:

{5, 5, 5, 5, 5, 5, 2, 1, 2, 4, 5, 5, 1, 3, 1, 1, 1, 5, 5, 1, 1, 1, 1, 1, 1, 1, 1, 1, 2, 2, 2, 2, 2, 3, 5, 5, 4, 5, 4, 5, 4, 5, 2, 2, 2, 3, 3, 3, 3, 3, 3, 3, 3, 3, 3, 3, etc. etc.}

He has calculated that there are exactly 918 possible partitions here, and that the aggregated data set contains 14,608 numbers in total. Is his analysis practical in any way? To what use could such a project be applied? Interestingly, by studying the relative occurrences of {1, 2, 3, 4, 5} in such abstract data set he discovered that their frequencies are inversely proportional to their size, so that a lone spy occurs 6,905 times, a cell consisting of a pair of spies occurs 3,228 times, a trio spy cell occurs 2,017 times, a cell of 4 spies occurs 1,408 times, and a cell of 5 spies occurs only 1,050 times. Moreover, he was able to reasonably fit these 5 values into an algebraic expression which he mathematically derived in the abstract and in general for all such types of partitions.

Could such highly abstract study be put into some concrete flesh-and-blood scenario so that real-life applications are made? Yes indeed! Imagine the headquarters of the American or Soviet intelligence departments during the height of the Cold War employing tens of thousands of spies. In one comprehensive spying program involving thousands of such 33-spy groups, each group is launched by a different officer, who randomly and independently partitions his 33-group into smaller cells of no more than 5 spies each. Each officer gives each possible partition equal chance (i.e. weight) of being applied to the 33 brave men and women about to depart on their patriotic mission. The statistical desk at the secret service headquarters is interested in the overall relative frequencies of the sizes of the cells occurring worldwide. Hence it was calculated that:

47% [*6,905 / 14,608*] of all the cells worldwide are of **1** spy.
22% [*3,228 / 14,608*] of all the cells worldwide are of **2** spies.
14% [*2,017 / 14,608*] of all the cells worldwide are of **3** spies.
10% [*1,408 / 14,608*] of all the cells worldwide are of **4** spies.
 7% [*1,050 / 14,608*] of all the cells worldwide are of **5** spies.

Certainly, the small is beautiful principle manifests itself decisively here, even though we have restricted the Integral Partition of 33 exclusively to small integers that are less than 6.

In a statistical sense, even for a single partition decision by a commanding officer responsible for only one such 33-spy group about to depart, one can inquire about the probability of any particular cell size, assuming the officer does not have any bias or preference for any particular size and equally chooses one partition among all the possibilities.

[8] Complete, Refined, and Irregular Equipartition Models

Three Equipartition models are discussed in this section, each based on some variation of Integer Partition. These three models are named Complete Equipartition, Refined Equipartition, and Irregular Equipartition.

Complete Equipartition is simply the entire set of all possible Integer Partitions of N, where all partitions are aggregated into one vast data set. This model allows for any integer from 1 to N to become a part of the N whole, without any limits or constraints. Indeed we insist that all the integers from 1 to N, small or big, must participate in the partition, and not only once, but as many times as possible. Complete Equipartition is called '**complete**' since it incorporates all possible breakups, completely, without leaving out any type of partition whatsoever. Complete Equipartition breaks up quantity N into very small integral parts such as 1 and 2; or into very big integral parts such as N - 2 and N – 1; and including the improper partition where N is actually not being broken up at all and instead it is being left intact as {N}. Since all possible partitions are given equal weights within the model, and are simply aggregated into that vast data set without any preferential adjustments, the prefix '**equi**' is included in the term 'Equipartition'.

For example, the Complete Equipartition of N = 27 is that vast data set containing among others {2, 5, 10, 10}, {1, 1, 25}, {1, 1, 1, 1, 1, 1, 1, 20}, {27}, {1, 26}, {1, 1, 25}, and so forth. This is written in full as {2, 5, 10, 10, 1, 1, 25, 1, 1, 1, 1, 1, 1, 1, 20, 27, 1, 26, 1, 1, 25, and so forth}. The entire data set of Complete Equipartition of 27 is way too long and vast to describe here.

In chapter 1 of this section, three examples of Complete Equipartition are discussed, namely those with N = 5, N = 7, and N = 13. Figures 3.1 and 3.2 clearly demonstrate the skewness of Complete Equipartition of 5. Figure 3.3 depicts the histogram of Complete Equipartition of 13, and that histogram is skewed as well, as it is consistently and monotonically falling to the right, except at the very end. At the high end of each Complete Equipartition, the partitions {1, N - 1} and {N} are the unique partitions involving integers N and N – 1, hence integers N and N – 1 are with equal frequency of 1, and so the histogram is flat on that small and insignificant portion.

Complete Equipartition is inherently skewed, and this is so regardless of the value of integer N. Yet, in spite of its innate skewness, Complete Equipartition is not Benford, regardless of the value of N, and this is so due to the extreme skewness associated with Complete Equipartition.

Complete Equipartition is even skewer than the Benford skewed configuration. For example, Complete Equipartition of N = 9 yields the vector of the count of the integers 1 to 9 occurrences {67, 26, 15, 8, 5, 3, 2, 1, 1}, and this translates into the vector of percents of occurrences of the integers - or rather 'digits' - of {52.3%, 20.3%, 11.7%, 6.3%, 3.9%, 2.3%, 1.6%, 0.8%, 0.8%}. Here integer or 'digit' 1 has the very high frequency of 52.3%, while in Benford's Law digit 1 has the relatively milder frequency of 30.1%.

Refined Equipartition refers to Integer Partitions of N where only {1, 2, 3, ... , n} are allowed in the partitions, and where n is much smaller than N. Since only partitions involving integers that are much smaller than N are allowed, this ensures that N is being broken into much smaller and **'refined'** parts, resulting in a significant fragmentation of the original quantity N. It should be noted that <u>all</u> integers from 1 to n are allowed to participate in the partition and are included, without any gaps or exclusions. Indeed we insist that all the integers from 1 to n must participate in the partition, and not only once, but as many times as possible.

The Equipartition Parable of a group of 33 spies - being separated into independent cells of no more than 5 spies each - is an example of Refined Equipartition. Here N = 33 and n = 5.

An article by Don Lemons in 1986 inspired and led Steven Miller to publish another related article in 2015 presenting a mathematically rigorous proof that Refined Equipartition endows the same quantitative portion from the entire quantity of the entire equipartition scheme to each integer - in the limit as N approaches infinity while n is kept fixed and finite so that n << N. Rephrasing Miller's assertion: The quantitative sum from each allowed integer in Refined Equipartition is a constant, granting each integer equal portion from the entire quantity embedded within that vast data set. Conceptually, Miller argues that even though a small integer is of low value, yet it has high frequency, so that its total quantity should be equivalent to the total quantity of a big integer of high value but of low frequency. In other words, that there exist perfect cancelation and offsetting effects between size and frequency.

Calculating quantitative sum for each integer in Refined Equipartition of 33 limited to 5 of the Equipartition Parable, we obtain: {(1)*(6905), (2)*(3228), (3)*(2017), (4)*(1408), (5)*(1050)}, namely {6905, 6456, 6051, 5632, 5250}. While total quantity per integer is not truly constant here, because Miller's constraints have not been fully met, yet the five portions are all nearly of about the same level, fluctuating around their average value of 6059, with relatively little variation between them. The quantitative configuration here is very similar to that of the oval shape in Figure 3.5 where Big, Medium, and Small all have about the same territory (i.e. the same total quantity). Therefore Refined Equipartition of 33 limited to 5 is very much in the spirit of the small is beautiful principle. Yet it certainly cannot be Benford with only 5 integers/digits, and where OOM = Log(5/1) = 0.70. Benford's Law applies only to data with high variability.

For a more detailed example, let us examine Refined Equipartition of 13 limited to 3. Here quantity 13 is broken into smaller integral parts that are not bigger than 3. This leads to the set of the following partitions involving only 1, 2, or 3 as parts:

{3, 3, 3, 3, 1}
{3, 3, 3, 2, 2}
{3, 3, 3, 2, 1, 1}
{3, 3, 3, 1, 1, 1, 1}
{3, 3, 2, 2, 2, 1}
{3, 3, 2, 2, 1, 1, 1}
{3, 3, 2, 1, 1, 1, 1, 1}
{3, 3, 1, 1, 1, 1, 1, 1, 1}
{3, 2, 2, 2, 2, 2}
{3, 2, 2, 2, 2, 1, 1}
{3, 2, 2, 2, 1, 1, 1, 1}
{3, 2, 2, 1, 1, 1, 1, 1, 1}
{3, 2, 1, 1, 1, 1, 1, 1, 1, 1}
{3, 1, 1, 1, 1, 1, 1, 1, 1, 1, 1}
{2, 2, 2, 2, 2, 2, 1}
{2, 2, 2, 2, 2, 1, 1, 1}
{2, 2, 2, 2, 1, 1, 1, 1, 1}
{2, 2, 2, 1, 1, 1, 1, 1, 1, 1}
{2, 2, 1, 1, 1, 1, 1, 1, 1, 1, 1}
{2, 1, 1, 1, 1, 1, 1, 1, 1, 1, 1, 1}
{1, 1, 1, 1, 1, 1, 1, 1, 1, 1, 1, 1, 1}

This yields the aggregated set of all the above partitions as follows:

{3, 3, 3, 3, 1, 3, 3, 3, 2, 2, 3, 3, 3, 2, 1, 1, 3, 3, 3, 1, 1, 1, 1, 3, 3, 2, 2, 2, 1, 3, 3, 2, 2, 1, 1, 1, 3, 3, 2, 1, 1, 1, 1, 1, 3, 3, 1, 1, 1, 1, 1, 1, 1, 3, 2, 2, 2, 2, 2, 3, 2, 2, 2, 2, 1, 1, 3, 2, 2, 2, 1, 1, 1, 1, 3, 2, 2, 1, 1, 1, 1, 1, 1, 3, 2, 1, 1, 1, 1, 1, 1, 1, 1, 3, 1, 1, 1, 1, 1, 1, 1, 1, 1, 1, 2, 2, 2, 2, 2, 2, 1, 2, 2, 2, 2, 2, 1, 1, 1, 2, 2, 2, 2, 1, 1, 1, 1, 1, 2, 2, 2, 1, 1, 1, 1, 1, 1, 1, 2, 2, 1, 1, 1, 1, 1, 1, 1, 1, 1, 2, 1, 1, 1, 1, 1, 1, 1, 1, 1, 1, 1, 1, 1, 1, 1, 1, 1, 1, 1, 1, 1, 1, 1, 1, 1, 1, 1}

The ordered set is then as follows:

{1, 2, 2, 2, 2, 2, 2, 2, 2, 2, 2, 2, 2, 2, 2, 2, 2, 2, 2, 2, 2, 2, 2, 2, 2, 2, 2, 2, 2, 2, 2, 2, 2, 2, 2, 2, 2, 2, 2, 2, 2, 2, 2, 2, 3, 3, 3, 3, 3, 3, 3, 3, 3, 3, 3, 3, 3, 3, 3, 3, 3, 3, 3, 3, 3, 3, 3, 3, 3, 3, 3}

The counts of integer 1, 2, 3 - namely their frequencies - are 102, 45, 27 respectively.
The entire data set of this Refined Equipartition contains 102 + 45 + 27 = 174 numbers in total.
Integers 1, 2, 3 are with frequency-percentages of 58.6%, 25.9%, 15.5% respectively.

While 3 << 13 is not actually true since 3 is not that much smaller than 13; and while N is still considered to be quite small with its value of 13, instead of approaching infinity or being some truly large number, nonetheless let us attempt to examine Miller's assertion in this case:

Total quantity of the entire data set of Refined Equipartition of 13 limited to 3 is obtained by simply summing up all the 174 integers, and this yields 273.

Total quantitative portion of any integer is simply (integer)*(frequency of the given integer), and here this yields:

Total quantity for integer 1:   (1)*(102) = 102
Total quantity for integer 2:   (2)*(45) = 90
Total quantity for integer 3:   (3)*(27) = 81

Total quantity of Equipartition: 273

While total quantity per integer is not truly constant here, because Miller's constraints have not been fully met, yet the three quantities are not that much different from each other, having little variation of only about 10 units above or below their average value of 91.

Finally, for properly constrained Refined Equipartition models where n << N and N → ∞, the observation that total quantity per integer is constant implies that frequency (or density) of an integer is inversely proportional to the value of the integer, and therefore frequency (or density) is of the form **k/x distribution**, known for its exact Benford behavior whenever defined range has an integral exponent difference between the max and min values [$LOG(max) - LOG(min) = Integer$], such as on integral powers of ten ranges. Hence Miller's assertion implies that Refined Equipartition models with particularly proper rangers are Benford. The following equations reinforce the above argument:

(total quantitative portion of the integer) = (integer)*(frequency of the integer)
(frequency of the integer) = (total quantitative portion of the integer)/(integer)
(frequency of the integer) = (constant for all the integers)/(integer)
Integer Frequency = Constant / Integer

Irregular Equipartition refers to Integer Partitions of N in which the parts are restricted and chosen from an arbitrary set of allowed integers. This arbitrary set does not necessarily increase nicely by one integer at a time, having possibly some gaps where some integers are skipped.

For example, for N = 25 representing the entire integral quantity to be partitioned, we allow only the arbitrary set of integers {1, 3, 6, 7, 11, 19} to be used in the partition. Integers 2 or 5 for example are not allowed to participate in the partition. Some possible partitions are:
{19, 6}, {19, 3, 3}, {11, 7, 3, 3, 1}, {6, 6, 6, 3, 3, 1}, and so forth.

Here, no mathematical results are offered for such messy and highly irregular Equipartition models, although these rare cases might emerge in some particular applications of partitions.

[9] Partition as a Set of Marks on the x-axis

It is essential to visualize each possible partition in Integer Partition of N as a sort of a scheme that places marks along the x-axis on the interval (0, N). These marks must be placed only upon the integers, avoiding the spaces between them. Figure 3.20 depicts such representation for the singular partition of 33 into {1, 1, 1, 1, 1, 1, 1, 2, 2, 2, 2, 3, 3, 3, 4, 5}. This is just one possible partition within the entire set of all 918 possible partitions for this Refined Equipartition of 33 limited to 5 of the Equipartition Parable.

Initially the integral parts are represented in order from 1 to 5 for convenience, first showing seven ones, then four twos, then three threes, a four, and finally a five - as seen in the upper panel of Figure 3.20. Such an order helps in organizing the parts carefully onto the x-axis. Yet, equipartition does not necessitate order of the parts in any way, so this is not the best or the only way of viewing partitions in our context where the focus is on the set of all possible partitions, or equivalently on random partitioning, and the implications about occurrences of relative quantities. Hence it would be better to randomize each singular occurrence of the parts, and this is shown in the lower panel of Figure 3.20 where the set of parts is shown in a randomized order as the equivalent partition {2, 5, 1, 3, 1, 1, 2, 2, 1, 1, 3, 4, 3, 2, 1, 1}.

Physical order of magnitude here is POM = Max/Min = 5/1 = 5, while order of magnitude is OOM = LOG(Max/Min) = LOG(5/1) = 0.7, and which are quite low.

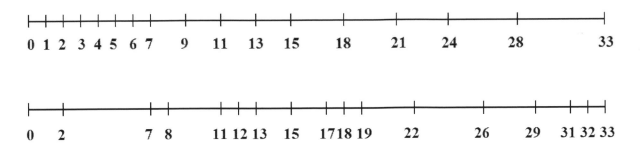

Figure 3.20: Integer Partition of 33 into {1, 1, 1, 1, 1, 1, 1, 2, 2, 2, 2, 3, 3, 3, 4, 5} along x-axis

Figure 3.21 depicts several other partition possibilities visualized as marks along the x-axis on the interval (0, 33). The last two partitions at the bottom are the two extremes, where either the set of the biggest possible pieces [integer 5 aided by integer 3] or the set of the smallest possible pieces [integer 1] are used to partition 33. It should be noted that it is very easy to obtain many more partitions here by simply starting out with a blank line between 0 and 33 and then randomly marking/poking marks along the integers as one sees fit. There is no need to calculate anything (the setup guarantees that it will all add up to 33 automatically) as long as we limit the gap between any two adjacent marks to the maximum length of 5.

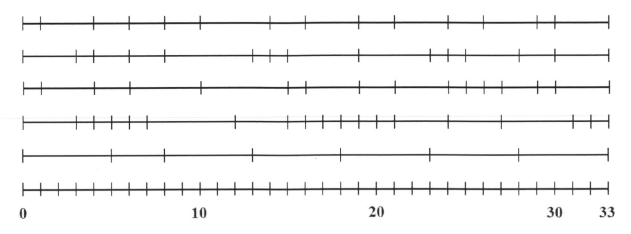

Figure 3.21: A Variety of Partitions of 33 into {1, 2, 3, 4, 5} Integral Parts along the x-axis

In each partition here of Figure 3.21 we are forcing the system into some significant fragmentation of the original quantity of 33 into much smaller parts. This is so due to the partial and limited compliance with the constraint n << N, although the statement 5 << 33 is not really true. In any case this partition still constitutes Refined Equipartition since it restricts the parts into smaller integers less than 6.

Even the usage of the combination of the largest parts possible near the bottom of Figure 3.21, namely {5, 3, 5, 5, 5, 5, 5} with six fives and one three, still results in significant fragmentation of the original 33 quantity into relatively much smaller parts.

It would take drawing 918 lines for Figure 3.21 in order to extend it fully until all possible partitions are visible, and for it to be properly called 'Equipartition', since by definition any such model must incorporate all possible partitions within the system.

Had we further restricted the partition of 33 to the set {2, 3, 5} say, as in Irregular Equipartition, then there would be even less variety in the ways distinct partitions can be formed, as only 3 sizes are allowed to participate in the partition, leading to fewer partitions.

Had we further restricted the partition of 33 to an even smaller set of {2, 3}, as in Irregular Equipartition, then there would be now even less variety in the ways distinct partitions can be formed, as only 2 sizes are allowed to participate in the partition, leading to even fewer partitions.

Had we further restricted the partition of 33 to the tiny set of {3} with only one integer, as in Irregular Equipartition, then there would be no variety whatsoever in the ways distinct partitions can be formed, as only 1 size is allowed now to participate in the partition, leading to one partition only. Indeed, this is the extreme case of **Singular Balanced Partition**, namely the case of a conserved quantity X partitioned into many <u>identical parts</u> yielding only <u>one size</u> exclusively. In this scenario, the parts are restricted exclusively to Q (which was 3 in the above example), assuming Q < X and where Q is a perfect divisor of X as in X/Q = I, where I is an integer. Here there is no need to give equal weights to numerous partitions because there is only one possible partition to consider in the entire process, namely {Q, Q, Q, ... I times} under the assumption that (Q + Q + Q + ... I times) = X.

Singular Balanced Partitions come with two flavors. The first flavor is of a conserved integral quantity X partitioned into one of its <u>integral</u> divisors. The second flavor is of a conserved real quantity X partitioned into any <u>real</u> number R < X such that X/R = I, where I is an integer, so that (R + R + R + ... I times) = X. It is noted that the prefix '**equi**' is omitted in the term 'Singular Balanced Partition' since only one partition is possible here and therefore there is no need to give **equal** weights to some non-existing multiple partitions.

Figure 3.22 depicts Singular Balanced Partition of the integral quantity 33 into identical integral parts of 3. Here there are no bigs, no smalls, no mediums, but only one uniform size of 3. Certainly the motto 'small is beautiful' is irrelevant here; no quantitative comparison or analysis of any sort can be made; and it is impossible to establish any relationship between this partition and Benford's Law in any sense. Physical order of magnitude here is POM = Max/Min = 3/3 = 1, and order of magnitude is OOM = LOG(Max/Min) = LOG(3/3) = LOG(1) = 0, namely as low as they could be! There is no variation here at all.

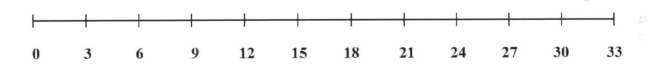

Figure 3.22: Singular Balanced Partition of 33 into Eleven Parts of 3's along the x-axis

Singular Balanced Partition is one extreme scenario where the allowed set of parts contains only one value Q, namely when the system is very inflexible, frugal, and totally unimaginative, not letting X be partitioned more liberally into all sorts of parts except for Q. The other extreme; the opposite pole of Singular Balanced Partition; is when the system is highly flexible, liberal and open to all sorts of values for the parts. So much so that the system allows all integral, fractional, rational, irrational, or whatsoever type of real numbers that exist between 0 and X! Here the allowed set of parts is {all real numbers on (0, X)}. Indeed, such a partition should be already very familiar with the reader, namely **Random Real Partition**! It is noted that the prefix '**equi**' is omitted in the term here since only a singular decisive random partition is needed and performed in this process - and which is sufficient to obtain the desired quantitative and near-Benford configurations in one fell swoop - given that plenty of parts are created. There is no need to give **equal** weights to some non-existing partitions.

Figure 3.23 depicts Random Real Partition of the quantity 33 into 32 parts of various real lengths, and which could be thought of as Random Pipe Breaking. Here there are very few big parts, some medium parts, and many small parts; all in the spirit of the small is beautiful principle. The relationship between this partition and Benford's Law is immediate and there is no need to repeat the experiment many times over and incorporate multiple partitions by constructing their aggregate set. This single set of resultant 32 parts in and of itself is not yet close enough to Benford because it does not contain enough parts, namely because the quantity (pipe) has not been broken thoroughly enough. To get directly to [nearly] Benford, all that is needed here is to break it more thoroughly into many more resultant parts (approximately over 5000 parts). Physical order of magnitude here is significantly larger as compared with the previous two examples of Figure 3.20 and Figure 3.22, coming at POM = Max/Min = 4.5/0.2 = 22.5, while order of magnitude is OOM = LOG(Max/Min) = LOG(4.5/0.2) = LOG(22.5) = 1.4.

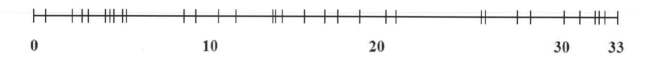

Figure 3.23: Random Real Partition of 33 into 32 real Parts along the x-axis

The choice to break the original quantity into 32 parts was somewhat deliberate, aiming to fit the number into the relationship $2^{INTEGER} = 32$ where INTEGER here is 5 and thus $2^5 = 32$ holds. This is so because of the possible [but erroneous] interpretation of Random Real Partition as Random Dependent Partition in the particular cases where [Number of Parts] $= 2^{INTEGER}$.

Figure 3.24 depicts the supposed or imagined 5 partition stages where quantity 33 is gradually and randomly being broken into 32 much smaller real quantities in the spirit of Random Dependent Partition - resulting in the same exact quantitative configuration and having identical parts as in the Random Real Partition case of Figure 3.23. Yet the two processes of Random Real Partition and Random Dependent Partition are distinct in nature, and they almost always lead to distinct resultant set of parts, in spite of the apparent successful superimposition of the two into a singular partition shown in Figure 3.24.

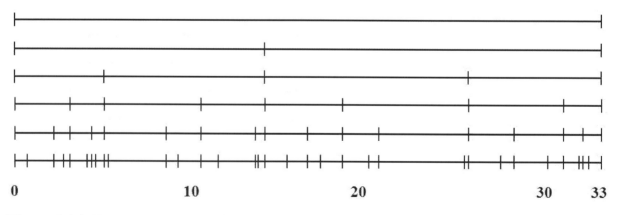

Figure 3.24: Erroneous Interpretation of Random Real Partition as Random Dependent Partition

Random Dependent Partition as in Random Rock Breaking involves taking a conserved real quantity X and breaking it into 2 real parts of random proportions chosen via the continuous Uniform(0, 1) - interpreted as the random ratio/percentage of the breakup. In the first stage, X is split into {X*Uniform(0, 1), X*(1 – Uniform(0, 1)). The process continues with the breaking up of each of these 2 parts separately but in the same random fashion, resulting in 4 parts altogether. Continuing breaking up the 4 into 8, the 8 into 16, the 16 into 32, and so forth, would result in the Benford configuration after about 10 or so such partition stages - depending on the desired level of compliance accuracy. Performing about 17 such stages would result in nearly perfect Benford configuration; and in the limit as the number of stages goes to infinity the logarithmic distribution is obtained exactly (in a mathematical rigorous way).

The complete success of arriving nearly perfectly (or exactly in the limit) at the Benford configuration for the process of Random Dependent Partition is contrasted with the slightly less successful story of Random Real Partition where Benford is found only approximately so, because the process is gradually metamorphosing into the Exponential Distribution which itself is only approximately Benford – assuming the creation of sufficiently large number of pieces/parts.

On the face of it, new marks in Random Dependent Partition are <u>dependent</u> on established older marks, because each new mark must to be placed between two older ones, while in Random Real Partition new marks are placed in a totally free and chaotic manner. Yet, a heuristic argument is made here, viewing the process of the Random Dependent Partition applying the Uniform(0, 1) and its resultant random x-axis placements of $(2^{STAGES} – 1)$ marks, as quite similar (yet not equivalent) to an <u>independent</u> generations of $(2^{STAGES} – 1)$ realizations from the Uniform(0, X) serving as marks on the x-axis – just as was seen in the corresponding cases of Figure 3.23 and Figure 3.24. Such a vista then lets us somehow unify in a limited sense the process of Random Real Partition with the process of Random Dependent Partition, because ultimately each scheme constitutes just a different description of events of the similar stochastic process which randomly places marks on (0, X) via the same distribution of the Uniform(0, 1)!

It is quite remarkable that these two random methods of placing marks on the x-axis via the same distribution of the Uniform(0, 1) would lead to somewhat distinct quantitative configurations for the resultant set of parts and therefore to distinct measures of compliance with Benford!

[10] Flexibility in the Selection of Parts Leads to Benford and High POM

Let us summarize and compare three different partition schemes:

(1) **Singular Balanced Partition** – due to its extremely rigid rule of how parts are selected, restricting them exclusively to only one size Q, there exists neither big nor small, but only one unique size for all the parts, no quantitative or digital comparisons can be made, and certainly there exists no relationship to Benford's Law whatsoever. Here order of magnitude is as low as can possibly be, namely 0.

(2) **Complete Equipartition** and **Refined Equipartition** – due to their limited flexibility in the rule of how parts are selected, allowing a particular diversity of integral sizes only, while excluding fractional and irrational sizes, these processes yield a skewed quantitative configuration for the set of all possible partitions aggregated as one vast data set, where the small is numerous and the big is rare. Only Refined Equipartition could be approximately Benford (or exactly so in the limit) provided that N is a very large number, that n is much smaller than N, and assuming that the range of allowed sizes - namely {1, 2, 3, ... n} - spans integral powers of ten, such as when n assumes the value of 10, 100, or 1000, and so forth. Here order of magnitude is well over 0, yet it is still a bit low.

(3) **Random Real Partition** – due to its extreme flexibility in the rule of how parts are selected, allowing for unaccountably infinite many real numbers on (0, X), namely any possible size whatsoever, this leads to an approximate Benford configuration in one fell swoop for just one single such random partition without any need to repeat the partition over and over again, and without the need to aggregate anything - assuming conserved quantity X is partitioned into very many real parts (approximately 5000 to 10000 pieces). Here order of magnitude is quite high, and especially so when numerous parts are created.

Conclusion: The more flexible a given partition process is in terms of how it selects its parts, the higher is resultant order of magnitude, the skewer is resultant set of parts, the more pronounced is the manifestation of the small is beautiful phenomenon, and possibly the closer it is to the Benford configuration.

[11] Testing Compliance of Random Real Partition with Benford's Law

Another perspective on Random Real Partition is its description as in the following scheme:

Generate N realizations from the continuous Uniform(0, X).
Order them from low to high, add 0 on the very left, and add X on the very right.
The data set is: $\{0, U_1, U_2, U_3, \ldots, U_{N-2}, U_{N-1}, U_N, X\}$.
Generate the difference data set out of the one above.
This data set is: $\{(U_1 - 0), (U_2 - U_1), (U_3 - U_2), \ldots, (U_{N-1} - U_{N-2}), (U_N - U_{N-1}), (X - U_N)\}$.
This data set is conjectured to be nearly Benford as N gets large. In practical terms it might be sufficient for N to be about 5000 to 10000 for a reasonable fit to Benford, and that nothing or not much is gained by increasing N from this level, since beyond around this level <u>saturation</u> sets in.

It should be noted that Monte Carlo simulation run of N realizations from the continuous Uniform(0, X) yields the same quantitative configuration for the difference data set had it been based on the continuous Uniform(0, 1), or on the continuous Uniform(0, Y), and so forth. This is so since these processes differ only in the sense of having a different scale, but structurally they perfectly correspond. As far as digital configuration is concerned, for Benford data sets the Scale Invariance Principle guarantees that digits distribution are the same in any scale, although for data sets which are not perfectly Benford, scale does indeed matter. Since Random Real Partition is only approximately Benford, therefore scale does matter here, but only slightly so.

Let us perform several Monte Carlo empirical tests regarding Random Real Partition and its compliance with Benford, applying the Uniform as the distribution generating the marks.

Uniform(0, 10) Partitioned via 25,000 Marks - {27.6, 16.9, 13.1, 10.3, 9.1, 7.2, 6.3, 5.3, 4.4}
Benford's Law for First Significant Digits   - {30.1, 17.6, 12.5, 9.7, 7.9, 6.7, 5.8, 5.1, 4.6}
SSD value is **9.2**, and such low value indicates that this partition is fairly close to Benford.
Here there are 25000/10 or 2500 marks per unit, and therefore this process corresponds to the Exponential Distributions with 2500 Lambda parameter value. In one computer simulation run for this Exponential Distribution, digits came as {27.2, 16.7, 13.3, 10.7, 9.2, 7.3, 6.3, 4.9, 4.4}.

Uniform(0, 800) Partitioned via 10,000 Marks - {32.2, 16.7, 11.3, 8.6, 7.5, 6.8, 6.0, 5.8, 5.1}
Benford's Law for First Significant Digits   - {30.1, 17.6, 12.5, 9.7, 7.9, 6.7, 5.8, 5.1, 4.6}
SSD value is **8.2**, and such low value indicates that this partition is fairly close to Benford.
Here there are 10000/800 or 12.5 marks per unit, and therefore this process corresponds to the Exponential Distributions with 12.5 Lambda parameter value. In one computer simulation run for this Exponential Distribution, digits came as {32.3, 16.5, 11.4, 8.7, 7.5, 6.8, 6.3, 5.4, 5.1}.

Uniform(0, 38) Partitioned via 35,000 Marks - {32.7, 17.9, 11.5, 8.7, 7.4, 6.2, 5.5, 5.1, 5.0}
Benford's Law for First Significant Digits   - {30.1, 17.6, 12.5, 9.7, 7.9, 6.7, 5.8, 5.1, 4.6}
SSD value is **9.7**, and such low value indicates that this partition is fairly close to Benford.
Here there are 35000/38 or 921.1 marks per unit, and therefore this process corresponds to the Exponential Distributions with 921.1 Lambda parameter value. In one computer simulation run for this Exponential Distribution, digits came as {32.8, 17.6, 11.8, 8.5, 7.4, 6.4, 5.6, 5.3, 4.7}.

Results are close to Benford, yet a more decisive result with lower SSD value (say less than 2) cannot be found here even if we partition an interval into many more parts. Saturation point is found somewhere around 5000, or 10000, or perhaps around 15000.

When Uniform(0, 38) for example is partitioned via only 1,000 marks, SSD values are somewhat higher. Six such Monte Carlo partition runs with 1,000 marks gave the following SSD values: 32.7, 52.6, 11.6, 26.8, 32.4, 21.1.

When Uniform(0, 38) for example is partitioned via only 100 marks, SSD values are significantly higher, since such partial and insufficient partition which yields only few parts is not effective enough and does not converge close enough to Benford and to the Exponential Distribution. Six such Monte Carlo partition runs with 100 marks gave the following SSD values: 114.8, 63.2, 77.1, 115.4, 43.5, 49.6.

[12] Models of Random Real Partition Applying a Variety of Distributions

One wonders what happens if another statistical distribution is substituted for the continuous Uniform! Could Random Real Partition be generalized to other distributions?! Surely the correspondence to the Exponential Distribution would be ruined in such substitution of distribution, but our quest is to get ever closer to Benford with much lower SSD.

Let us perform several Monte Carlo empirical tests on other (non-Uniform) statistical distributions to examine the possibility of improving and generalizing this random process.

Normal(19, 4) Partitioned via 35,000 Marks - {28.5, 17.4, 13.2, 10.5, 8.3, 6.8, 6.0, 5.0, 4.3}
Benford's Law for First Significant Digits - {30.1, 17.6, 12.5, 9.7, 7.9, 6.7, 5.8, 5.1, 4.6}
SSD value is **4.0**, and this lower value (in comparison to the Uniforms) indicates that applying the Normal in partitions may be in general closer to Benford than the application of the Uniform.

Exponential(2.4) Partitioned via 20,000 Marks - {29.2, 17.4, 13.1, 9.7, 8.2, 7.0, 5.7, 5.4, 4.4}
Benford's Law for First Significant Digits   - {30.1, 17.6, 12.5, 9.7, 7.9, 6.7, 5.8, 5.1, 4.6}
SSD value is **1.4**, and this extremely low value (in comparison to the Uniforms and the Normals) indicates that applying the Exponential in partitions may be in general closer to Benford than the applications of the Uniforms or the Normals.

Could the superior result of the partitioning via the Exponential be explained in terms of the fact that the Exponential distribution itself is quite close to Benford? Let us examine then random partitions with the applications of the highly logarithmic distributions of the Lognormal and k/x to get a clue.

k/x on (1, 10) Partitioned via 30,000 Marks - {30.0, 18.0, 12.2, 9.4, 7.8, 6.9, 5.7, 5.2, 4.7}
Benford's Law for First Significant Digits - {30.1, 17.6, 12.5, 9.7, 7.9, 6.7, 5.8, 5.1, 4.6}
SSD value is **0.4**, and such exceedingly low value indicates that applying the k/x distribution gets us extremely close to Benford!

Lognormal(9.3, 1.7) Partitioned via 35,000 M. - {30.0, 17.6, 12.4, 10.0, 8.0, 6.6, 6.0, 5.0, 4.4}
Benford's Law for First Significant Digits   - {30.1, 17.6, 12.5, 9.7, 7.9, 6.7, 5.8, 5.1, 4.6}
SSD value is **0.2**, and such exceedingly low value indicates that applying the Lognormal gets us extremely close to Benford! [Note on Notation: Lognormal with shape = 1.7 and location = 9.3.]

Yet, intuitively it's very clear that the logarithmic behavior of the distribution being utilized to randomly partition the interval has nothing to do with the resultant logarithmic behavior of the parts. But if this is the case, then what could account for the correlation between logarithmic-ness of the partitioning distribution itself and the logarithmic-ness of the resultant set of parts as was observed in the simulations above? The straightforward answer to this dilemma is that it is not the logarithmic-ness of the partitioning distribution that induces better logarithmic results for the set of parts, but rather its skewness! Figure 3.25 depicts several random realizations from a highly (negatively) skewed distribution, resulting in many small parts on the right where density is high but only a few big parts on the left where density is low. Skewness implies that not all regions on the x-axis are equally likely, some regions have higher densities and some have lower densities. As a consequence, the regions with higher densities are those producing numerous marks which are (relatively speaking) narrowly crowded together and parenting many small parts. On the other hand, regions with lower densities, especially long tails of distributions, are those producing fewer marks which are (relatively speaking) widely spread apart from each other, and parenting only a few big parts. Clearly, skewness of the generating distribution induces more pronounced quantitative differentiation in the generated set of parts; and since quantitative differentiation is the driving force of digital logarithmic-ness, hence partitioning along skewed distributions induces more skewness and better result for the parts.

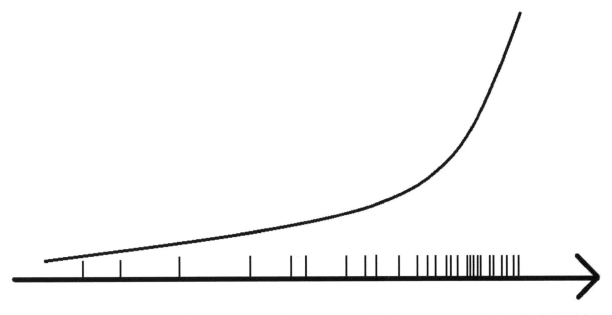

Figure 3.25: It is not Logarithmic-ness but Skew-ness which Improves Results in Real Partitions

Let us verify the above argument at least in one concrete case. The distribution PDF(x) = k*x^3 defined over (1, 50) [*where k = 4/(50^4 – 1) = 0.00000064*] is highly skewed, resembling a great deal the curve in Figure 3.25. Its digital configuration is totally non-Benford. This is so mostly because it has a rising histogram, where the small is rare and the big is numerous. Yet, according to the argument above, its highly skewed density implies that partitioning (1, 50) along it should lead to a strong Benford configuration for the parts! Let us then check this result empirically:

k*x^3 on (1, 50) Partitioned via 25,000 Marks - {30.8, 17.5, 12.0, 9.1, 8.0, 6.8, 5.6, 5.4, 4.8}
Benford's Law for First Significant Digits   - {30.1, 17.6, 12.5, 9.7, 7.9, 6.7, 5.8, 5.1, 4.6}
SSD value is **1.3**, and this very low value indicates that partitioning along highly skewed distributions leads to stronger Benford results for the parts (in comparison to symmetrical ones).

<u>Note:</u> Here the Cumulative Function is (k/4)*(x^4 – 1) over the range (1, 50), and it is set equal to the Uniform(0, 1) in Monte Carlo computer simulations in order to generate realizations. The random number function in MS-Excel called RAND() generates values uniformly distributed on (0, 1). Hence, in the expression RAND() = (k/4)*(x^4 – 1) we solve for x to obtain the relationship x = 4th root of [RAND()*(4/k) + 1], x being a random realization from k*x^3 distribution.

Yet, there seems to be a stubborn refusal on the part of the Normal and especially on the part of the Uniform towards letting Random Real Partition achieving a near perfect convergence to the logarithmic configuration, and no matter how many marks are being placed on the x-axis, SSD never seems to manage to get below 6 or 5.

The failure to achieve complete Benfordness when the Uniform is chosen to serve as the distribution responsible for placing the marks on the x-axis can be neatly explained via the intimate connection between the Uniform and the Exponential distributions. This well-known result in Mathematical Statistics states that distances between [Uniformly] random markings along the length of one spacial dimension are distributed as in the Exponential Distribution. The Exponential Distribution is specified by the single parameter called Lambda. Lambda is the event rate, namely the average number of events per unit time, or the average number of spacial occurrences per unit length. In the context of Benford's Law, first digits of the Exponential Distribution are known to be close to LOG(1 + 1/d) but not close enough, no matter what value is assigned to parameter Lambda, with SSD fluctuating roughly between 5 and 13.

[13] Deterministic Dependent Partition Applying Fixed Ratios

The partition process described in the manuscript submitted to the publisher of Kossovsky's book in November 2013, and the publishing by the mathematician Steven Miller of an article about Random Dependent Partition in Dec 2013 (as well as a previous article in Nov 2011) refer essentially to the same process. Discussions about the process can be found in Kossovsky (2014) Chapter 92 titled "Breaking a Rock Repeatedly into Small Pieces is Logarithmic", and Miller et al (2013) "Benford's Law and Continuous Dependent Random Variables"; the latter containing a rigorous mathematical proof as well as the correct statement regarding fixed deterministic ratio of partitions.

In Kossovsky (2014) the description is of a piece of rock of a given weight being repeatedly broken into binary smaller pieces. In Miller Steven et al (2013) the description is of an original one-dimensional linear stick of length L being repeatedly broken into binary smaller segments. Surely, both descriptions, of rocks and sticks, representing weights and lengths, are simply particular manifestations of the same generic idea of repeated random divisions of a given conserved quantity into smaller and smaller ones.

Interestingly, Miller includes the case of a deterministic fixed p ratio [and its complement $(1 - p)$] breakup, such as in, say, 20% - 80%, or 40% - 60%, instead of utilizing random ratios via the Uniform(0, 1) distribution. The fixed ratio result is constrained to cases where $LOG((1 - p)/p)$ cannot be expressed as a rational N/D value, N and D being integers. When $LOG((1 - p)/p)$ is a rational number no convergence is found. Miller has provided rigorous mathematical proofs for both scenarios, for the random case, as well as for the deterministic case. There are two factors which render Miller's deterministic case less relevant to real-life physical data sets. The first factor is the extremely slow rate of convergence in the fixed deterministic case, which necessitates thousands if not tens of thousands of stages, in contrast to the random case which rapidly converges extremely close to the logarithmic after merely, say, 10 or 13 cycles! It may be that there exist some very long deterministic decomposition processes in nature which involve such huge number of stages, but one would conjecture that these must be quite rare in nature even if they exist at all, and that they are not the typical data sets that the scientist, engineer, or the statistician encounters. The second factor is the rarity with which decompositions in nature are conducted with such precise, fixed, and orderly ratio, and perhaps this never occurs at all. Mother Nature is known to behave erratically and chaotically when she feels weak and unable to hold her compounds intact anymore, passively letting them decompose slowly and gradually, and be partitioned in a random way, using totally random ratios. She is even more chaotic when she rages and in anger spectacularly explodes her constructs into bits and pieces rapidly in quick successions; and to expect her to deliberately, calmly, and steadily apply continuously the same fixed p ratio is unrealistic. Expecting to find in nature a decomposition process that (1) comes with an enormous number of stages, and (2) that it steadily keeps the same deterministic fixed p ratio throughout, is being doubly unrealistic.

As one simulation example, a rock weighing 33 kilograms is repeatedly broken in 13 stages into $2^{13} = 8192$ pieces by randomly deciding on the breakup proportions via the Uniform(0, 1).

Breaking a 33-kilo Rock in 13 Stages - {29.9, 17.2, 12.6, 9.7, 8.1, 6.7, 5.9, 5.6, 4.3}
Benford's Law for First Order Digits - {30.1, 17.6, 12.5, 9.7, 7.9, 6.7, 5.8, 5.1, 4.6}

SSD value is 0.6, and such extremely low value indicates that Random Rock Breaking process is extremely close to Benford. The 3 smallest pieces and the 3 biggest pieces after the 13th stage are: {0.0000000000079, 0.0000000002123, 0.0000000004750, … , 0.42, 0.44, 1.45}. Order of magnitude seems to be incredibly large, calculated as LOG(1.45/0.0000000000079) = LOG(1.83*10^{11}) = 11.3, and (on the face of it) this guarantees a near perfect logarithmic behavior. But this is deceiving! The upper 1% and lower 1% whiskers (extreme outliers) are attempting to make us believe that the spread of the data is huge, but in fact it is not as dramatic as it seems. Avoiding the exaggeration of the whiskers and concentrating only on the core 98% of the data, we obtain: 1%-percentile = 0.0000000259, and 99%-percentile = 0.0645, hence a much more realistic order of magnitude is LOG(0.0645/0.0000000259) = LOG(2494364) = 6.4, and which is still considered rather unusually high in Benford's Law, guaranteeing an excellent fit to the logarithmic. CPOM = $(P_{90\%} / P_{10\%})$ = (0.0068572/0.0000025) = 2697; it's quite high! Figure 3.26 depicts the histogram of almost all the resultant pieces which lie on the short interval (0, 0.07). It shows severe skewness where the small is numerous and the big is rare. The choice of a logarithmic vertical scale enables us to see the overall data structure clearly, but unfortunately it masks the dramatic fall in the histogram, namely its severe skewness.

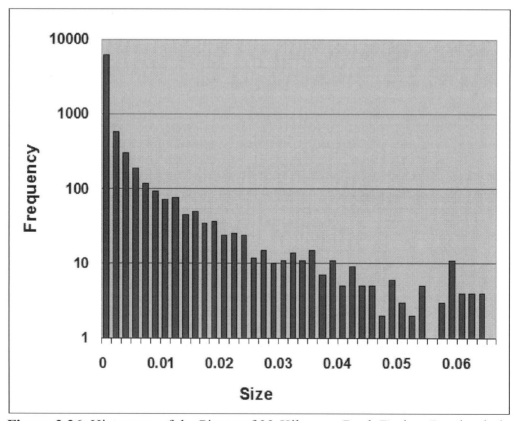

Figure 3.26: Histogram of the Pieces of 33-Kilogram Rock Broken Randomly in 13 Stages

Further scrutinizing the set of 8192 randomly obtained pieces after the 13th stage reveals that the original 33-kilogram rock has been thoroughly divided into totally distinct parts, so that the set of 8192 pieces does not contain any duplicated quantity! In other words, the resultant set of 8192 pieces contains 8192 distinct sizes! This fact should certainly be considered as a strong 'pro-partition' feature of the Random Rock Breaking model, explaining why it gets so fast and so close to Benford in one fell swoop, after only very few stages.

From numerous empirical experimentations (simulations), it is concluded that the rate of convergence for Random Rock Breaking is quite rapid, and after 10 stages, having merely 2^{10} or 1024 pieces, SSD is almost always below 10.0. After 14 stages, SSD is almost always below 1.0.

It is essential to visualize how quantities here evolve algebraically; and this is accomplished by writing the process carefully stage by stage. Assuming the original weight of the rock is 1 kilogram, U_J being the Jth realization from the Uniform(0, 1) in the whole simulation scheme, and $(1 - U_J)$ being its complement:

$\{1\}$

$\{U_1, (1 - U_1)\}$

$\{U_2{*}(U_1),\ \ (1 - U_2){*}U_1,\ \ U_3{*}(1 - U_1),\ \ (1 - U_3){*}(1 - U_1)\}$

$\{U_4{*}(U_2{*}U_1),\ \ \ (1 - U_4){*}(U_2{*}U_1),\ \ U_5{*}(1 - U_2){*}U_1,\ \ \ (1 - U_5){*}(1 - U_2){*}U_1,$
$\ \ U_6{*}U_3{*}(1 - U_1),\ \ \ (1 - U_6){*}U_3{*}(1 - U_1),\ \ U_7{*}(1 - U_3){*}(1 - U_1),\ \ \ (1 - U_7){*}(1 - U_3){*}(1 - U_1)\}$

And so forth to higher stages.

Clearly, it would be exceedingly rare to find identical values (pieces) here. For example, it is possible in principle to have $U_4{*}(U_2{*}U_1) = U_6{*}U_3{*}(1 - U_1)$, but this would require a very particular set of choices of U_J's and which is extremely unlikely and rare, having zero probability in a formal mathematical sense. The Uniform(0, 1) contains unaccountably infinite real numbers!

The above sequence of algebraic expressions demonstrates that Random Rock Breaking can be thought of also as a multiplicative process albeit with strong dependencies between the terms. Since multiplication processes in general are known to be highly skewed quantitatively and to have strong logarithmic digital tendencies, we expect (and get) skewness and Benford behavior here. This assertion relies on the fact that the arithmetical terms of the process involve numerous multiplicands, and this fact is more than sufficient to obtain a powerful logarithmic tendency and skewness for the products - given that multiplicands come with high order of magnitude. The use of the Uniform(0, 1) which ['supposedly'] possesses infinitely large order of magnitude guarantees that the process can easily overcome any possible challenge from the algebraic dependencies between the terms. Order of magnitude here is naively calculated as LOG(1/0) = LOG(Infinite) = Infinite. Yet, CPOM is (0.9)/(0.1) = 9, and not infinite! As discussed in Section 2, multiplication processes involving random variables with high order of magnitude lead to a strong and rapid convergence to the logarithmic. By its very nature, Random Rock Breaking model cannot use any other random variable with low order of magnitude such as for example

Uniform(5, 7), because it needs to break a whole quantity of a particular rock weight into two fractions, and this can only be achieved via Uniform(0, 1) which is of high order of magnitude.

Let us now turn our attention to Miller's model of a deterministic fixed ratio of breakup. It is essential to visualize how quantities evolve here algebraically; and this is accomplished by writing the process carefully stage by stage. Assuming the original weight of the rock is 1 kilogram; p is the fixed deterministic ratio; and s = (1 − p) is its complement:

{1}

{p, s}

{pp, ps, sp, ss}

{ppp, pps, psp, pss, spp, sps, ssp, sss}

{pppp, ppps, ppsp, ppss, pspp, psps, pssp, psss, sppp, spps, spsp, spss, sspp, ssps, sssp, ssss}

{ppppp, pppps, pppsp, pppss, ppspp, ppsps, ppssp, ppsss, psppp, pspps, pspsp, pspss, psspp, pssps, psssp, pssss, spppp, spppp, sppsp, sppss, spspp, spsps, spsss, spsss, ssppp, sspps, sspsp, sspss, ssspp, sssps, sssssp, ssssp, sssss}

And so forth to higher stages.

This relates to the Binomial Distribution. Clearly there are many repeating terms here. In the 2nd stage, **ps** = **sp**. In the 3rd stage, **pps** = **psp** = **spp**, as there are 3 ways to order such a product, where s is on the right, or in the center, or on the left. Also: **pss** = **sps** = **ssp**. In the 4th stage, **ppps** = **ppsp** = **pspp** = **sppp**, as well as **ppss** = **psps** = **pssp** = **spps** = **spsp** = **sspp**.

In how many ways can $(P^r)(S^p)$ be arranged? First, let us define n = r + p, namely the number of stages. The number of ways they can be arranged is given by the Binomial Coefficient:

$$\binom{n}{r} = \frac{n!}{(r)!\,(n-r)!}$$

Applying the binomial coefficient for ppps = ppsp = pspp = sppp, namely $(P^3)(S^1)$:

$$\binom{4}{3} = \frac{4!}{(3)!\,(1)!} = \frac{4*3*2*1}{(3*2*1)(1)} = 4$$

Applying the binomial coefficient for ppss = psps = pssp = spps = spsp = sspp, namely $(P^2)(S^2)$:

$$\binom{4}{2} = \frac{4!}{(2)!\,(2)!} = \frac{4*3*2*1}{(2*1)(2*1)} = 6$$

Hence there are 4 ways to arrange ppps, ppsp, pspp, sppp.

Hence there are 6 ways to arrange ppss, psps, pssp, spps, spsp, sspp.

As a demonstration of the incredibly slow convergence rate in Miller's deterministic case, a rock weighing 33 kilograms is repeatedly broken in 13 stages into $2^{13} = 8192$ pieces by deterministically deciding on the breakup fixed ratio of p = 0.7 and s = (1 – p) = 0.3.

Breaking 33-kilo Rock, 13 Stages, 70% - 30% - {38.7, 4.4, 15.7, 15.7, 1.0, 3.5, 0.0, 20.9, 0.0}
Benford's Law First Order Significant Digits - {30.1, 17.6, 12.5, 9.7, 7.9, 6.7, 5.8, 5.1, 4.6}

Digit distribution is nowhere near Benford as yet after only 13 stages, and SSD value is 658.1.

Convergence is exceedingly slow! Deterministic Rock Breaking process requires a truly huge number of stages in order to converge to Benford.

Scrutinizing the above set of 8192 deterministically obtained pieces after the 13th stage reveals that the original 33-kilogram rock hasn't been divided into numerous sizes, and that the set of 8192 pieces contains numerous duplicated quantities. So much so, that there exists only 14 distinct sizes within the entire set of 8192 pieces! The details of these 14 sizes are as follow:

Size	Occurrences
0.0000053	1
0.0000123	13
0.0000286	78
0.0000668	286
0.0001560	715
0.0003639	1287
0.0008491	1716
0.0019812	1716
0.0046228	1287
0.0107865	715
0.0251685	286
0.0587266	78
0.1370287	13
0.3197337	1

For example, there are 1716 pieces of size 0.0019812. Each such piece springs from:

0.0019812 = 33*0.7*0.7*0.7*0.7*0.7*0.7*0.7*0.3*0.3*0.3*0.3*0.3*0.3, or from
0.0019812 = 33*0.3*0.7*0.7*0.7*0.7*0.7*0.7*0.3*0.3*0.7*0.3*0.3*0.3 or from
0.0019812 = 33*0.3*0.3*0.3*0.7*0.7*0.7*0.7*0.7*0.7*0.7*0.3*0.3*0.3
etc. etc.

and in general
$0.0019812 = 33*(0.7^7)(0.3^6).$

167

In how many ways can $(0.7^7)(0.3^6)$ be arranged? The answer is 1716. Here $7 + 6 = 13$, signifying that there are 13 stages in this process.

The Binomial Coefficient here is:

$$\binom{13}{7} = \frac{13!}{(7)!\,(6)!} = \frac{6227020800}{5040 * 720} = \frac{6227020800}{3628800} = 1716$$

Such meager set of only 14 distinct sizes for the resultant set of 8192 pieces, with so many repetitions, does not bode well for Benford and skewness! Indeed the process hasn't even begun to converge here.

Such meager set of only 14 distinct sizes for the resultant set of 8192 pieces, with so many repetitions, reveals that the original 33-kilogram rock hasn't been thoroughly divided, and that this process of partition is incomplete.

Such meager set of only 14 distinct sizes for the resultant set of 8192 pieces, with so many repetitions, and the fact that the sizes are much smaller than 33, remind us so much of Refined Equipartition here, much more so than of Random Rock Breaking! [*well, only at this early 13th stage, but it's not so later.*]

Within the perspective of Refined Equipartition, these two facts also explain why the process hasn't even begun to arrive at the Benford configuration yet. In Refined Equipartition it was absolutely necessary to consider the set of all possible partitions to arrive at a Benford-like configuration, but here only a single partition is considered, and this explains why this process isn't yet Benford.

Interestingly, values within this set of 14 sizes increase steadily by a multiplicative factor of 2.33 (almost), pointing to a bit more similarity here in a sense with Refined Equipartition model which increases the sizes steadily with the additive value of 1.

[14] Chaotic Rock Breaking

The connection or relevance of the Refined Equipartition model to physical real-life data sets is often questioned by those who know Mother Nature well and are familiar with the way she works, and it is posited that she would probably never bother to delicately break her quantities carefully only along sets of parts with exact integral relationships, and that she would definitely not restrict herself only to much smaller and refined pieces (namely n << N). Indeed, this line of thought suggests that also Random Real Partition and Random Dependent Partition are not the typical processes that she likes doing or even capable of performing.

Let us imagine a real-life assembly-line with workers and management as in typical large corporations, attempting to physically perform the Random Dependent Partition process on a very long metal pipe. First the workers decide on the first random location where the pipe is to be cut, followed by actual cutting, and then the designations of "1" and "2" tags for the newly created pieces are made. This is followed by the orderly cutting of piece "1" and then piece "2" using random location within each piece. Without such designations and tags, there exists the possibility that by mistake one piece is cut twice leaving the other piece intact. Next, the workers designate the newly created four pieces as "1", "2", "3", and "4". At this stage, piece designation becomes even more crucial to avoid confusion and mistakes, and to remember which pieces were already cut and which pieces are awaiting their turn. It is highly doubtful that Mother Nature is capable or even interested in such serious and rigid type of work.

Consideration of how Random Real Partition would take place in such real-life assembly-line setup also leads one to think that this is not the type of work that would interest Mother Nature much. In this case, hundreds or rather thousands of workers are set up at random points along the long metal pipe with saws, ceramic mills, diamond drills, and other such cutting tools, all ready to simultaneously cut the pipe when the order is given. Upon hearing the first ear-piercing whistle of the foreman, workers stand to attention ready to do the cutting. The moment the second ear-piercing whistle is heard, they all cut rapidly, forcefully, and simultaneously, while Mother Nature is looking at them benevolently from above, feeling pity for the workers for all their highly coordinated effort and forced concentration. This description is called Scenario A. It should be noted that there is no compelling reason to assume that in Random Real Partition the breakups should be 'performed' simultaneously. Indeed, there is no need whatsoever to introduce the time dimension here, although without timing and detailed description of how Random Real Partition is performed, the concept is not a partition 'process' per se, but rather mere abstraction of how a given quantity may be broken into many smaller parts. This is why one cannot ask Mother Nature to 'perform' such a partition in the abstract, unless she is told what to do precisely, stage by stage. What could be suggested here (to be called Scenario B) is the successive gentle markings one by one of random points along the long metal pipe, followed by the actual cutting in a random fashion all of these marks by a lone, underpaid, exploited, and stoic worker, who cuts the long pipe for hours on end, one mark at a time (chosen randomly).

Obviously, this discussion about Mother Nature is metaphorical, and surely there exist in nature some particular decomposition processes that perfectly match the mathematical models of Random Real Partition and Random Dependent Partition, and even perhaps Refined Equipartition approximately, but what is conjectured here is that the typical decomposition process in nature is highly chaotic, totally lacking structure.

How would temperamental Mother Nature go about breaking a rock or a pipe her way, leisurely, chaotically, and consistent with her free-spirit attitude and her strong dislike of regimentation?

Her first act in the process is the breaking of the original rock of weight X into two parts randomly via the continuous Uniform(0, 1) to decide on the proportions of the two fragments. There is no need whatsoever to designate any pieces with any tags thereafter, since order of breakups does not matter to her in the least. Her second act is the totally relaxed and random selection of any one of the two pieces, followed by its fragmentation into two parts randomly via the continuous Uniform(0, 1), resulting in 3 pieces. Her third act is the totally relaxed and random selection of any one of the three pieces, followed by its fragmentation into two parts randomly via the continuous Uniform(0, 1), resulting in 4 pieces. This continues on and on for sufficiently large number of stages to obtain the Benford configuration.

Obviously, after each stage, the number of existing pieces increases by one. Also, it is obvious that the total quantity of the entire system (overall sum - overall weight of all the pieces) is conserved throughout the entire process. The specific description of physical balls or rocks of uniform mass density being broken, and the focus on the weight variable, is an arbitrary one of course, and the generic model is of pure quantities and abstract numbers.

Schematically the process is described as follow:

1) Initial Set = {X}

2) Repeat C times:

    Choose one value at random, remove it from the set of values, then split it as in {Uniform(0, 1), 1 − Uniform(0, 1)}, then place these two values back into the set.

3) Final Set of (C + 1) pieces is Benford, assuming C > 1000 approximately.

This is coined as 'Chaotic Rock Breaking'. From the point of view of Mother Nature, this is the most natural and straightforward way to randomly break a rock into small fragments. She closes her eyes; picks up randomly one piece at a time; breaks it in a random fashion; throws the two pieces back into the pile; and leisurely repeats this procedure over and over again until she gets tired or bored, or until she notices that the Benford configuration has already been achieved almost perfectly – and which immediately extinguishes her motivation to do any further work.

The table in Figure 3.27 depicts digital results for 23 distinct Monte Carlo computer simulations of Chaotic Rock Breaking with varying number of resultant pieces (i.e. varying number of fragmentation acts), all starting with 100-kilogram rock, and using the random breakup ratio of the Uniform(0, 1).

In the first row of the table, 100-kilogam rock has been broken 50 times, resulting in 51 pieces, and here digital configuration is not Benford, as indicated by the very high 207.0 value for SSD. Figure 3.27 is not about a single process and 23 snapshots taken at different times, but rather about 23 independent and totally different Monte Carlo simulation runs (i.e. the computer simulation starts anew from the beginning with a single whole 100-kilogram rock for each of the 23 different rows.)

# of Pieces	Digit 1	Digit 2	Digit 3	Digit 4	Digit 5	Digit 6	Digit 7	Digit 8	Digit 9	SSD
51	21.6	25.5	9.8	3.9	9.8	5.9	9.8	5.9	7.8	207.0
101	30.7	11.9	9.9	6.9	13.9	8.9	4.0	5.0	8.9	109.9
201	29.9	18.9	12.9	14.4	4.5	7.0	5.5	4.0	3.0	40.2
301	31.2	16.9	9.6	13.6	7.3	8.0	5.6	4.0	3.7	29.5
401	30.7	19.5	12.7	8.5	7.5	3.7	5.5	7.0	5.0	17.9
501	32.1	19.4	11.6	8.6	8.0	5.8	6.0	4.0	4.6	11.4
601	31.1	15.0	11.5	10.8	7.3	6.7	7.5	5.7	4.5	13.8
701	29.8	18.5	13.7	9.3	9.1	5.7	4.6	4.1	5.1	7.8
801	31.6	17.4	12.1	8.6	8.7	6.7	5.9	5.2	3.7	5.0
901	30.4	17.2	13.8	9.4	6.1	7.2	6.0	6.7	3.2	9.8
1001	29.3	18.1	11.3	10.1	9.5	6.7	5.5	5.9	3.7	6.5
1301	29.0	18.6	12.8	9.4	9.8	6.1	5.6	4.7	4.1	6.6
1501	32.1	17.6	12.1	9.6	7.5	5.6	5.4	4.9	5.3	6.4
1701	30.5	18.6	12.4	10.2	8.2	6.8	5.0	4.1	4.3	3.3
2001	30.1	16.4	13.4	9.8	7.4	6.4	5.6	5.6	5.3	3.4
2301	30.6	16.2	12.4	9.2	7.6	7.4	6.8	5.3	4.4	4.0
2501	29.9	17.0	12.8	8.9	9.0	6.9	5.6	5.2	4.8	2.6
3001	29.5	17.5	12.6	10.4	8.2	6.3	5.7	4.7	5.1	1.5
3501	30.2	18.1	11.5	9.9	7.8	6.8	5.8	5.7	4.3	1.7
4001	29.3	17.7	12.4	10.4	7.9	6.5	6.3	4.6	4.8	1.7
10001	30.3	18.0	12.7	9.3	7.7	6.6	5.9	4.9	4.6	0.5
15001	29.4	17.7	12.5	9.9	7.7	6.9	5.8	5.3	4.6	0.7
20001	30.0	17.2	12.4	9.9	8.0	6.8	6.0	5.1	4.8	0.3

Figure 3.27: Digital Results for 23 Chaotic Rock Breaking Processes Via Uniform(0, 1)

The chart in Figure 3.28 depicts the scatter plot of SSD versus the number of pieces for the simulation runs of the upper 20 rows of Figure 3.27. A logarithmic scale is used for the vertical axis for better visualization and clarity. Given that a value of SSD less than 10.0 is tolerated, and that any data set with SSD less than this arbitrary cutoff point of 10.0 is considered as Benford; then from these Monte Carlo simulation results one could conclude that Chaotic Rock Breaking with over 1000 fragmentations (pieces) is definitely Benford. When comparing this result with the Random Rock Breaking process where approximately 10 stages are needed at a minimum to arrive close enough to the Benford digital configuration with SSD below 10.0, and the implied 2^{10} or 1024 pieces, it seems that both, Chaotic and Random, have the same or very similar rates of convergence. Perhaps the adjective 'Chaotic' has stronger connotation with lack of order and total unpredictability than adjective 'Random' has, hence the motivation behind the coining of the terms 'Random Rock Breaking' and 'Chaotic Rock Breaking' reflecting the two distinct levels of randomness, yet they both seem to converge just as rapidly.

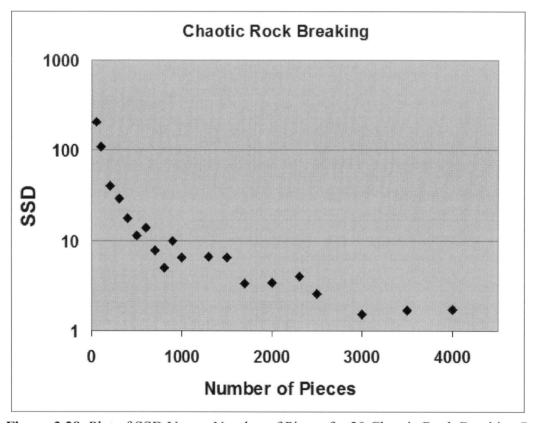

Figure 3.28: Plot of SSD Versus Number of Pieces for 20 Chaotic Rock Breaking Processes

It is beneficial to visualize how quantities here evolve algebraically; and this is accomplished by writing one possible random scenario stage by stage. Assuming the original weight of the rock is W kilogram, U_J being the Jth realization from the Uniform(0, 1) in the whole simulation scheme, and $(1 - U_J)$ being its complement:

$\{W\}$ W is broken

$\{U_1 * W, \ (1 - U_1) * W\}$ then only $U_1 * W$ is broken

$\{U_2 * U_1 * W, \ (1 - U_2) * U_1 * W, \ (1 - U_1) * W\}$ then only $(1 - U_1) * W$ is broken

---

$\{U_2 * U_1 * W, \ (1 - U_2) * U_1 * W, \ U_3 * (1 - U_1) * W, \ (1 - U_3) * (1 - U_1) * W\}$ then $U_2 * U_1 * W$

$\{U_4 * U_2 * U_1 * W, \ (1 - U_4) * U_2 * U_1 * W, \ (1 - U_2) * U_1 * W, \ U_3 * (1 - U_1) * W, \ (1 - U_3) * (1 - U_1) * W\}$

And so forth, producing many more broken pieces.

Clearly, after sufficient number of such fragmentation stages (approximately 1000), on average each term (i.e. each piece) contains numerous multiplicands, and therefore the set of pieces can be thought of as emerging from a multiplication process with high order of magnitude, and thus leading to Benford. Roughly speaking, order of magnitude of the Uniform(0, 1) is calculated as LOG(1/0), and theoretically or potentially this is infinitely large. In reality it is finite and not as large, because realizations or simulations from the Uniform(0, 1) never truly get near 0; outliers are misleading and exaggerating variability; and Core Physical Order of Magnitude (CPOM) is by far the most appropriate measure here.

The common denominator in all three processes of Random Real Partition (Scenarios A and B), Random Rock Breaking, and Chaotic Rock Breaking, is that each process simply places marks randomly everywhere on the x-axis, not discriminating against any particular sub-sections within the relevant x-axis range. Yet, intriguingly, Random Rock Breaking and Chaotic Rock Breaking processes come with digital results that vary considerably from Random Real Partition [i.e. both converging fully to Benford in the limit]! If one attempts to visually follow 'in real time' the sequence of marks being formed onto the x-axis for all three processes, then it might be possible in some cases to distinguish between them and to know which process is actually being run. In Random Real Partition Scenario A, the observer sees all the marks being established simultaneously in one epic instance, although as discussed earlier, Random Real Partition is not clearly defined as a 'process' per se, involving the time dimension. In Random Rock Breaking the observer sees a whole new set of marks established between each older set of marks, in orderly and highly regimented stages. In Chaotic Rock Breaking as well as in Random Real Partition Scenario B, the observer sees new marks being established, one at a time, totally randomly and chaotically, without any apparent structure or regimentation, and without any regards to currently existing older marks, as if the process totally lacks memory of all the previous stages – yet these two processes lead to distinct quantitative and digital results!

The whole motivation in introducing the idea of Chaotic Rock Breaking is to allow Mother Nature to act naturally, totally lacking structure and order. Yet curiosity prompts us to investigate hybrid models of such chaotic fragmentation using the predictable, fixed, and deterministic ratio p. In other words, for the same process of Chaotic Rock Breaking, instead of random ratio of partition as in the Uniform(0, 1), a fixed ratio p is used throughout the entire process. Randomness is still preserved in these hybrid models because the piece to be broken is always being chosen randomly. Surely Mother Nature cannot remember nor concentrate hard enough to make sure that the same p ratio is applied in all partitions; and expecting this from her is really not realistic. In any case, let us explore empirical results from such hybrid models. The table in Figure 3.29 gives the results for the deterministic fixed ratio 85% - 15% of Chaotic 100-kilogram Rock Breaking. Figure 3.29 is not about a single process and various snapshots taken at different times, but rather about independent and totally different Monte Carlo simulation runs with varying lengths (i.e. the computer simulation starts anew from the beginning with a single whole 100-kilogram rock for each of the 12 different rows.)

Obviously, for this hybrid model with deterministic fixed 85% - 15% ratio, convergence to Benford is much slower as compared with the pure Chaotic Rock Breaking model. Moreover, it is not even known whether or not a much longer run well over 33092 pieces would manage to get SSD below 1 say. In other words, it is not even certain that full convergence in the hybrid 85% - 15% case can ever be achieved. Simulations well beyond 33000 prove somewhat time-consuming, and surely they should be explored in the future.

# of Pieces	Digit 1	Digit 2	Digit 3	Digit 4	Digit 5	Digit 6	Digit 7	Digit 8	Digit 9	SSD
101	35.6	17.8	14.9	5.9	8.9	5.9	5.9	1.0	4.0	69.3
201	26.4	17.4	15.9	7.5	10.0	8.5	3.5	6.0	5.0	44.2
301	32.6	21.6	10.0	8.0	8.3	4.7	6.6	3.0	5.3	41.4
501	28.1	19.2	13.8	7.8	8.6	8.4	4.2	6.2	3.8	19.2
801	28.1	18.9	14.7	7.9	9.7	6.4	4.1	5.4	4.9	20.3
1001	27.7	18.7	15.5	10.2	8.0	7.3	4.9	4.8	3.0	20.0
2001	30.3	18.3	13.3	6.5	9.4	7.2	4.7	5.1	5.1	15.0
4001	28.8	18.0	14.5	7.8	9.6	7.6	3.9	5.5	4.1	17.1
7001	28.5	18.5	14.2	7.7	10.0	6.9	4.3	5.6	4.3	17.0
12001	28.2	17.4	14.6	8.0	9.9	7.1	4.4	6.3	4.1	18.5
25001	28.9	17.9	14.2	8.3	9.0	7.4	4.6	5.9	3.8	10.5
33092	28.8	17.7	14.2	9.0	8.9	7.0	5.0	5.5	3.8	7.1

Figure 3.29: Digital Results of Hybrid Chaotic Rock Breaking with Fixed 85% - 15% Ratio

It stands to reason that this Hybrid Chaotic Rock Breaking model with 85% - 15% fixed ratio converges to Benford by far faster than Miller's Deterministic Rock Breaking with 85% - 15% fixed ratio. As was indeed seen in the previous chapter, breaking 33-kilo rock in 13 stages, with fixed ratio of 70% - 30%, and yielding 8192 pieces, was not even close to Benford. This is so because Hybrid Chaotic Rock Breaking model still contains the random element within itself, in the sense that each piece about to be broken is chosen randomly, while Deterministic Rock Breaking on the other hand is a purely deterministic model, and there is nothing random about it.

The table in Figure 3.30 gives digital results for a variety of deterministic fixed ratios for the Hybrid Chaotic Rock Breaking model. All simulation runs start with 100-kilogram rock. There is a bit of a mystery here why some fixed p rates come out very close to Benford, while others show just the Benfordian tendency but without any obvious strong convergence. Certainly Miller's irrationality constraint of LOG((1 - p)/p) does not apply here at all since this is a very different mathematical model than Deterministic Rock Breaking, and even though the ratio is deterministic, yet the process selects the next piece about to be broken in a random manner. It is possible that ultimately in the limit when the number of pieces goes to infinity, or in a practical way as Monte Carlo simulations are run for a truly large number of pieces on a more powerful computer, full or partial convergence to Benford could be found here equally and uniformly for the entire variety of these deterministic fixed ratios.

Fixed p Ratio	# Pieces	Digit 1	Digit 2	Digit 3	Digit 4	Digit 5	Digit 6	Digit 7	Digit 8	Digit 9	SSD
0.5000000	5001	30.2	9.7	20.7	8.0	2.2	8.0	12.6	0.0	8.7	256.3
0.2715402	5001	27.5	14.8	14.5	13.3	2.7	12.2	0.2	11.1	3.8	156.6
0.2777000	5001	31.4	18.2	12.1	7.4	8.8	7.2	6.1	3.2	5.5	13.5
0.3368796	2001	30.7	16.0	11.8	11.0	5.9	6.1	7.4	5.6	5.3	12.8
0.4254751	2001	27.0	22.8	10.7	6.5	8.7	7.0	4.3	3.1	9.7	83.1
0.0520000	2001	29.2	19.9	11.0	9.8	6.6	6.1	9.3	4.3	3.5	24.7
0.2524733	1001	29.9	18.8	14.5	8.6	7.2	5.9	5.2	5.7	4.3	8.5
0.2684584	1001	30.0	18.5	12.5	11.9	8.0	5.6	4.1	9.3	0.2	46.3

Figure 3.30: Digital Results of Hybrid Chaotic Rock Breaking with a Variety of Fixed Ratios

[15] Random Minimum Breaking

Instead of randomly selecting the next piece to be partitioned as in Chaotic Rock Breaking, a process is envisioned where at each stage the minimum of all currently existing pieces is selected and partitioned randomly via the Uniform(0, 1). In other words, the minimum is constantly being partitioned in a random fashion, over and over again. Here, resultant data of this process rapidly approaches extreme quantitative skewness, and Benford digital behavior is confirmed. Surely, by selecting the minimum at each stage of partition, skewness is constantly being increased in the system. The fact that the smallest piece is partitioned into two pieces [thus increasing the number of the small in the system] while big pieces are left intact [leaving the number of the big in the system unchanged] implies that at each stage greater skewness is achieved.

In one Monte Carlo simulation run of 500-kilogram rock, with 400 stages of splitting up the minimum via the Uniform(0, 1), the following digital configuration was obtained:

Breaking Min of 500-kilo Rock in 400 Stages - {29.5, 17.3, 15.8, 10.3, 5.0, 5.0, 5.8, 6.5, 5.0}
Benford's Law First Order Significant Digits - {30.1, 17.6, 12.5, 9.7, 7.9, 6.7, 5.8, 5.1, 4.6}

Digit distribution is quite close to Benford, with SSD value of 24.9.

The biggest 18 values of the ordered resultant data set after the 400th stage are as follows:

{378.0, 63.6, 29.4, 18.9, 7.3, 2.0, 0.72, 0.08, 0.03, 0.00159, 0.00046, 0.000046, 0.00001125, 0.00000504, 0.000000518, 0.000000236, 0.000000125, 0.00000000258}

The final minimum after the 400th stage is the exceedingly low value of $8.28*10^{-306}$!

Log histogram for the resultant data set after the 400th stage is nearly uniform, implying that resultant data set is of the deterministic flavor in Benford's Law. The logarithm of successive elements of the ordered resultant data set decreases by 0.77 on average.

The logarithm of the biggest 18 values of the ordered resultant data set after the 400th stage are as follows:

{2.58, 1.80, 1.47, 1.28, 0.87, 0.29, -0.14, -1.07, -1.54, -2.80, -3.34, -4.34, -4.95, -5.30, -6.29, -6.63, -6.90, -8.59}

Surely, Mother Nature would adamantly refuse to perform this tedious process. She avoids at all cost such regimented and intense partition style where the next minimum must be carefully ascertained at each stage.

[16] Random Maximum Breaking

When the maximum is repeatedly selected and partitioned randomly via the Uniform(0, 1), resultant data set approaches the Uniform Distribution in the limit as the number of such partitions goes to infinity. Such total absence of quantitative skewness precludes Benford digital behavior. Here, the maximum is constantly being partitioned in a random fashion, over and over again. Surely, by selecting the maximum at each stage of partition, skewness is constantly being decreased in the system. The fact that the biggest piece is partitioned into two pieces [thus increasing the number of the big in the system] while small pieces are left intact [leaving the number of the small in the system unchanged] implies that at each stage greater uniformity and evenness is achieved.

In one Monte Carlo simulation run of 100-kilogram rock, with 4000 stages of splitting up the maximum via the Uniform(0, 1), the following digital configuration was obtained:

Breaking Max of 100-kilo Rock in 4000 Stages - {22.0, 23.0, 22.9, 21.5, 2.2, 2.3, 1.9, 2.0, 2.3}
Benford's Law First Order Significant Digits   - {30.1, 17.6, 12.5, 9.7, 7.9, 6.7, 5.8, 5.1, 4.6}

Digit distribution is not of the Benford configuration at all, as indicated by the very large SSD value of 424.9.

Quantitatively, resultant data set after the 4000th stage is uniformly distributed on the interval between 0.000007 and 0.049550. Out of 4001 pieces, 3225 pieces fall uniformly on the interval (0.01, 0.05), with only 776 pieces falling below 0.01. This explains why first digits are nearly evenly distributed between digits 1, 2, 3, and 4.

The smallest 7 values and the biggest 7 values, of the ordered resultant data set after the 4000th stage, are as follows:

{0.000007, 0.000010, 0.000012, 0.000032, 0.000045, 0.000051. 0.000053, . . .
. . . 0.049492, 0.049497, 0.049506, 0.049527, 0.049534, 0.049546, 0.049550}

Surely, Mother Nature would adamantly refuse to perform this tedious process. She avoids at all cost such regimented and intense partition style where the next maximum must be carefully ascertained at each stage.

[17] Mathematical Model for Equipartition

Let us introduce standard notations for the Equipartition model. We begin with the Equipartition Parable of a group of 33 spies being separated into independent cells of maximum 5 spies each.

Quantity to be partitioned: $X = 33$

Allowed set of parts: $\{x_j\} = \{1, 2, 3, 4, 5\}$ [index j running from 1 to 5]

Writing down some particular partitions could prove quite tedious and too long at times. For example: **{1, 1, 1, 1, 1, 1, 1, 1, 1, 1, 1, 1, 1, 1, 1, 1, 1, 1, 1, 1, 2, 2, 2, 2, 2, 3}** is too long and ink-consuming. Since there are usually repeating terms in Equipartition, it would be much easier and more efficient to express them more concisely by simply counting the repetitions n_j of each part x_j. For the partition above, there are 20 ones, 5 twos, 1 three, 0 fours, and 0 fives, hence $\mathbf{n_j = \{20, 5, 1, 0, 0\}}$.

Using Dot Product notation, the fact that any given partition here must add up to 33 can be succinctly expressed as $\mathbf{33 = \{20, 5, 1, 0, 0\}*\{1, 2, 3, 4, 5\}}$.

The other partitions

$\{5, 5, 5, 5, 5, 5, 2, 1\}$
$\{2, 4, 5, 5, 1, 3, 1, 1, 1, 5, 5\}$
$\{1, 1, 1, 1, 1, 1, 1, 1, 1, 1, 2, 2, 2, 2, 2, 3, 5, 5\}$
$\{4, 5, 4, 5, 4, 5, 2, 2, 2\}$
$\{3, 3, 3, 3, 3, 3, 3, 3, 3, 3, 3\}$

are written in the format $X = \{n_j\}*\{x_j\}$, where $\{x_j\}$ is a fixed vector, while $\{n_j\}$ varies:

$33 = \{1, 1, 0, 0, 6\}*\{1, 2, 3, 4, 5\}$
$33 = \{4, 1, 1, 1, 4\}*\{1, 2, 3, 4, 5\}$
$33 = \{10, 5, 1, 0, 2\}*\{1, 2, 3, 4, 5\}$
$33 = \{0, 3, 0, 3, 3\}*\{1, 2, 3, 4, 5\}$
$33 = \{0, 0, 11, 0, 0\}*\{1, 2, 3, 4, 5\}$

And which can be expanded and written explicitly as:

$33 = 1*1 + 1*2 + 0*3 + 0*4 + 6*5$
$33 = 4*1 + 1*2 + 1*3 + 1*4 + 4*5$
$33 = 10*1 + 5*2 + 1*3 + 0*4 + 2*5$
$33 = 0*1 + 3*2 + 0*3 + 3*4 + 3*5$
$33 = 0*1 + 0*2 + 11*3 + 0*4 + 0*5$

In general, a conserved positive integral quantity X is partitioned exclusively into positive integral pieces $\{x_j\}$.

These pieces are a <u>fixed</u> set of N positive integral quantities $\{x_j\}$ and are called the '**part set**'. The part set is fixed within any given Equipartition model, and all possible partitions are written with respect to this fixed set, but each Equipartition model has its own distinct part set.

$\{x_j\} = \{ x_1,\ x_2,\ x_3,\ \dots,\ x_N \}$. Typically $\{x_j\}$ is a set of consecutive integers starting at 1, as:

$$\{x_j\} = \{ 1, 2, 3, \dots, N \}.$$

For any particular partition, each x_j could be repeated T times and so $n_j = T$, or it may occur only once and so $n_j = 1$, or it may not occur at all and so $n_j = 0$.

For any particular partition, X is written as a linear combination of $\{x_j\}$ by way of $\{n_j\}$ as:

$X = n_1 * x_1 + n_2 * x_2 + n_3 * x_3 + \dots + n_N * x_N$

Or in Dot Product notations: $X = \{n_j\} * \{x_j\}$.

The '**number of parts set**', namely $\{n_j\}$, of N non-negative integers is of course <u>not fixed</u>, but rather it <u>varies</u> according to the particular partition chosen. Obviously the set $\{n_j\}$ is restricted to those linear combinations that add up to X.

The partition of X is then expressed as a linear combination of $\{x_j\}$ and $\{n_j\}$:

$$X = \sum_{j=1}^{N} n_j * x_j$$

The variable vector $\{n_j\}$ specifies a particular partition of X.

It is preferable that the part set $\{x_j\}$ starts at 1, namely that $x_1 = 1$. This is so in order to ensure that a partition will always exist for every X.

The part set $\{x_j\}$ should be conveniently written as monotonically increasing set, namely that $x_{j+1} > x_j$ for all j.

Obviously, the part set should consist only of distinct values, namely that $x_j \neq x_i$ for all $j \neq i$.

In the final description of this abstract model, all $\{n_j\}$ partitions are given equal weights. This is the reason for the prefix '**equi**' in the term 'Equipartition'.

The term 'configurational entropy' in Thermodynamics refers to the assumption that all possible system configurations are equally likely. In the same vein, here all possible partitions $\{n_j\}$ are considered as equally-likely, and this is consistent with the above assignment of equal weights to all partitions. Physics inspires us, so we borrow from physics the principle regarding equality of all possible configurations. Fortunately and perhaps surprisingly, imitating nature in this abstract mathematical model leads to Benford under certain conditions! Yet, this is **not** regarded as a statement of some physical law of nature about real-life actual partitions, and once this is stated mathematically it can be run and simulated successfully on the computer without direct reference to physical reality or any laws of nature.

Equipartitions fall into 3 distinct categories, Complete Equipartition, Refined Equipartition, and Irregular Equipartition.

Complete Equipartition is one in the style of Integer Partitions of Number Theory with a smooth and simple part set $\{x_j\} = \{1, 2, 3, \dots, X\}$. Here $N = X$ and $x_N = X$. The part set $\{x_j\}$ increases monotonically and nicely by one integer at a time from 1 all the way to X. It is called 'complete' since it incorporates all possible breakups.

Refined Equipartition is one where the part set $\{x_j\}$ contains only integers that are much smaller than X thus ensuring that X is being broken only into much smaller and refined parts, resulting in a significant fragmentation of the original X quantity.
Formally stated: $N \ll X$, or as: $x_N \ll X$. Here the largest (last) element in the part set is much smaller than X. Also (for some results) it is thought of as a limiting process where x_N (and by implication N) is fixed, while X tends to infinity. It should be noted that the part set $\{x_j\}$ in refined equipartition increases nicely and monotonically one integer at a time from 1, so that there are no gaps. Mathematical applications of general results from restricted Integer Partition in Number Theory are of course possible even for this limited/refined partition.

Irregular Equipartition is one in which the part set $\{x_j\}$ does not increase nicely by one integer at a time; so that it has some gaps where some integers between 1 and x_N are 'missing'. There exist a small number of applications for such irregular partitions, but usually there is nothing to gain by increasing the part set in such irregular manner. In any case, Irregular Equipartitions shall be omitted for the rest of the whole discussion. Letting all the integers between 1 and x_N participate in the breakup of quantity X ensures a smooth and more thorough partition. Here, no mathematical results are offered for this messy irregular equipartition, although indeed some of the mathematical results and expressions given here may apply, or at least may constitute good approximations for Irregular Equipartitions as well.

Since Irregular Equipartitions are excluded from our discussion, the part set $\{x_j\}$ starts from 1, and increases monotonically by 1, until it reaches N, as it imitates the positive integers in a finite way; hence X can be expressed as:

$$X = n_1*1 + n_2*2 + n_3*3 + ... + n_N*N$$

Or equivalently, the partition of X can be expressed more concisely as the linear combination:

$$X = \sum_{j=1}^{N} n_j * j$$

The focus of the mathematical analysis here is on the configuration of the aggregated data set of all possible partitions, namely on the relative frequency or repetition of each x_j within that vast data set. Indirectly, we can learn about the relative frequencies of x_j by summing n_j separately for each size j over all partitions, which is simply the total number of occurrences of x_j within the entire equipartition model. An alternative measure is calculating $AVG(n_j)$, namely the average number of parts n_j of size x_j.

A single partition of X into part set $\{x_j\}$ is not Benford in the least. Only the aggregation of all possible partitions leads to Benford under certain conditions. For this aggregation, once the list of all possible partitions is available, one is interested in counting how many times x_1 occurs, and how many x_2 occurs, and so forth, namely the grand histogram of all x_j incorporating all partitions as one vast data set. It should be noted that if the value of n_1 is 48 say for one particular partition, then this implies that x_1 occurs 48 times in this particular partition. In order to arrive at the aggregated x_j histogram incorporating occurrences from all partitions, one needs to simply sum (over all partitions) those n_1 occurrences to obtain the grand count of x_1, and then sum (over all partitions) those n_2 occurrences to obtain the grand count of x_2, and so forth. Indeed, $AVG(n_j)$ is actually this exact sum of x_j re-scaled (i.e. divided) by the number of all possible partitions, hence $AVG(n_j)$ conveys the relative proportions of x_j as well.

P is defined as the number of all possible partitions for any particular equipartition scheme. n_{kj} refers to n_j of the k-th partition; with index k running from 1 to P. AVG (n_j) is then:

$$\sum_{k=1}^{P} (n_{kj})/P = AVG(n_j)$$

The number of times each x_j occurs within the <u>entire</u> equipartition scheme is:

$$\sum_{k=1}^{P} n_{kj} = [\# \ of \ occurrences \ of \ x_j \ in \ entire \ scheme \ ]$$

$$AVG(n_j) * P = [\# \ of \ occurrences \ of \ x_j \ in \ scheme \ ]$$

$$AVG(n_j) * [\# \ of \ Partitions] = [\# \ of \ occurrences \ of \ x_j \ in \ scheme \ ]$$

For either Complete Equipartitions or Refined Equipartitions (but not for Irregular Equipartitions), mathematical analysis in Miller (2015) leads to an exponential-like discrete distribution for AVG(n_j) with the following expression:

$$AVG(n_j) = \frac{1}{e^{\lambda x_j} - 1}$$

Where the constant λ – called the Lagrange multiplier – is uniquely determine via the constraint:

$$X = \sum_{j=1}^{N} \frac{x_j}{e^{\lambda x_j} - 1}$$

Let us derive this last equation of total quantity X which determines the value of λ by combining the expression of the above definition of AVG (n_j) together with Miller's result for AVG (n_j):

$$AVG\left(n_j\right) = \sum_{k=1}^{P} (n_{kj})/P$$

$$AVG\left(n_j\right) = \frac{1}{e^{\lambda x_j} - 1}$$

Equating the two right hand sides we get:

$$\sum_{k=1}^{P} (n_{kj})/P = \frac{1}{e^{\lambda X_j} - 1}$$

Multiplying both sides by x_j we get:

$$\sum_{k=1}^{P} x_j(n_{kj})/P = \frac{x_j}{e^{\lambda X_j} - 1}$$

Summing both sides over index j in order to aggregate all sizes, we get:

$$\sum_{j=1}^{N} \sum_{k=1}^{P} x_j(n_{kj})/P = \sum_{j=1}^{N} \frac{x_j}{e^{\lambda X_j} - 1}$$

Switching the order of the indices, we get:

$$\frac{1}{P} \sum_{k=1}^{P} \sum_{j=1}^{N} x_j n_{kj} = \sum_{j=1}^{N} \frac{x_j}{e^{\lambda X_j} - 1}$$

Since original quantity X is being conserved in any given partition, the inner summation is X:

$$\sum_{j=1}^{N} x_j n_{kj} = \text{Dot Product of } \{x_j\} \text{ and } \{n_j\} = (\text{Total Quantity in a Partition}) = X$$

Hence:

$$\frac{1}{P} \sum_{k=1}^{P} X = \sum_{j=1}^{N} \frac{x_j}{e^{\lambda X_j} - 1}$$

Finally, evaluating the summation on the left side of the last equation, and applying the summation rule $\sum_{k=1}^{P} \mathrm{Constant} = P * \mathrm{Constant}$, we get:

$$\frac{1}{P}\sum_{k=1}^{P} X = \frac{1}{P}(P * X) = X = \sum_{j=1}^{N} \frac{x_j}{e^{\lambda x_j} - 1}$$

Clearly, the above expression $\dfrac{1}{e^{\lambda x_j} - 1}$ for AVG(n_j) is skewed in favor of the small (e.g. n_1), discriminating against the big (e.g. n_N). In other words, the expression is in a sense inversely proportional to x_j in a complicated way (appearing within an exponent in the denominator).

The above expression for AVG(n_j) reminds us of the continuous Exponential Distribution $\lambda e^{-\lambda x}$ defined on the infinite real range of $(0, +\infty)$, and which is known for being fairly close to Benford regardless of the value of its parameter λ, but the expression above is only for the integral values of the part set $\{x_j\}$, namely for the discrete (and very finite) set $\{1, 2, 3, \dots, N\}$, and therefore it is not possible at all to draw any conclusions whatsoever about possible logarithmic behavior here even in the approximate. Moreover, for $\{x_j\} = \{1, 2, 3\}$ say, digits 4 to 9 never get a chance to lead in base 10 (although the possibility exists that the frequency vector might fit Benford's Law base 4).

Yet, such result where quantities are skewed in favor of the small raises our hopes and induces us to explore the possibility that Benford's Law could be found here. Indeed, additional mathematical arguments in Miller (2015) for Refined Equipartitions (but not for Complete Equipartitions) lead to much superior result given that $x_N \ll X$ (or equivalently $N \ll X$), and formally in the limit as X goes to infinity while N (or equivalently x_N) remains fixed:

$$\mathrm{AVG}(n_j) \approx \frac{(X/N)}{x_j}$$

This result implies a monotonically decreasing set of values for the average of n_j. Here the expression for the average of n_j is inversely proportional to x_j!

The above expression can also be written as:

$$(x_j) * \mathrm{AVG}(n_j) \approx \frac{X}{N}$$

Let us combine two results; the first result [definition rather] was discussed earlier:

$$AVG(n_j)*[\# \text{ of Partitions}] = AVG(n_j)*[P] = [\# \text{ of occurrences of } x_j \text{ in scheme}]$$

The second result is common sense and quite obvious:

$$[\text{total quantity for size } x_j \text{ in scheme}] = x_j*[\# \text{ of occurrences of } x_j \text{ in scheme}]$$

Together these relationships yield:

$$[\text{total quantity for size } x_j \text{ in scheme}] = x_j*[AVG(n_j)*[P]] = [x_j*AVG(n_j)]*[P]$$

From Miller's result above $[x_j*AVG(n_j)] \approx [X/N]$, hence we finally get:

$$[\text{total quantity for size } x_j \text{ in scheme}] \approx \frac{XP}{N}$$

Since P, X, N are all fixed values and independent of j for a given Refined Equipartition scheme, it follows that total quantity for each size x_j is a constant, namely the same for all sizes.

Thus, an appealing interpretation of Miller's result is that Refined Equipartition endows the same quantitative portion from the entire quantity of the entire equipartition scheme of XP to each size of the N existing sizes, namely (XP)/N. This is so since there are N sizes of x_j, and each partition contains the quantity X within itself as its own sum, therefore XP is the quantity of all possible partitions in the entire scheme, namely the total quantity of that vast data set of the Refined Equipartition model. In other words: the entire quantity of the entire equipartition scheme XP is divided fairly and equally among the N sizes of x_j.

This result is gotten only if we thoroughly break up the conserved quantity X into very small and highly refined pieces – as is the case in Refined Equipartitions. Complete Equipartitions on the other hand do not lead to the above result because they also include 'crude', 'partial' or 'improper' partitions, where fairly large pieces are left intact.

Finally, the observation that variable $AVG(n_j)$ is inversely proportional to x_j implies that it is of the k/x distribution form, albeit in a discrete sense , not in a continuous sense. Since the continuous k/x density is known for its exact Benford behavior in cases when range falls exactly between 1 and an integral power of ten, namely $[1, 10^N)$, N being an integer greater than 0, such as for example on $[1, 10)$, $[1, 100)$, $[1, 1000)$, and so forth, it follows that Refined Equipartition could possibly be Benford-like whenever part set $\{x_j\}$ consists of corresponding integral ranges, such as $\{1, 2, 3, 4, 5, 6, 7, 8, 9, 10\}$, or perhaps more appropriately on $\{1, 2, 3, 4, 5, 6, 7, 8, 9\}$, since real x-axis sub-range $[9, 10)$ is 'covered' by integer 9.

To empirically test the first general equipartition result for AVG(n_j) in one concrete case, a computer program is run to produce <u>Complete Equipartition</u> for X = 25. Here the part set is { x_j } = {1, 2, 3, … , 23, 24, 25}. The computer does not assume anything and does not apply any mathematical formulas, except that it gives equal weights to all possible partitions. It simply displays all possible partitions, one by one, converts them into {n_j} vector format, and then calculates AVG(n_j) directly. There are 1958 possible partitions here.

Empirical results - computer generated AVG(n_j):
{3.75, 1.65, 0.96, 0.63, 0.44, 0.31, 0.23, 0.17, 0.13, 0.09, 0.07, 0.05, 0.04, 0.03, 0.02,
0.015, 0.011, 0.008, 0.006, 0.004, 0.003, 0.002, 0.001, 0.001, 0.001}

Theoretical expression of $1/(e^{\lambda X_j} - 1)$ for AVG(n_j):
{3.59, 1.57, 0.92, 0.60, 0.41, 0.30, 0.22, 0.16, 0.12, 0.09, 0.07, 0.06, 0.04, 0.03, 0.03,
0.020, 0.016, 0.012, 0.009, 0.007, 0.006, 0.004, 0.004, 0.003, 0.002}

This result shows a good fit of empirical to the theoretical expression of AVG(n_j). Here 0.24583 was the calculated value of the Lagrange multiplier λ. Figure 3.31 depicts AVG(n_j) for integers 1 to 12. There is nothing here resembling Benford's Law of course except for the conceptual observation that small is beautiful. One could examine the first digits distribution of these 25 integers, but nothing of significance would come out of such a study.

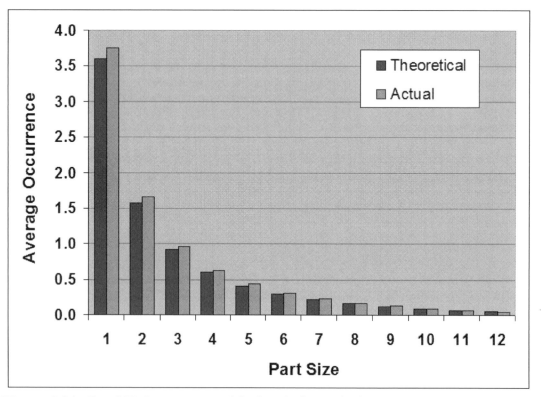

Figure 3.31: Good Fit between Empirical and Theoretical in Complete Equipartition of 25

Another empirical test is performed regarding the two expressions for AVG(n_j).
This computer program produces <u>Refined Equipartition</u> for X = 33 and part set
{ x_j } = {1, 2, 3, 4, 5}, namely that of the Equipartition Parable. The computer does not assume anything and does not apply any mathematical formulas, except that it gives equal weights to all possible partitions. It simply displays all possible partitions, one by one, converts them into {n_j} vector format, and then calculates AVG(n_j) directly. There are 918 possible partitions here.

Empirical results - computer generated AVG(n_j) = {7.52, 3.52, 2.20, 1.53, 1.14}
Theoretical expression of **1/($e^{\lambda X_j}$ - 1)** for AVG(n_j) = {7.50, 3.52, 2.20, 1.54, 1.15}

This result shows an excellent fit between the empirical result and the theoretical expression. Here 0.1251 was the calculated value of the Lagrange multiplier λ. The 2nd theoretical expression (X/N)/x_j assumes that N $\ll$ X, and although this is principally Refined Equipartition, yet this condition is not completely satisfied here since 5 is not really much less than 33, consequently the fit of empirical to the theoretical is not as strong for this result.

Empirical results - computer generated AVG(n_j) = {7.52, 3.52, 2.20, 1.53, 1.14}
Theoretical expression of **(X/N)/x_j** for AVG(n_j) = {6.61, 3.30, 2.20, 1.65, 1.32}

Converting the above empirical results of AVG(n_j) into actual occurrences of x_j is straightforward via the use of the expression [occurrences of x_j] = AVG(n_j)*[# of partitions] = AVG(n_j)*[918]. Hence for this Refined Equipartition: 1 occurs 6905 times, 2 occurs 3228 times, 3 occurs 2017 times, 4 occurs 1408 times, and 5 occurs only 1050 times.
The proportions of the occurrences of {1, 2, 3, 4, 5} are {47%, 22%, 14%, 10%, 7%}.
Total Quantity Per Size = [Occurrences of x_j in scheme]*[x_j] = {6905, 6456, 6051, 5632, 5250}

There is nothing here resembling Benford's Law base 10 of course, except for the conceptual observation that small is beautiful. One could examine the first digits distribution of these 5 integers (which are just the integers themselves), but absolutely nothing of significance would come out of this study. Digits 6 to 9 never occur in any case. Yet, the nature of the part set here is approximately logarithmic-like because it is roughly of the k/x distribution form. A comparison to Benford's Law base 6 does not yield a very good fit, and it disappoints a bit:

Equipartition, X = 33, part set {1, 2, 3, 4, 5} - {47%, 22%, 14%, 10%, 7%}
Benford's Law Base 6 for the First Digits   - {39%, 23%, 16%, 13%, 10%}

Conceptually explaining such skewed result of AVG(n_j) here is quite straightforward. It is simply the direct consequence of the model itself. Here total quantity is X = 33, and the part set is {x_j} = {1, 2, 3, 4, 5}. Since n_1 refers to x_1 which contains very little quantitative value (i.e. quantity 1) hence it often needs to be frequent and repetitive in order for 33 to be obtained; while n_5 refers to x_5 which contains much bigger quantitative value (i.e. quantity 5) hence it could often be quite infrequent and scarce and 33 may still be obtained.

Indeed, the underlying explanation why average n_j is monotonically decreasing springs from a very profound, universal, and yet extremely simple principle regarding how a conserved quantity can be partitioned. The obvious principle or observation here is that: 'One big quantity is composed of numerous small quantities', or: 'Numerous small quantities are needed to merge into one big quantity'. Hence partitioning a fixed conserved quantity into parts can be done roughly-speaking in two extreme styles, either via a breakup into many small parts, or via a breakup into few big parts. A more moderate style would be to have a mixture of all kinds of sizes, consisting of many small ones, some medium ones, and a few big ones. This conceptual outline explains why average n_j is skewed quantitatively.

Since $AVG(n_1)$ represents the expected number of small things - it is big.
Since $AVG(n_N)$ represents the expected number of big things - it is small.

For an additional illustration of the ideas and results of this chapter, another concrete numerical example shall be given where total quantity $X = 5$, and the part set $\{x_j\} = \{1, 2, 3, 4, 5\}$.
This scenario is of the Complete Equipartition type, violating the constraint $N \ll X$.
In any case, the quantitative lesson this scenario teaches us is generic and universal in all partitions. Clearly n_1 can assume many values within its relatively large possible range of 0 to 5, given that the rest of n_j values are such that it all adds up to 5. And clearly n_5 can assume only two values within its relatively smaller possible range of 0 to 1, given that the rest of n_j values are such that it all adds up to 5. The exhaustive list of all 7 possible partitions written in Dot Product notations as $\{n_j\} * \{x_j\}$ is:

$\{5, 0, 0, 0, 0\} * \{1, 2, 3, 4, 5\} = 5$
$\{3, 1, 0, 0, 0\} * \{1, 2, 3, 4, 5\} = 5$
$\{2, 0, 1, 0, 0\} * \{1, 2, 3, 4, 5\} = 5$
$\{1, 2, 0, 0, 0\} * \{1, 2, 3, 4, 5\} = 5$
$\{1, 0, 0, 1, 0\} * \{1, 2, 3, 4, 5\} = 5$
$\{0, 1, 1, 0, 0\} * \{1, 2, 3, 4, 5\} = 5$
$\{0, 0, 0, 0, 1\} * \{1, 2, 3, 4, 5\} = 5$

$AVG(n_j) = \{12/7, \ 4/7, \ 2/7, \ 1/7, \ 1/7\}$
$AVG(n_j) = \{1.71, \ 0.57, \ 0.29, \ 0.14, \ 0.14\}$

The relationship (occurrences of x_j) = $AVG(n_j) *$(# of partitions) is quite obvious here, namely (occurrences of x_j) = $AVG(n_j) *$(7), thus $AVG(n_j)$ = (occurrences of x_j)/(7).

It is essential to dispel any possible mistaken perception that in Complete Equipartitions each size possesses approximately the same portion of overall quantity, namely the false perception that x_j*(occurrences of x_j) for all j are roughly equal, they are not! Calculating this here we get: for size 1 overall quantity is 1*12; for size 2 overall quantity is 2*4; for size 3 overall quantity is 3*2; for size 4 overall quantity is 4*1; for size 5 overall quantity is 5*1; so that the vector regarding overall quantities for the five sizes is {12, 8, 6, 4, 5}, and which is decisively not equal, but rather uneven, as the small size of 1 earns 12 quantitative units while the big size of 5 earns only 5 quantitative units. This quantitative configuration is of extreme skewness, excessively favoring the small over the big, and this is indeed the case in all Complete Equipartitions. Benford digital configuration is not valid here.

The standard measuring rod (benchmark) with which all quantitative configurations should be compared to is the equitable configuration where all the sizes share equally and fairly the same portion of overall quantity, so that x_j*(occurrences of x_j) for each size j is $\approx$ (XP)/N as in Refined Equipartitions, or more generally stated for all types of partitions and equipartitions: when quantitative portions for all the sizes are the same.

NOTE: Since this is Complete Equipartition of 5, only **$1/(e^{\lambda X_j} - 1)$** expression for AVG(n_j) is valid, while $(X/N)/x_j$ expression is not appropriate here. The two comparisons are:

Empirical calculations for AVG(n_j) = {1.71, 0.57, 0.29, 0.14, 0.14}
Theoretical $1/(e^{\lambda X_j} - 1)$ for AVG(n_j) = {1.63, 0.62, 0.31, 0.17, 0.10}
Theoretical $(X/N)/x_j$ for AVG(n_j) = {1.00, 0.50, 0.33, 0.25, 0.20}

As a concrete logarithmic example, total conserved quantity is $X = 10{,}000{,}000$; the part set is $\{x_j\} = \{1, 2, 3, \ldots, 997, 998, 999\}$. This is the Refined Equipartition case; hence both mathematical expressions for $AVG(n_j)$ can be applied. Lagrange multiplier λ is calculated as 0.00009749151. Needless to say, this is not performed on the computer due to the incredibly large number of possible partitions here numbering in the trillions of trillions and more. Normal personal computers cannot even begin to handle such fantastically huge number of combinatorial possibilities. Hence ('empirical') computer corroboration of the theory is not possible in this case. Rather the theoretical expressions $1/(e^{\lambda X_j} - 1)$ and $(X/N)/x_j$ are taken on faith and used with the aid of the computer to arrive at numerical results. Applying the expression $\mathbf{1/(e^{\lambda X_j} - 1)}$ we get:

$AVG(n_1) = \mathbf{10257}$
$AVG(n_2) = \mathbf{5128}$
$AVG(n_3) = \mathbf{3419}$

… etc.

$AVG(n_{997}) = \mathbf{9.80}$
$AVG(n_{998}) = \mathbf{9.79}$
$AVG(n_{999}) = \mathbf{9.78}$

The relationship [# of occurrences of x_j in scheme] = $AVG(n_j)*$[# of Partitions] leads to the relative percentages of x_j without the need to find out the actual value of the # of Partitions.

1st digits of 999 x_j values, $1/(e^{\lambda X_j} - 1)$ - {32.4, 17.7, 12.2,   9.3, 7.5, 6.3, 5.5, 4.8, 4.3}
Benford's Law for First Digits order   - {30.1, 17.6, 12.5,   9.7, 7.9, 6.7, 5.8, 5.1, 4.6}
SSD value is 6.1, and such low value indicates that this process is very close to Benford. The choice of the 1 to 999 range was deliberate in order to have it spanning integral powers of ten.

Here $N \ll X$, namely that $999 \ll 10{,}000{,}000$, therefore the theoretical expression $AVG(n_j) = (X/N)/x_j$ can be applied. Results are as follows:

$AVG(n_1) = \mathbf{10010}$
$AVG(n_2) = \mathbf{5005}$
$AVG(n_3) = \mathbf{3337}$

… etc.

$AVG(n_{997}) = \mathbf{10}$
$AVG(n_{998}) = \mathbf{10}$
$AVG(n_{999}) = \mathbf{10}$

1st digits of 999 x_j values, $(X/N)/x_j$ - {32.3, 17.6, 12.2,   9.3, 7.6, 6.4, 5.5, 4.8, 4.3}
Benford's Law for First Digits order - {30.1, 17.6, 12.5,   9.7, 7.9, 6.7, 5.8, 5.1, 4.6}
SSD value is 5.3, and such low value indicates that this process is very close to Benford.

The Achilles' heel of Refined Equipartition model is the highly unusual deterministic flavor of the main result $\text{AVG}(\,n_j\,) \approx (X/N)/x_j$. It implies that resultant x_j data is constantly and consistently logarithmic throughout its entire range when measured on mini sub-intervals between integral powers of ten such as (1, 10), (10, 100), and so forth, and that Digital Development Pattern does not exist. As mentioned in chapters 15 and 19 of section 1, practically all random data sets come with Digital Development Pattern, almost without any exception. Surely data derived from exponential growth series is of the k/x deterministic flavor, but this is rather rare, such as bank account frozen for decades or centuries steadily earning interest without any withdrawals or deposits, or growing bacteria in the laboratory with many weeks or months of hourly colony recordings. For this reason the applicability of Refined Equipartition to real-life data in the natural world is highly doubtful. Indeed, such probable lack of connection to real-life data is in perfect harmony with another lack of connection to the highly random nature of real-life partition processes which almost never occur along those neat integral lines of Refined Equipartition, and this can be nicely explained by those who intimately know Mother Nature and the way she works. She would probably never bother to carefully break her entities and quantities carefully only along part sets with exact integral relationships. She is simply too busy and is always rushing, building and destroying in a hurry, along whatsoever parts that seem the easiest or the fastest to use, including fractional and real messy part sets. Moreover, those statisticians expecting her to carefully abstain from breaking X along large pieces and to stick only to much smaller and refined pieces (i.e. the constraint of N << X for Refined Equipartition) are doubly naïve and do not really know how Mother Nature typically works. She does not like to be constrained in any way, and she is usually quite crude, rarely refined.

Random Real Partition, Random Dependent Partition, and Chaotic Rock Breaking, on the other hand, have that logarithmic flavor compatible with random distributions and real-life data sets, because these three processes come with a decisive Digital Development Pattern. Therefore these three random partition models are better candidates to explain data sets related to partitions in the natural world, more so than Refined Equipartition which lacks that nearly universal pattern of development. This is also highly consistent with the flexible and crude way these three random partition models are performed, allowing for any fractional and real value for the parts, as well as allowing for any size whatsoever, including very large and crude pieces. These three random partition models are thus much more compatible with the chaotic and crude manner Mother Nature does her daily work (especially Chaotic Rock Breaking as discussed in chapter 14) and could serve as more appropriate models for real-life partition-related data.

Summarizing the differences between these three partition processes and Refined Equipartition:

Refined Equipartition:
Not a realistic model, naïvely expecting Mother Nature to partition so carefully along integral parts, as well as so refinely by avoiding big pieces. Strangely not showing any Digital Development Pattern which accompanies almost all random data.

Random Real Partition, Random Dependent Partition, and Chaotic Rock Breaking:
Realistic models, allowing Mother Nature to behave more naturally and to partition along any integral, fractional, or real parts, including the permission to crudely break along big pieces. Showing a decisive Digital Development Pattern which is so typical in real-life random data.

[18] Balls Distributed inside Boxes

Oded Kafri (2009) "Entropy Principle in Direct Derivation of Benford's Law".

Conceptually, we partition a big pile of balls into separate mini piles to reside inside boxes. Kafri's discussion and ideas lead to two distinct yet very similar descriptions:

Model A: All possible permutations of **L** undistinguishable balls inside **X** distinguishable boxes, aggregated as one large data set. In the same vein as the term 'configurational entropy' of Thermodynamics, here all possible permutations are considered as equally likely and thus are given equal weights. For example, for 5 undistinguishable balls inside 3 distinguishable boxes, the data set to be crowned as Benford is the number of balls in a box, namely the collection of all the (21*3 = 63) numbers in the left panel of Figure 3.32, zero included. This yields the skewed proportions of {28.6%, 23.8%, 19.0%, 14.3%, 9.5%, 4.8%} for the quantities {0, 1, 2, 3, 4, 5}, but which is not close enough to Benford Base 7. Kafri is suggesting further ignoring all the zeros, but a comparison to Benford Base 6 is still not satisfactory. Perhaps a scheme which averages numerous results with certain X and L parameters, together with some limiting process might yield Benford exactly. What starkly differentiates Kafri's model from Equipartition model is the insistence on considering the boxes as distinguishable, thus introducing order as a factor into the scheme. For Kafri, partitioning 5 into {3, 1, 1} or {1, 3, 1} or {1, 1, 3} signify three different post-partition states, whereas in Equipartition these three partitions are identical as the model disregards order altogether. That the small always outnumbers the big in Kafri's model can be seen clearly in Figure 3.32, and this is true for all L and X values whenever X > 2. The only exception is for X = 2 which yields equal proportions for all sizes.

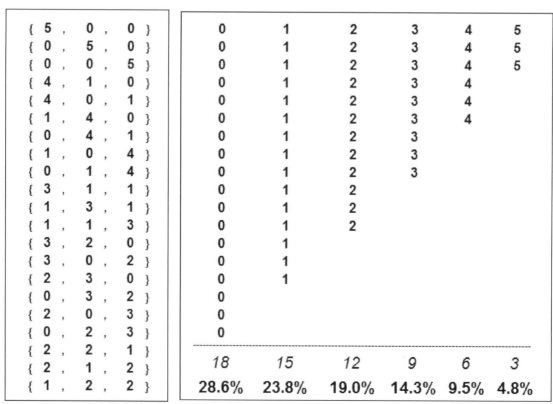

Figure 3.32: All Possible Permutations of 5 Undistinguishable Balls in 3 Distinguishable Boxes

Model B: The physical random throw of L actual (and thus distinguishable) balls into X distinguishable boxes, and where the principle of entropy is not being applied in any way. Instead, this is thought of as a statistical/stochastic process where each ball is randomly landing in any one of the boxes, so that all the boxes have equal probability in attracting a thrown and flying ball. The variable to be crowned as Benford is the number of balls in a box as above. For example, for 5 balls thrown into 3 distinguishable boxes, Monte Carlo simulations yield the proportion {13.2%, 32.9%, 32.9%, 16.4%, 4.1%, 0.4%} for the occurrences of {0, 1, 2, 3, 4, 5}.

In Kossovsky (2014) it is shown these two descriptions of Kafri's scheme, namely model A and model B, regarding the existence of L balls inside X boxes, are truly two distinct scenarios yielding two distinct numerical and quantitative results.

Interestingly, Kafri's Balls and Boxes scheme could be viewed conceptually not only as partitions but also as consolidations. One vista is to consider an initial glued group of balls undergoing partitions, where many balls end up in potentially different boxes experiencing separations from each other. An alternative vista is to consider the throwing or placing of disconnected and dispersed individual balls into boxes as consolidations, where some balls end up together touching each other and sharing a box.

Modification of Kafri's model type A, fitting it into Refined Equipartition in a sense, should bring it around to the Benford configuration. In order to achieve that, it is necessary to add four new features (the first two features are actually part of Kafri's original suggestion):

To ensure that a genuine refined partition is occurring here, there must be some superficial restriction on the maximum number of balls residing within any given box, and such a limit must be made well below L. Such a limit on the number of balls in any given box would eliminate not only the possibility of one box containing all L balls, but would also eliminate the possibility of one box containing too many balls in general. This feature ensures that balls are well-broken and nicely spread out. The maximum number of balls per box is designated as B, and the constraint $B \ll L$ is added to ensure that this corresponds to the Refined Equipartition constraint $N \ll X$.

The elimination of all empty boxes (zeros) from the calculations.

The condition $X = L$ is postulated due to two arguments. (1) $X < L$ - namely more balls than existing boxes - would not let us spread the balls in the most extreme way possible, namely having all the ball alone separated, and original L quantity totally broken up into 1's. We want to ensure that extreme fragmentation is possible. (2) $X > L$ - namely more boxes than existing balls - would introduce redundant boxes into the system; boxes that are never going to host any balls, even if we break the L pile of balls into 1's. In conclusion: if the possibilities of $X < L$ as well as $X > L$ are eliminated, it then follows that $X = L$.

Order is eliminated from the model. All the balls are undistinguishable and all the boxes are undistinguishable. The configuration {1, 1, 3} signifies that two boxes are with one ball each, and that another box is with three balls, and this can also be equally written as {3, 1, 1}. No designations such as 'left box', 'center box', or 'right box' are made.

[19] Division of Logarithmic Data along Small/Medium/Big Sizes

The main result of Refined Equipartition is that $AVG(n_j) \approx (X/N)/x_j$. This implies that all the sizes obtain equal quantitative portions, namely that [total quantity for size x_j] $\approx (XP)/N$, so that all the sizes share the same portion fairly from the overall quantity of the system. The oval-shaped area in Figure 3.5 enables us to clearly visualize this property of the partition-resultant data where all sizes share fairly and equally in the bounty of the overall quantity of the system. The quest in this chapter is to examine real-life logarithmic data sets and abstract logarithmic distributions and to empirically determine whether or not the equitable and fair quantitative configuration of Figure 3.5 is indeed the norm for Benford-obeying data sets and distributions, as opposed to the alternative configurations of Figure 3.6 and Figure 3.7 where only the Small or only the Big dominate all other sizes.

Hence, each logarithmic data set or distribution under consideration shall be divided in its entirety into 3 size categories - designated as Small, Medium, and Big, and then quantitative portions for each of the three sizes shall be calculated. Hopefully, these 3 sizes would be found to share equally and fairly between them overall quantity of the entire system, namely that each of the 3 sizes should earn approximately 33% of the overall quantity in the entire system.

In order to divide a given data set into Small, Medium, and Big categories, it is necessary to decide on two border points, Border Point A and Border Point B, reasonably chosen within the range of the particular data under consideration. Consequently, all data points falling to the left of Border Point A are deemed as Small; all data points falling between Border Point A and Border Point B are deemed as Medium; and all data points falling to the right of Border Point B are deemed as Big. In other words, Border Points A and B are the threshold or cutoff points differentiating between the three sizes. This arrangement can be visualized in Figure 3.33 which depicts the classification of Small/Medium/Big via these two Border Points. The reason for the slightly lesser range allocated to the Medium shall be explained shortly. By convention, any data points falling exactly on Border Points A are deemed as Medium; and any data points falling exactly on Border Points B are deem as Big.

For smoothly spread data, it is only from the knowledge of the range of the data under consideration that we construct these two border points, and almost without utilizing any information regarding relative concentrations of values within the range provided by the data's density or histogram. For data with severely irregular spread, where the portions of the data on the margins are highly diluted so that the edges are considered outliers, these two border points are constructed in a way that attempts to ignore and omit these outliers from the calculations.

The most obvious way to choose Border Point A and Border Point B is by simply dividing the entire range, from its minimum to its maximum, equally into 3 sub-intervals, each having the same length, namely (Max – Min)/3. Yet, in order to avoid excessive influence from outliers on the very margins of data, a more reasonable approach here is to start at the 1st percentile point and to end at the 99th percentile point, utilizing these percentile points as the two edges of the core 98% range of data. Figure 3.34 depicts the arrangement of the two Border Points A and B.

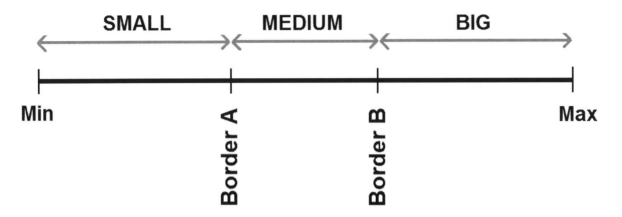

Figure 3.33: Classification of Small, Medium, and Big via Two Border Points

The 1% percentile point, or the 1st percentile point, denoted as $P_{1\%}$, is the value below which 1% of the ordered data may be found.

The 99% percentile point, or the 99th percentile point, denoted as $P_{99\%}$, is the value below which 99% of the ordered data may be found.

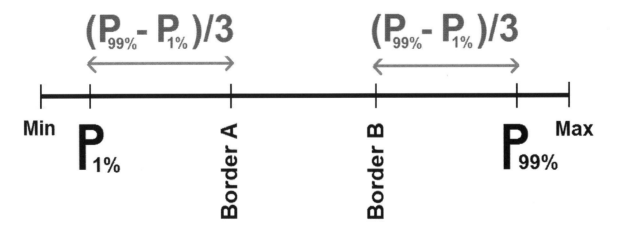

Figure 3.34: The Creation of Two Border Points via the Exclusion of Top 1% and Bottom 1%

In what might seem as a paradox, once Border Point A and Border Point B are chosen with the deliberate exclusion of the outliers, the entire set of numbers in the data are now incorporated into the designation Small/Medium/Big, including even those outliers on the margins. In other words, after excluding the margins from the determination of Border Points A and B, the points falling between Min and $P_{1\%}$ are to be included at the end and are allocated to Small, and the points falling between $P_{99\%}$ and Max are to be included at the end and are allocated to Big. This seemingly contradictory attitude towards outliers is motivated by the realization of the potential for significant adverse effects on the Border Points A and B due to outliers; as opposed to the insignificant and very mild effects of adding only very few extra data points (outliers) to Small and Big. By including the outliers in the classification scheme at the end, we ensure that nothing within the entire data set is omitted or neglected. The diagrams of Figure 3.33 and Figure 3.34 define the two border points as follows:

Border Point A = $P_{1\%}$ + $(P_{99\%} - P_{1\%})/3$
Border Point B = $P_{99\%}$ − $(P_{99\%} - P_{1\%})/3$

In one concrete numerical example, demonstrating the absolute necessity of avoiding the utilization of the maximum, minimum, and possible outliers in the construction of the two Border Points, a data set is imagined having 30,000 points which are approximately uniformly distributed on (5, 35), plus a single outlier value of 155. Allowing this outlier to influence decisions regarding Border Points A and B by partitioning the entire range from the minimum 5 to the maximum 155 into 3 equal sections of $(155 - 5)/3 = (150)/3 = 50$ width each, would cause the Small to be defined over (5, 55), and thus having the Small artificially earn nearly 100% of overall data portion. By using the 1st and the 99th percentiles, we arrive at a much more reasonable partitioning scheme along (5, 15.1) for the Small, [15.1, 24.9) for the Medium, and [24.9, 155) for the Big. It should be noted how Medium earns less range than either Big or Small.

For this data set, $P_{1\%} = 5 + 0.01*(35 - 5) = 5.3$, and $P_{99\%} = 5 + 0.99*(35 - 5) = 34.7$. It follows that Border Point A = $P_{1\%} + (P_{99\%} - P_{1\%})/3 = 5.3 + (34.7 - 5.3)/3 = 5.3 + 9.8 = 15.1$, and that Border Point B = $P_{99\%} - (P_{99\%} - P_{1\%})/3 = 34.7 - (34.7 - 5.3)/3 = 34.7 - 9.8 = 24.9$.

Medium is full of envy; it is bitterly complaining that its share on the entire range is less than that allocated to either Small or Big; and that it is being discriminated against. Subsequently, the algorithmist points out to Medium that differences are really tiny; that its loss is no more than 2% of the overall range, and finally Medium reluctantly accepts the arrangement so as not to appear as obstructionist, still whispering to itself all sorts of old grievances and ridiculous accusations against the Small and especially against the Big, such as in the alleged maltreatments and discriminations of middle children in large families.

By including the outliers at the end of the classification scheme we ensure that the entire data set is presented, and part of the motivation for this is the partition-vista of the data, which induces the desire that the partition would apply to all segments and parts of the data, even on the margins.

Perhaps the noun 'Composition' would be more appropriate for this vista than the noun 'Partition'. Surely this vista is readily and nicely interpreted in the case of population data as the total population of the entire country. This vista is also readily and nicely interpreted in the case of county or province area data as the total area of the entire country. This vista is trivially interpreted in the case of data on the time intervals between earthquakes in a given year as simply one year time interval. But how could one possibly interpret for example the sum of all the distances from the Solar System to a large collection of stars?! And how could one possibly interpret for example the sum of the prices of all the items on sale in a big catalog of a large retail company?! In any case, this partition-vista of data shall be taken here, and therefore all outliers and data points on the margins are incorporated into the scheme. We wish to have the sum of all the parts equals exactly to the whole.

As an example of how all the quantitative results pertaining to this algorithm for a variety of logarithmic data sets and distributions are going to be presented below [namely the format and style of the presentation of the quantitative analysis], the imaginary data set supposedly generating the oval shape in Figure 3.5 will be presented as in Figure 3.35. Here the areas represent the quantitative portions for each of the 3 sizes, Small, Medium, and Big. The numbers within each area represent the number of parts/pieces/points. It should be emphasized that the numbers inside the areas do not represent quantitative portions, but rather the number of parts/pieces/points for a given size. In Figure 3.5 there are 220 small parts in the middle, 35 medium parts on the right, and only 7 big parts on the left. Since Figure 3.5 was drawn with the 'artistic intention' of endowing 1/3 of total oval area to each of the 3 sizes equally, this fact is reflected here a bit more clearly and exactly showing that each size obtains 1/3 of the total circular area.

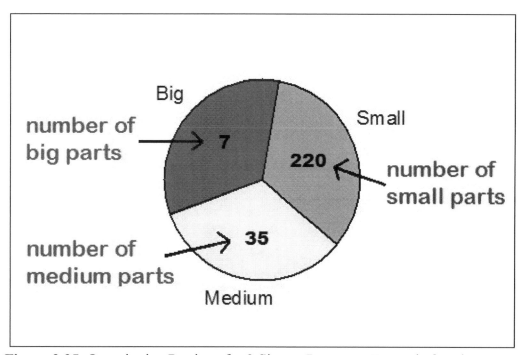

Figure 3.35: Quantitative Portions for 3 Sizes – Prototype Example for Figure 3.5

Figure 3.36 depicts quantitative results for Monte Carlo computer simulations obtaining 35,000 realizations from the Lognormal(9.3, 1.7). Quantitative portions are:

Small = 50.8%   **Medium = 15.8%**   **Big = 33.4%**

Figure 3.37 depicts quantitative results for the US Census Data on 2009 Population count of its 19,509 incorporated cities & towns. See Kossovsky (2014) Chapter 12. Quantitative portions are:

Small = 36.7%   **Medium = 14.8%**   **Big = 48.5%**

Figure 3.38 depicts quantitative results for the 19,452 Global Earthquakes occurring in 2012. The data here is of the time in the units of seconds between successive earthquakes. See Kossovsky (2014) Chapter 11 for more information. Quantitative portions are:

Small = 46.0%   **Medium = 33.9%**   **Big = 20.1%**

Figure 3.39 depicts quantitative results for the prices of 14,914 items on sale in the catalog of Canford Audio PLC. See Kossovsky (2014) Chapter 44. Quantitative portions are:

Small = 53.3%   **Medium = 16.9%**   **Big = 29.7%**

Figure 3.40 depicts quantitative results for the 8,192 Pieces in a 13-Stage Random Rock Breaking process in one Monte Carlo simulation run. Quantitative portions are:

Small = 34.3%   **Medium = 15.3%**   **Big = 50.4%**

Figure 3.41 depicts quantitative results for 30,000 values of the simulated Symmetrical Triangle on (1, 5) serving as exponents. The realizations from the simulations of the Triangle density are used as exponents of base 10; namely that the final data set is $10^{\text{Symmetrical Triangle}}$.
See Kossovsky (2014) Chapter 64 for discussions about the topic. Quantitative portions are:

Small = 47.6%   **Medium = 24.0%**   **Big = 28.4%**

Figure 3.42 depicts quantitative results for the US Market Capitalization as of Oct 9, 2016, encompassing 2,883 companies. The largest 6 companies were omitted in this analysis since they appear as extreme outliers here. Quantitative portions are:

Small = 41.0%   **Medium = 12.7%**   **Big = 46.2%**

Figure 3.43 depicts quantitative results for 20,000 values of the simulated chain of five Uniform(0, b) Distributions, where parameter b is chosen randomly from yet another such Uniform Distribution, while ultimate Uniform is with a fixed parameter b = 5666. Schematically written as Uniform(0, Uniform(0, Uniform(0, Uniform(0, Uniform(0, 5666))))).

Small = 48.9%   **Medium = 26.3%**   **Bi g= 24.8%**

Figure 3.44 depicts quantitative results for the prices of 8,079 items on sale in the catalog of MD Helicopters Inc. for the year 2012. See link at http://www.mdhelicopters.com/v2/index.php. MD Helicopters Inc. manufactures and retails helicopters and parts. Quantitative portions are:

Small = 32.9%   **Medium = 15.8%**   **Big = 51.3%**

Figure 3.45 depicts quantitative results for 30,000 simulated values of the perfectly and uniquely Benford k/x distribution defined on the interval (1, 10). Quantitative portions are:

Small = 32.4%   **Medium = 32.0%**   **Big = 35.5%**

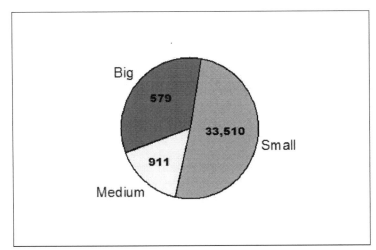

Figure 3.36: Quantitative Portions – Lognormal(9.3, 1.7)

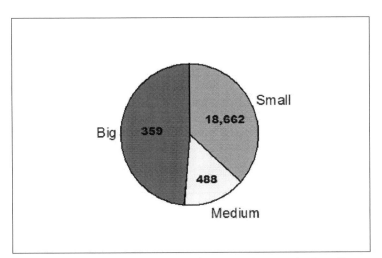

Figure 3.37: Quantitative Portions – US Population Census in 2009

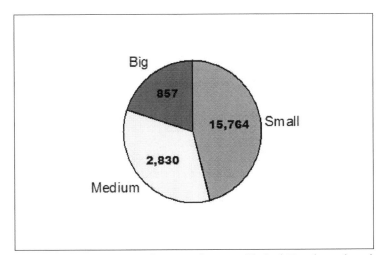

Figure 3.38: Quantitative Portions – Global Earthquakes in 2012

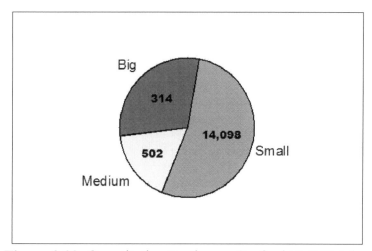

Figure 3.39: Quantitative Portions – Canford PLC Catalog

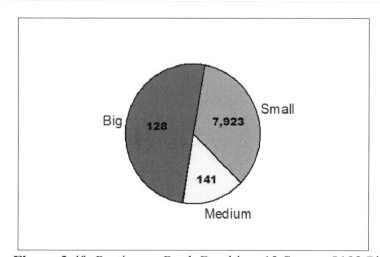

Figure 3.40: Portions – Rock Breaking, 13 Stages, 8192 Pieces

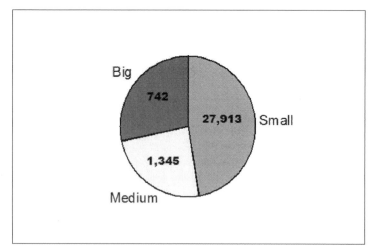

Figure 3.41: Portions – Log is Symmetrical Triangle on (1, 5)

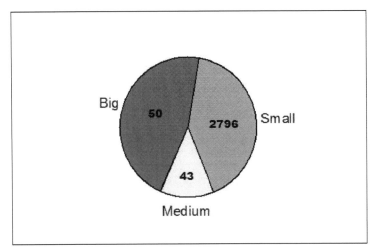

Figure 3.42: Portions – US Market Capitalization - Oct 9, 2016

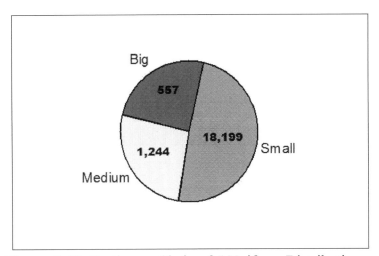

Figure 3.43: Portions – Chain of 5 Uniform Distributions

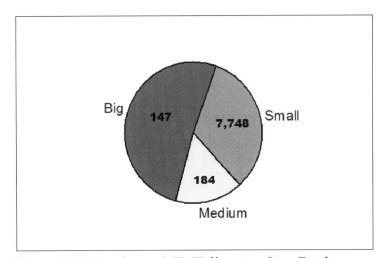

Figure 3.44: Portions – MD Helicopters Inc. Catalog

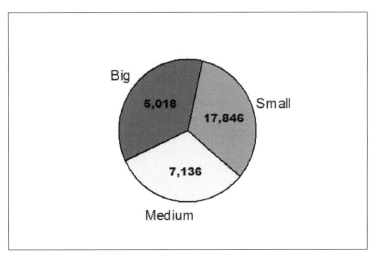

Figure 3.45: Portions – The Perfectly Benford k/x over (1, 10)

The first common feature noted here in all of these ten logarithmic data sets and distributions, is the fact that (for the most part) no particular size dominate quantitative-wise, as if the three sizes always sincerely wish to share equally and fairly between them overall quantity of the entire system, namely 33% for each of the three sizes, although reality often intervenes and imposes upon them to give a bit to, or to take a bit from, one size or another, and even then it's never by much. The lowest quantitative portion for any size in the entire set of ten is for the most part 15% (except once in the case of US Market Capitalization which is 12.7%). The highest quantitative portion for any size in the entire set of ten is 53.3%. We never encounter here extreme situations where one size dominates with say 60% or 70% of quantitative portion, or where one size diminishes and obtains only say 10% or less.

The second common feature noted here in all of these ten logarithmic data sets and distributions, is the fact that the number of Small, Medium, and Big parts consistently reflects the small is beautiful phenomenon; and that in almost all of these ten data sets and distributions the Small is more numerous than the Medium, and the Medium is more numerous than the Big. There is almost no exception here to this strict and nearly universal rule [well, only for US Market Capitalization data set, the Big with 50 points has a very slight advantage over the Medium with 43 points, being a tiny bit more numerous, yet the Big and the Medium combined which earn in total 93 points pales in comparison to the Small which earns 2796 points!]

[20] Division of Non-Logarithmic Data along Small/Medium/Big Sizes

Applying the above algorithm in the division of data into Small/Medium/Large sizes for a group of 5 non-logarithmic data sets and distributions shows decisive deviations not only from the small is beautiful principle, but also from the approximate or near equality in quantitative portions seen in all of the 10 logarithmic data sets and distributions of the previous chapter. Here for non-logarithmic data sets and distributions, one size often strongly dominates all other sizes, obtaining 60% or even 70% of overall quantity.

Here is a descriptive summary of the quantitative results for these 5 data sets and distributions:

Figure 3.46 depicts quantitative results for 25,000 realizations of Monte Carlo computer simulations from the Uniform Distribution on (5, 78,000). Surely, since the density here is flat, uniform, and horizontal, each size contains about the same number of points (a tiny bit less for Medium since it is defined as a bit shorter than either Small or Big). Hence, none is more frequent than the others; rather all three sizes occur with almost equal frequency. The ramification regarding quantitative portions is quite straightforward, as there are 8,342 Small points lying between the Minimum of 5 and Border Point A of 26,279; there are 8,220 Medium points lying between Border Point A of 26,279 and Border Point B of 51,737; and there are 8,438 Big points lying between Border Point B of 51,737 and the Maximum of 78,000, hence each size obtains substantially different quantitative portion depending on the qualitative territory its points lie in, and this produces the relative results where Big > Medium > Small. Quantitative portions are:

Small = 11.2%   **Medium = 32.8%**   **Big = 56.0%**

Figure 3.47 depicts quantitative results for 25,000 realizations of Monte Carlo computer simulations from the Normal(177, 40). Here Medium benefits substantially from the fact that most of the data points and also most of the quantities are around the center, namely around the mean/median/mode, as in all Normal distributions. Here Medium is by far the most frequent size in terms of having the largest number of data points. Medium also earns the most in terms of quantitative portions. Quantitative portions are:

Small = 15.0%   **Medium = 56.6%**   **Big = 28.5%**

Figure 3.48 depicts quantitative results for 25,000 realizations of Monte Carlo computer simulations from the x^3 distribution over (1, 50) discussed earlier in this section in chapter 12. Here the curve rises sharply and constantly, hence not only the expression 'big is exceedingly beautiful' is true here, but there is also a consistent advantage here to bigger sizes over smaller sizes in both measures, namely in the number of data points, as well as in quantitative portions, where Big > Medium > Small. Quantitative portions are:

Small = 4.7%   **Medium = 22.6%**   **Big = 72.7%**

Figure 3.49 depicts quantitative results for the data set pertaining to areas of the 3,143 Counties in the US. See Kossovsky (2014) Chapter 45 for more information. Here Small is excessively more frequent than the other sizes, over and above its usual level of frequency according to Benford's Law, and especially regarding quantitative portions, which are:

Small = 59.4%   **Medium = 13.2%**   **Big = 27.3%**

Figure 3.50 depicts quantitative results for the State of Oklahoma <u>payroll</u> data of the Department of Human Services for the 1st quarter of 2012. This data can be found on their website https://data.ok.gov/Finance-and-Administration/State-of-Oklahoma-Payroll-Q1-2012/dqi7-zvab. Only those 2189 rows from the column 'Amount' pertaining to the Department of Human Services are considered. See Kossovsky (2014) Chapter 138, Page 592 for more information. Here the medium is the most significant in both measures, namely in the number of data points, as well as in quantitative portions, where Medium > Small and where Medium > Big. Here quantitative portions are:

Small = 19.0%   **Medium = 66.7%**   **Big = 14.4%**

In conclusion: a clear distinction is seen between the 10 <u>logarithmic</u> data sets and distributions and the 5 <u>non-logarithmic</u> ones. The non-logarithmic data sets strongly deviate from the small is beautiful principle regarding number of points per size, as well as allowing one size to strongly dominate (quantitative-portion-wise) all the other sizes, well above its supposed 33% allocation.

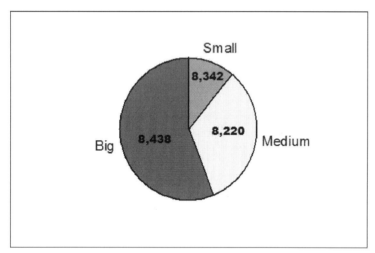

Figure 3.46: Portions – Uniform Distribution on (5, 78000)

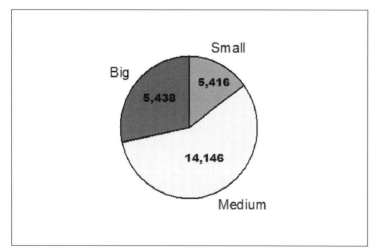

Figure 3.47: Quantitative Portions – Normal(177, 40)

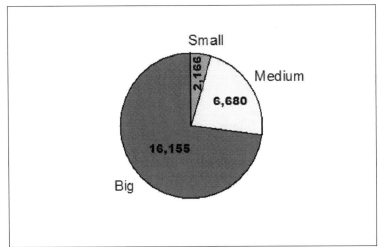

Figure 3.48: Quantitative Portions – x^3 Distribution over (1, 50)

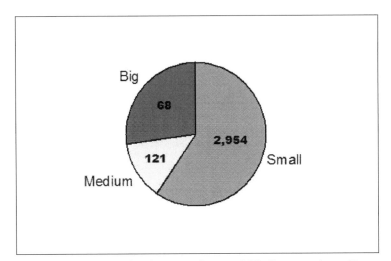

Figure 3.49: Quantitative Portions – US County Area Data

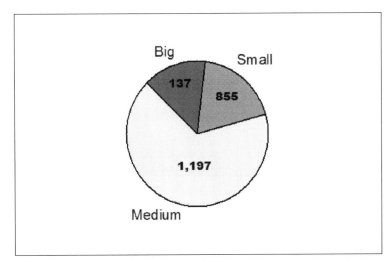

Figure 3.50: Quantitative Portions – Oklahoma Payroll Data

[21] The Constancy of Quantitative Portions in the k/x Case

Figure 3.45 of the perfectly Benford k/x distribution over (1, 10) comes the closest to that ideal, equitable, and fair 33.3% division for all the three sizes. Medium earns a tiny bit less than either Small or Big, but this is due only to its slightly lesser range as defined in the algorithm which endows the extra section of top 1% to Big and the extra section of bottom 1% to Small, and nothing extra to Medium. One must bear in mind though that k/x has that rare deterministic flavor in Benford's Law, totally lacking Digital Development Pattern. The distribution k/x is a unique case in the field Benford's Law. Indeed, from the definition of the k/x distribution we can deduce one of its essential properties, namely that by doubling quantity x we reduce its frequency (density height) by half. Surely for the transition such as: $x \to 2x$ the density of k/x always diminishes by half as in: $k/x \to k/(2x) = (k/x)/2$. As a consequence, sums of quantities (approximately - rectangle area times some middle x value) are constant everywhere on the x-axis for all sub-intervals having an identical width on the x-axis.

Let us consider two adjacent regions under the k/x curve, a left region and a right region, both of equal length on the x-axis, as seen in Figure 3.51. Here, the left region has the advantage of having relatively higher density, but it also has the disadvantage of laying over lower x values. The right region has the disadvantage of having relatively lower density, but it has the advantage of laying over higher x values. As it happened, one factor offsets the other factor, and it all cancels out exactly, resulting in equal sums for the left and for the right regions. Let us prove this quantitative equality in the generic sketch of k/x distribution illustrated in Figure 3.51.

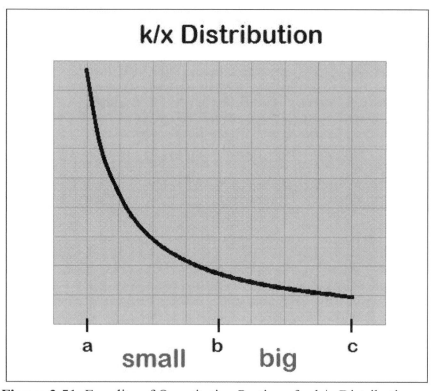

Figure 3.51: Equality of Quantitative Portions for k/x Distribution

When a continuous random distribution probability density is considered, it is not possible to sum 'amounts' in the usual manner by adding discrete values on the histogram; there are none here. To better illustrate the point, one has to trace back the path we always take going from a histogram of discrete values to an abstract continuous density probability curve (best fitting the actual data) by way of dividing the height of each rectangle of x-axis unity width by the total number of data points in the entire data set. Hence we define a fixed imaginary integer N representing the number of all the values within some imaginary data set perfectly fitting the density curve in question, namely fitting k/x defined over (a, c).

The product of an infinitesimal or tiny rectangular area under the k/x curve times N, should then represent the number of 'discrete values' falling within that infinitesimal sub-interval. If that product is then further multiplied by the value of the x-axis at the bottom of the infinitesimal rectangular, it yields the sum of all amounts within that tiny rectangular area. Algebraically:

Sum within infinitesimal sub-interval = xN[infinitesimal area]

It shall be assumed that (b – a) = (c – b), namely that the two Small and Big sub-intervals are of equal length. Let us first sum the quantitative portion of Small falling on the sub-interval (a, b). The overall sum of all the infinitesimal areas' sums within the sub-interval (a, b) is given by:

$$\text{Sum for (a, b)} = \sum xN[\text{mini area}] \quad \textit{[for all rectangles from a to b].}$$

Turning it into a definite integral, we get:

$$\text{Sum for (a, b)} = \int_a^b xNf(x)dx$$
$$\text{Sum for (a, b)} = \int_a^b xN(k/x)dx$$

The x term cancels out, and we are left with:

$$\text{Sum for (a, b)} = \int_a^b Nkdx$$
$$\text{Sum for (a, b)} = Nk(b - a) = Nk(\text{Length of subinterval})$$

Namely the same constant value of Nk[Length of sub-interval] for any sub-interval of comparable length. Hence the same result is obtained for Big falling on the sub-interval (b, c).

The setup of this proof surely resonates as something quite familiar in statistical methods. In the discrete case, the idea of summation is always employed in the definition of the average, as in [Avg] = [Sum]/[Number of Values]. Here too, the above expression $\int xNf(x)\, dx$ corresponds to the generic definition of the average $\int xf(x)\, dx$ *[over entire range]* in mathematical statistics, except that extra N term. But since N represents [Number of Values], it follows that the expression $\int xNf(x)\, dx$ directly signifies sum.

[22] Closer Quantitative Scrutiny of Random Logarithmic Data Sets

In light of the very consistent result in the rare case of the <u>deterministic</u> k/x distribution showing equality of quantitative portions by size, the critical question that arises here regards all the other much more common <u>random</u> data cases; namely what effect does Digital Development Pattern have on quantitative portions by size for logarithmic real-life data sets and distributions. Empirically, from the nine logarithmic random data sets and distributions of Figures 3.36 - 3.44, it seems that Small often obtains somewhat higher quantitative portions than the expected 33.3% of the purely Benford deterministic k/x case. The averages of quantitative portions by size for these nine random data sets and distributions are: **Small = 43%, Medium = 20%, Big = 37%**.

In order to gain better understanding about quantitative allocation by size for logarithmic data under the influence of Digital Development Pattern, it is necessary to let go of the parsimonious and simple division of data into merely 3 sizes of Small, Medium, and Big, and instead examine quantitative portions of numerous sizes by dividing the range of data into much smaller sub-intervals. Even dividing the range into only 9 categories say, 'very small', 'somewhat small', 'just small', 'almost medium, 'somewhat medium', 'just medium', 'almost big', 'somewhat big', 'very big' sizes, wouldn't do! It is necessary to examine a very large number of sizes, much larger than merely 9.

In Kossovsky (2014) Chapter 81, the concept of Leading Digits Inflection Point (LDIP) is discussed - exclusively in terms of its effects on digital behavior. LDIP is defined as the maximum or top point on the histogram of the logarithm-transformed data where it's temporary flat and horizontal. Also, 10^{LDIP} is the corresponding LDIP on the x-axis itself for the raw data. It is precisely at LDIP that the data or curve behaves purely logarithmically as in the k/x distribution case. LDIP is a <u>digital turning point</u>, where to the left of it digital configuration is milder than Benford, and to the right of it digital configuration is more extreme and skewer than Benford. Since Digital Development Pattern revolves around LDIP, the conjecture that arises naturally in the context of relative quantities is that LDIP also determines quantitative portions locally for mini sub-intervals on the x-axis; namely that LDIP is also a <u>quantitative turning point</u>.

Two factors lead to the more precise statement of the conjecture:

(1) The well-known fact that in almost all logarithmic (and even in some non-logarithmic) random data sets, the histogram 'very briefly' rises on the extreme left part of the x-axis for the lowest values in the beginning of the data set. The obvious implication regarding relative quantities is that quantitative portions on mini sub-intervals on the left-most part of the data increase as x increases, so that there the big contains more quantitative portion than the small. Surely, around the left-most part of the data, as x advances to higher values under taller heights of the local histogram, quantitative portions rise accordingly. All this occurs to the left of LDIP.

(2) Precisely at LDIP, the data behaves purely logarithmically as in the k/x distribution case. This fact implies that around LDIP quantitative portions on local mini sub-intervals are equal, as is the case for the k/x distribution.

Hence the conjecture extrapolates these two facts, and states that anywhere to the left of LDIP bigger sizes obtain higher quantitative portions than smaller sizes, while anywhere to the right of LDIP smaller sizes obtain higher quantitative portions than bigger sizes. Let us now turn to empirical observations in order to verify this theoretical conjecture. Indeed, closer empirical scrutiny applying numerous sizes (instead of merely 3) reveals a very consistent pattern in all logarithmic random data sets and distributions, where quantitative portions are dramatically different on the region to left of LDIP as compared with the region to right of LDIP. For the region to the right of LDIP, sub-intervals contain less and less quantitative portions as x gets larger. The exact opposite occurs for the region to the left of LDIP.

Figure 3.52 depicts the typical way quantitative portions occur in almost all random logarithmic data and distributions. Figure 3.52 utilizes 35,000 realizations from computer simulation runs of the Lognormal(9.3, 1.7). A total of 82 quantitative portions for 82 sizes are calculated here in steps of 1000 units of length on the x-axis, beginning from the origin 0 and ending at 82,000. Hence sums are calculated on [0, 1000), [1000, 2000), [2000, 3000), … , [80000, 81000), [81000, 82000). For example, the set of all the numbers falling within [0, 1000) are added (i.e. their sum is calculated) yielding the quantitative value of 1,554,816, and this is divided by the total quantitative sum of the entire 35,000 simulated Lognormal values which is 1,629,854,915, pointing to the ratio (1,554,816)/(1,629,854,915) or 0.095% as the quantitative portion.

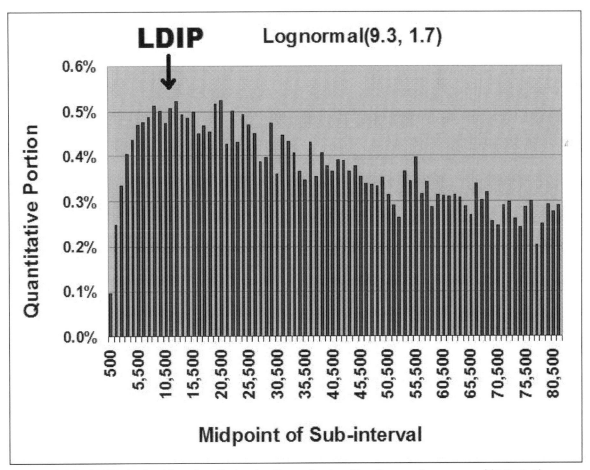

Figure 3.52: LDIP is the Turning Point in Quantitative Portions – Lognormal(9.3, 1.7)

In Kossovsky (2014) Chapter 81, Page 347, LDIP of the Lognormal is derived, yielding the generic expression [e] $^{\text{LOCATION PARAMETER}}$ for the LDIP on the x-axis. Applying this expression here we obtain [e] $^{9.3}$ = [2.718281] $^{9.3}$ = 10,938 as the LDIP on the x-axis. This calculation shows an excellent fit between the theoretical LDIP and the empirical observation of its location on the x-axis where a decisive quantitative reversal is occurring, as seen in Figure 3.52. It should be noted that the apparent falling long tail on the right in Figure 3.52 does indeed exist, and extended empirical examinations show that quantitative portions continue to fall further as additional mini sub-intervals are considered there for higher values of x well beyond 82,000. Hence in a sense, there is a strong resemblance (in overall-shape) between the typical histogram of random logarithmic data itself and its related chart of quantitative portions with the % unit for the vertical axis. Both appear to rapidly and impulsively rise strongly for a very short interval on the left in the beginning, only to then deeply regret this rushed act and to totally reverse themselves and fall gradually, slowly, and almost monotonically, on a much longer interval to the right.

A quantitative scheme with very few sizes applying very wide sub-intervals per size, typically encloses or 'swallows' LDIP within the smallest (first) size, and this totally masks the typical brief rise in quantitative portions around the left-most part of the data as was seen in Figure 3.52.

Figures 3.53 and 3.54 - which also pertain to the Lognormal(9.3, 1.7) - depict the typical way quantitative portions occur in almost all random logarithmic data sets and distributions when only very few sizes are created, applying very wide sub-intervals in the definition of sizes, and resulting in LDIP residing comfortably well within the smallest (first) size.

In the setup of Figures 3.53 and 3.54, the entire part of the range left of LDIP is being absorbed and easily swallowed within the smallest (first) size. The entire range of all 35,000 realizations of Lognormal(9.3, 1.7) is divided into much wider sub-intervals in long steps of 25,000 units of length on the x-axis, beginning from the origin 0 and ending at 375,000.

A total of 15 quantitative portions for 15 sizes are calculated here. Hence sums are calculated on [0, 25,000), [25,000, 50,000), … , [325,000, 350,000), [350,000, 375,000). For example, the set of the 23,987 numbers falling within [0, 25,000) are added (i.e. their sum is calculated) yielding the quantitative value of 182,794,046, and this is divided by the total quantitative sum of the entire 35,000 simulated Lognormal values which is 1,629,854,915, pointing to the ratio (182,794,046)/(1,629,854,915) or 11.2% as the quantitative portion. Also here, the set of the 4,457 numbers falling within [25,000, 50,000) are added (i.e. their sum is calculated) yielding the quantitative value of 158,330,171, and this is divided by the total quantitative sum of the entire 35,000 simulated Lognormal values which is 1,629,854,915, pointing to the ratio (158,330,171)/(1,629,854,915) or 9.7% as the quantitative portion.

Since the right edge of the first [0, 25,000) sub-interval is well beyond LDIP value of 10,938, no quantitative rise can be seen here whatsoever, and successive sub-intervals (i.e. sizes) consistently obtain lower and lower quantitative portions.

Note: Figure 3.54 depicts the relative [not the absolute/actual] quantitative portions for these 15 sizes. Total quantitative portion between 0 and 375,000 is only 64.5%, and the rest being 35.5% falls farther to the right. For example: area of 'Small 1' appears as $(11.2\%)/(64.5\%) = 17.4\%$.

Size Name	From	To	Quantitative Portion	Number of Data Points
Small 1	0	25,000	11.2%	23,987
Small 2	25,000	50,000	9.7%	4,457
Small 3	50,000	75,000	7.6%	2,021
Small 4	75,000	100,000	6.3%	1,184
Small 5	100,000	125,000	4.9%	719
Medium 1	125,000	150,000	4.3%	513
Medium 2	150,000	175,000	3.3%	331
Medium 3	175,000	200,000	2.9%	251
Medium 4	200,000	225,000	2.7%	208
Medium 5	225,000	250,000	2.5%	171
Big 1	250,000	275,000	2.4%	149
Big 2	275,000	300,000	1.9%	109
Big 3	300,000	325,000	2.0%	107
Big 4	325,000	350,000	1.4%	67
Big 5	350,000	375,000	1.4%	61

Figure 3.53: Table of Quantitative Portions for 15 Sizes - Lognormal(9.3, 1.7)

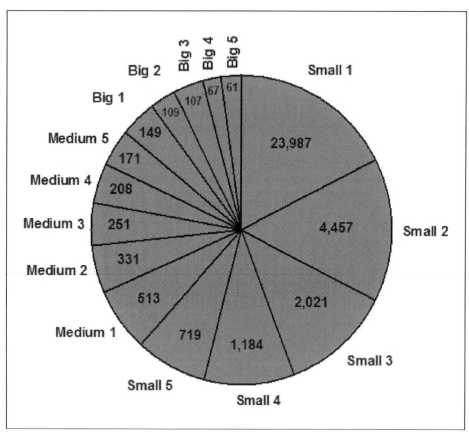

Figure 3.54: Chart of Quantitative Portions for 15 Sizes - Lognormal(9.3, 1.7)

Figure 3.55 is another demonstration of the typical way quantitative portions occur in almost all random logarithmic real-life and physical data sets. Figure 3.55 pertains to the data set of all 19,451 time intervals (in seconds) between successive global earthquakes in the year 2012. A total of 82 quantitative portions for 82 sizes are calculated in steps of 100 units of length on the x-axis, beginning from zero (which is very near the minimum value of 0.01) and ending at 8,200. Hence sums are calculated on [0, 100), [100, 200), [200, 300), ... , [8000, 8100), [8100, 8200). For example, the set of all the numbers falling within [0, 100) are added (i.e. their sum is calculated) yielding the quantitative value of 71,101, and this is divided by the total quantitative sum of the entire 19,451 time intervals which is 31,617,664, pointing to the ratio (71,101)/(31,617,664) or 0.22% as the quantitative portion.

Note: The total quantitative sum of the entire 19,451 time intervals is 31,617,664. This value closely corresponds to the theoretical/calculated 60*60*24*366 = 31,622,400 seconds for the entire year, since 2012 was a leap year (given one extra day) with 366 days in total.

The approximate value of LDIP for the global 2012 earthquake data set [according to the conjecture] can be found by visually reading the highest point of the chart of Figure 3.55 where 'curve' is momentarily flat and horizontal. The value of 1550 approximately seems like a reasonable choice.

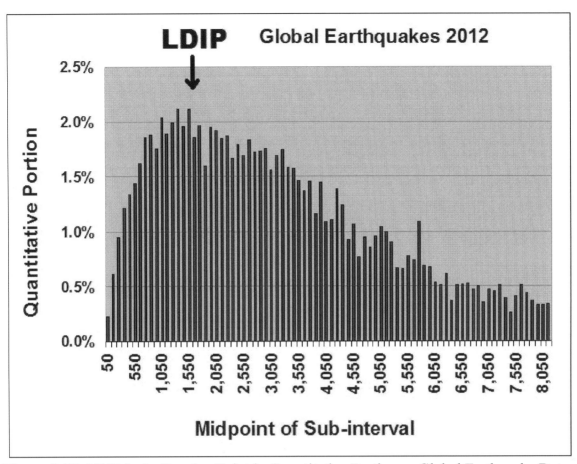

Figure 3.55: LDIP is the Turning Point in Quantitative Portions – Global Earthquake Data

Let us now figure out the approximate value of LDIP for the global 2012 earthquake data set by visually reading the highest point on its log histogram where 'curve' is momentarily flat and horizontal, as depicted in Figure 3.56. The choice of 3.2 log value seems like a reasonable one.

The LDIP on the x-axis itself should then be approximately $10^{3.2} = 1585$, and this value corresponds nicely with the quantitatively-estimated LDIP shown in Figure 3.55.

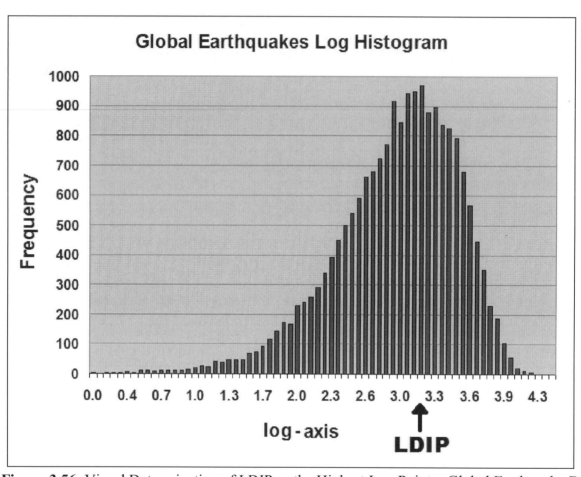

Figure 3.56: Visual Determination of LDIP as the Highest Log Point – Global Earthquake Data

Quantitative portions for the global 2012 earthquake data set applying only 3 sizes via the algorithm in chapter 19 yielded: Small = 46.0%, Medium = 33.9%, and Big = 20.1%. Figure 3.55 which applies 82 sizes provides a more detailed description of how quantitative portions are occurring here. According to the algorithm in chapter 19, Border Point A is 2764, and Border Point B is 5516. Therefore the definition of Small classifies those values which are less than 2764 as such. It follows that LDIP value of approximately 1585 is well within the range of the category of Small, and this fact gives rise to the expectation that quantitative proportions are such that Small > Medium > Big, as indeed shown by the result of the algorithm.

[23] General Derivation of Log Density for Any Random Distribution

In order to rigorously prove the above conjecture, namely that Leading Digits Inflection Point is also the decisive quantitative turning point for any random distribution f(x), it is necessary first to derive the generic expression of the density distribution of the log-transformed g(y), namely g(log(x)), and then to demonstrate that the x-axis locations of its rise and fall corresponds exactly to the x-axis locations of the rise and fall of local mini quantitative portions.

Let f(x) be a continuous and differential function over the range R, serving as the probability density function for variable X.

Let g(y) be the probability density function for the base B logarithmically-transformed X variable, namely for $Y = LOG_B X$.

The Distribution Function Technique shall be applied using similar notations as in Freund's book "Mathematical Statistics", sixth edition, chapter 7.3.

The Distribution Function (Cumulative Distribution Function) of Y, denoted as G(y), is then:

$G(y) = Probability[\ Y < y\ ] = Probability[\ LOG_B X < y\ ].$

Taking base B to the power of both sides of the last inequality we obtain:

$G(y) = Probability[\ B^{LOG_B(X)} < B^y\ ] = Probability[\ X < B^y\ ]$, hence

$$G(y) = \int_{-\infty}^{B^y} f(x)\,dx$$

Applying the fundamental relationship in mathematical statistics between f(x) and F(x), namely:

$$f(x) = \frac{d(F(x))}{dx}$$

$$Probability\ Density\ Function(x) = \frac{d(Cumulative\ Function(x))}{dx}$$

We then differentiate G(y) with respect to **y**, and obtain:

$$g(y) = \frac{d}{dy} \int_{-\infty}^{B^y} f(x)dx$$

Letting **u(y) = B^y**, we then substitute **u** as a function of **y** for B^y and obtain:

$$g(u(y)) = \frac{d}{dy} \int_{-\infty}^{u} f(x)dx$$

The Chain Rule states that:

If G(y) = f(u(y)), then G'(y) = f'(u(y))*u'(y)

and in Leibniz's notation:

$$\frac{dG}{dy} = \frac{dG}{du} * \frac{du}{dy}$$

[Taking Leibniz's notation to extreme, as if on faith, the **du** in the numerator 'cancels out' the **du** in the denominator, a procedure which highly annoys Calculus, provoking it to protest against the demeaning reduction of its rich complexity into simple Algebra.]

We obtain:

$$g(u(y)) = [\frac{d}{du} \int_{-\infty}^{u} f(x)dx] * [\frac{du}{dy}]$$

$$g(u(y)) = [\frac{d}{du} \int_{-\infty}^{u} f(x)dx] * [\frac{d}{dy} B^y]$$

The Second Fundamental Theorem of Calculus states that:

$$\frac{d}{dx}\int_{a}^{x} f(t)dt = f(x)$$

Hence g(y) - the Probability Density Function of variable y - is reduced to:

$$g(y) = f(B^{y}) * \frac{d}{dy} B^{y}$$

From the basic rules of derivatives of exponential functions we get:

$$g(y) = f(B^{y})*\ln(B)*B^{y}$$

Using base 10 would imply that B = 10; recalling the fact that **f** stands for the probability density function for the original variable x itself; and recalling the fact that $Y = LOG_{B}X$, namely that $B^{Y} = x$; we finally obtain:

$$g(y) = f(B^{y})*\ln(B)*B^{y}$$

$$g(y) = f(x)*\ln(10)*x$$

$$g(\log(x)) = \ln(10)*x*f(x)$$

Probability Density Function of $Log(x) = \ln(10)*x*f(x)$

[24] Leading Digits Inflection Point is also Quantitative Turning Point

Leading Digits Inflection Point is defined as the location on the log-axis where the probability density function g(log(x)) curve is at its maximum, namely where g(log(x)) is momentarily flat, horizontal and uniform, neither rising nor falling. Extensive empirical evidences strongly indicate that the nature of random and statistical data (Benford and non-Benford) in extreme generality is characterized by log-gradualism; that log histogram almost never starts suddenly nor ends abruptly around some high initial or final value; rather it starts and ends very low near the log-axis itself; rising up to LDIP; then falling back towards the log-axis; mimicking an upside-down U curve; and that it nearly always shows a marked curvature, as it concaves down with negative 2nd derivative around the core of the data. This implies that there is typically only one unique maximum; namely only one LDIP.

In chapter 21 on the quantitative portions in the k/x distribution case, the generic expression for quantitative portion within any sub-interval (a, b) for <u>any</u> random distribution was given as:

$$\text{Sum } for(a,b) = \int_a^b N * x * f(x) \, dx$$

Clearly the two expressions, the one for quantitative portions as well as the one for log density, both revolve around the product **x*f(x)**, except for the irrelevant constants N and ln(10). The fact that the expression for quantitative portions involves a definite integral, while the expression for log density does not involve definite integrals is also irrelevant, since the former can be written locally simply as [b − a]*N*x*f(x) where b is only tiny bit greater than a. Hence, to the left of LDIP where x*f(x) is rising, quantitative portions are also rising, and to the right of LDIP where x*f(x) is falling, quantitative portions are also falling. Therefore the derivation concerning the generic expression of log density of the previous chapter constitutes a rigorous proof that LDIP is also a quantitative turning point, apart from serving also as a digital turning point.

[25] Leading Digits Inflection Point is Number System Invariant

The definition of Leading Digits Inflection Point as the location on the log-axis where the curve is at its maximum clearly necessitates a positional number system as well as a particular base B. Without a positional number system no discussion about logarithms or digits and their distributions can take place. Remarkably, the position of LDIP on the x-axis is unchanged under a base transformation. Surely the log density of any given distribution appears differently in different bases, yet when LDIP on that B-base log-axis is translated into its related location on the x-axis, it always points to the same point regardless of the value of the base B. One doesn't need any rigorous mathematical proof or complex arguments in order to demonstrate this result, as it follows directly from the principles of the General Law of Relative Quantities as outlined in the 7th Section of Kossovsky (2014). Since it was shown earlier that LDIP is also the quantitative turning point, it then follows that it is base-invariant as well as number-system-invariant. Histograms and continuous density curves are invariant under a base transformation as well as being totally independent on the number system in use. The only changes that occur when a base is transformed or when a totally new number system is put into use (say Roman or Egyptian Numerals) are the symbols below the x-axis and to the left of the y-axis in the Cartesian Plane, but not the [quantitative meaning of the] actual set of tuples constituting the curve. LDIP is also the turning point in relative quantitative portions measured judiciously on tiny sub-intervals, and this quantitative portion is a primitive and basic concept, not necessitating any digits, base, or even a number system. Quantitative portion contribution from each data point in the data set is the confluence of the x-axis and the y-axis positional interplay; being the product [length of the x-value from origin]*[height of the y-axis].

[26] Curved Two-Dimensional Random Partition Process

During a conversation with Don Lemons about the relationship between random partition processes and Benford's Law, the author proposed the most primitive perhaps or the most simplistic two-dimensional random partition of (say) an oval-shaped tablecloth, soft fabric, or a cardboard, where the cutting is performed in the freest possible way, without any restrictions. Figure 3.57 depicts an example of an original oval shape partitioned along random curved paths.

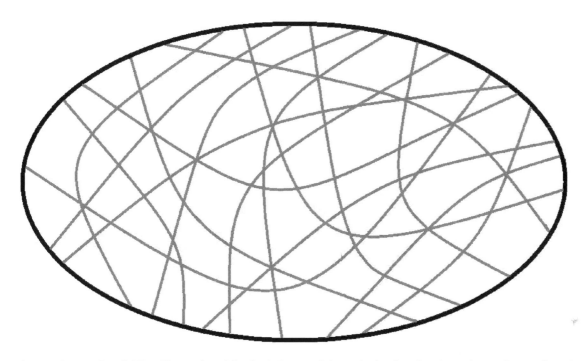

Figure 3.57: Oval-like Shaped Tablecloth is Partitioned via the Cutting along Curved Paths

A large and strong scissor is imagined, being allowed to curve around while cutting without constraints. This is done by cutting from any edge to any other edge, while turning around any which way one sees fit, and without stopping in the middle. The overall oval shape stays intact after each such cutting act, to let the next cutting act be performed also on an oval-shaped tablecloth, until the very end of the long cutting process, and then when no more cutting is desired, when the scissor is put away for good, the still-standing fragile oval with its physically cut pieces is dismantled, after which the oval shape is no longer intact, and its pieces are gathered into one big pile, followed by the recording of the values of the areas of all the pieces. The author conjectures that such restriction-less, natural, and spontaneous partition should lead to Benford even with only very few resultants pieces. Due to the complex two-dimensional curve-like nature of this partition model, the enormous challenge here is in setting up any Monte Carlo computer simulations in order to verify the conjecture, and especially in providing a rigorous mathematical proof.

[27] Linear Two-Dimensional Random Partition Process

Don Lemons has suggested modifying the above oval-like curved partition process, performing it in a linear way instead, so as to obtain a process that could be more systematic, simpler, and by far easier to simulate and to provide a rigorous mathematical proof. This modified model is of strictly linear partitions, both vertically and horizontally, and with the original cardboard being a square or a rectangle, as opposed to a round oval shape.

This linear partition model performed on a square or on a rectangle, is done by creating random marks via the continuous Uniform on the y-axis vertically, in addition to creating random marks via the continuous Uniform on the x-axis horizontally, followed by the partitioning the original square or rectangle along these two sets of marks, resulting in the creation of numerous mini rectangles, and where the sum of all the mini areas is equal to the area of the original square or rectangle. The areas of these mini rectangles are measured and recorded to verify the Benford digital configuration for these values.

As an illustrative example of the linear model - visualized in Figure 3.58 - a square area of 7 meters width by 7 meters height is to be randomly partitioned via Monte Carlo simulations of 24 random horizontal marks and 24 random vertical marks using the continuous Uniform(0, 7) for each dimension to generate independent random realizations.

The actual cutting along those **24** random marks scattered throughout the x-axis horizontal dimension creates **25** horizontal width-compartments.

The actual cutting along those **24** random marks scattered throughout the y-axis vertical dimension creates **25** vertical height-compartments.

Together, these two-dimensional partitions create 25*25 or **625** random areas which are almost always rectangles. Figure 3.58 depicts the results of one Monte Carlo simulation of such partition.

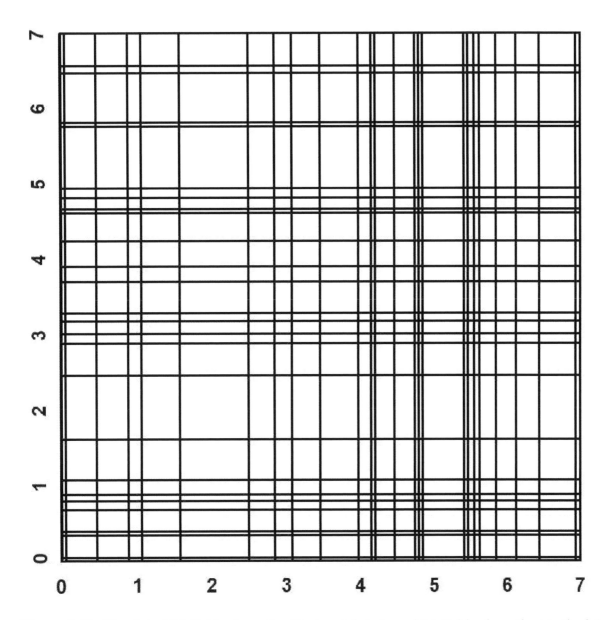

Figure 3.58: Simulated 24 Grids along the Horizontal Axis and 24 Grids along the Vertical Axis

The following is the set of these 24 simulated random marks along the **horizontal** dimension, sorted low to high, plus the two boundary values of 0 and 7:

{**0.000**, 0.023, 0.412, 0.862, 1.043, 1.660, 2.523, 2.826, 3.102, 3.447, 3.966, 4.130, 4.139, 4.366, 4.787, 4.857, 4.869, 5.416, 5.421, 5.489, 5.610, 5.821, 6.164, 6.450, 6.986, **7.000**}

The following is the set of the 25 <u>differences</u> in the above set, representing the 25 **width** values seen in Figure 3.58:

{0.023, 0.389, 0.450, 0.181, 0.617, 0.863, 0.303, 0.277, 0.345, 0.519, 0.164, 0.008, 0.227, 0.422, 0.070, 0.012, 0.546, 0.005, 0.068, 0.121, 0.211, 0.343, 0.286, 0.536, 0.014}

The following is the set of these 24 simulated random marks along the **vertical** dimension, sorted low to high, plus the two boundary values of 0 and 7:

{**0.000**, 0.032, 0.311, 0.346, 0.645, 0.794, 0.876, 1.065, 1.601, 2.450, 2.928, 3.006, 3.196, 3.323, 3.729, 3.903, 4.264, 4.664, 4.678, 4.848, 4.926, 5.755, 5.818, 6.528, 6.671, **7.000**}

The following is the set of the 25 <u>differences</u> in the above set, representing the 25 **height** values seen in Figure 3.58:

{0.032, 0.279, 0.035, 0.299, 0.148, 0.082, 0.189, 0.536, 0.849, 0.478, 0.078, 0.190, 0.127, 0.406, 0.174, 0.361, 0.400, 0.014, 0.170, 0.078, 0.829, 0.063, 0.710, 0.143, 0.329}

These two sets of differences are multiplied [i.e. each element in one set is multiplied by each element of the other set], yielding the set of 635 products representing the areas of the 635 mini rectangles created by the partition process.

First digits proportions are fairly close to Benford, even for this limited 25 by 25 partition:

Random 625 Rectangles - {30.1%, 18.2%, 11.2%, 11.2%, 7.7%, 5.9%, 5.9%, 4.2%, 5.6%}
Benford's Law 1st Digits - {30.1%, 17.6%, 12.5%,   9.7%, 7.9%, 6.7%, 5.8%, 5.1%, 4.6%}

Figure 3.59 depicts the bar chart of the digital comparison to Benford's Law. The low Sum of Squares Deviation (SSD) value of 7.0 for the set of resultant 625 rectangles indicates that this digit configuration is in good conformity with Benford's Law. Figure 3.60 depicts the histogram of the areas of resultant 625 rectangles, confirming the general principle in random partition processes that predicts decisive quantitative skewness where the relatively small is much more numerous than the relatively big.

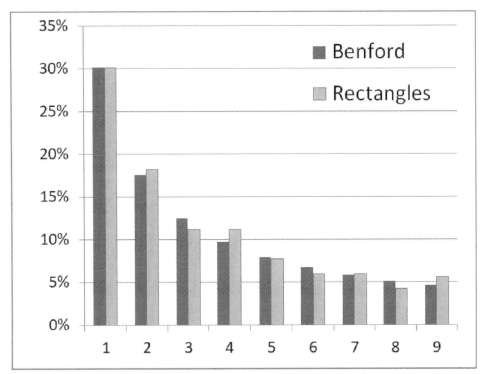

Figure 3.59: Good Fit to Benford - Simulated 625 Random Rectangles

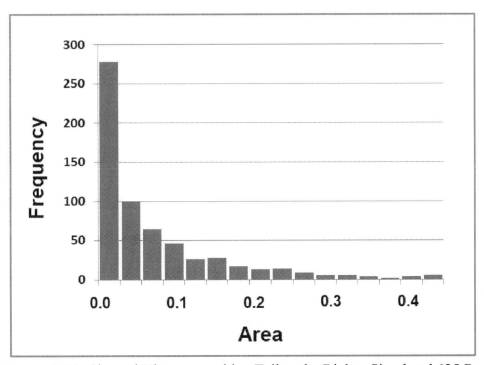

Figure 3.60: Skewed Histogram with a Tail to the Right - Simulated 625 Rectangles

Performing 8 additional Monte Carlo simulations of the above 25 by 25 linear partition scheme yielded a variety of results, mostly with slightly higher SSD values. Yet, quite surprisingly though, the average digit configuration of these 8 simulations came out extremely close to Benford, with the very low SSD value of 1.9! And this is so in spite of the fact that this modest 25 by 25 partition process involves only 48 actual cutting acts, and yields only 625 parts! Figure 3.61 depicts these 8 simulation results, including the digit-by-digit average of these 8 results at the bottom line (in bold font). Figure 3.62 depicts the excellent fit to Benford of the average of these 8 simulations.

Digit 1	Digit 2	Digit 3	Digit 4	Digit 5	Digit 6	Digit 7	Digit 8	Digit 9	SSD
33.3%	16.5%	11.0%	8.6%	8.3%	7.4%	5.0%	5.8%	4.2%	**16.5**
28.8%	16.3%	11.5%	9.6%	7.7%	7.5%	7.8%	5.6%	5.1%	**9.8**
30.9%	18.2%	10.4%	10.4%	8.8%	6.9%	5.0%	4.8%	4.6%	**7.5**
28.3%	17.8%	13.4%	10.7%	8.5%	6.6%	4.5%	5.9%	4.3%	**7.9**
28.2%	19.4%	12.2%	9.8%	9.8%	5.0%	4.5%	6.9%	4.5%	**18.2**
29.8%	17.9%	10.2%	9.8%	9.3%	5.6%	5.9%	6.4%	5.1%	**10.3**
31.8%	16.2%	12.2%	9.9%	7.8%	6.2%	6.7%	5.1%	4.0%	**6.7**
27.7%	18.4%	11.8%	9.9%	9.1%	6.6%	7.2%	5.3%	4.0%	**10.8**
29.8%	**17.6%**	**11.6%**	**9.8%**	**8.7%**	**6.5%**	**5.8%**	**5.7%**	**4.5%**	**1.9**

Figure 3.61: 8 Distinct Simulations of 25 by 25 Partitions – Average is Nearly Benford

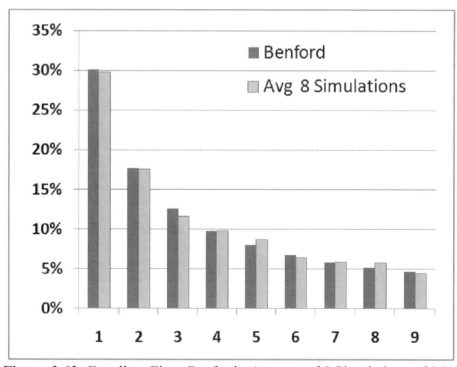

Figure 3.62: Excellent Fit to Benford - Average of 8 Simulations of 25x25 Partitions

Yet, this result actually should not be surprising. Indeed, these 8 distinct simulations could be imagined or thought of as a single complex partitioning of a long 56 by 7 original rectangle made of cardboard or tablecloth, containing within itself 8 squares (of dimension 7 by 7 each), and where each square undergoes 25 by 25 linear partition process independently of its adjacent squares. For such a vista, the satisfactory result with the very low value of 1.9 for SSD is very much expected due to the thorough and very refine partition process of this long rectangle, yielding numerous mini rectangular parts, namely yielding 8*625 or 5000 mini rectangles, all of which naturally should be very close to Benford. All this is in the spirit of the generic principle learnt in this entire Section 3 of this book, namely that almost any partition process that is performed in a random fashion without too many constraints and yielding numerous pieces or parts converges to the Benford digital configuration and shows a marked preference for the small at the expense of the big.

Superior and practically perfect results are gotten by increasing the number of simulated grids to 499 by 499 resulting in 500*500 or 250,000 random rectangles, and where compliance with Benford's Law is as perfect as can be gotten for any random data type.

Random 250,000 Rectangles - {29.9%, 17.7%, 12.5%, 9.7%, 8.0%, 6.7%, 5.8%, 5.1%, 4.6%}
Benford's Law First Digits   - {30.1%, 17.6%, 12.5%, 9.7%, 7.9%, 6.7%, 5.8%, 5.1%, 4.6%}

Figure 3.63 depicts the bar chart of the digital comparison to Benford's Law. The exceedingly low Sum of Squares Deviation (SSD) value of 0.1 for the set of resultant 250,000 rectangles indicates that this digit configuration is practically in complete conformity with Benford's Law!

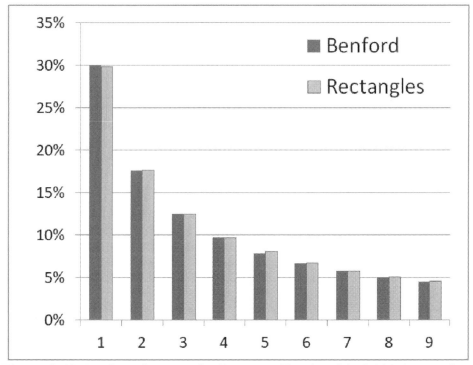

Figure 3.63: Perfect Fit to Benford's Law - Simulated 250,000 Rectangles

[28] Two Vistas of the Linear Random Partition Process

The mathematical-arithmetical vista of the linear partition model is of two distinct one-dimensional random markings placements. Hence, assuming plenty of markings on the x-axis along the horizontal dimension, the horizontal width-compartments (i.e. differences between the markings) are distributed as in the Exponential Distribution [as discussed earlier in this Section 3, chapters 5 and 11]. The same can be said about the vertical y-axis dimension. In addition, since the value of each mini rectangular area is obtained via the multiplication of its width by its height, therefore the model can be viewed as involving the random multiplicative process of two Exponential Distributions [see the end of chapter 8 in Section 2 for an example of the superb Benford convergence for such multiplication].

The model is said to stand on two legs: (1) two distinct one-dimensional markings placement processes leading to two Exponentials, and (2) a multiplicative process. Both legs are known to tend to the Benford digital phenomenon in and of themselves. Therefore, the confluence of these two processes reinforces the effects of each on digits configuration leading to full convergence to Benford. Chapter 95 titled 'Hybrid Causes Leading to Logarithmic Convergence' in Kossovsky (2014) explores this frequently occurring manifestation of the Benford phenomenon.

The physical partition vista of the linear partition model is of the singular partitioning of a rectangular-shaped tablecloth, soft fabric, or a cardboard, where the cuttings are performed strictly in a linear way along the horizontal and vertical dimensions – and without any reference to multiplicative involvement.

Conceptually, for the physical partition vista, this partition process is much more effective in its 2nd stage as compared with its 1st stage. In the 1st stage, random horizontal cuts are made, signifying a normal one-dimensional partition process which can only lead to the Exponential configuration at most. But in the 2nd stage where random vertical cuts are made, each such single vertical line from end-to-end performs multiple partitions because it cuts across many already-existing horizontal ones! This is why the process of partition accelerates in the 2nd stage, and this may explain why we are able to fully converge to Benford after the 2nd stage, instead of obtaining only the approximate Benford digital configuration of the Exponential.

The superb digital results obtained via the 500 by 500 partition, with its resultant 250,000 random rectangles, and where compliance with Benford's Law was practically perfect, can never be matched by simply partitioning one-dimensional linear length, say a pipe, via 250,000 partitions, since such one-dimensional process leads only to the Exponential Distribution and its approximate and not exact Benford digital configuration (due to the saturation that always sets in beyond 5000 to 10000 partitions, where a near perfect fit to the Exponential is gotten, and where further improvement in digital configuration is not possible, as results continue to be Exponential no matter how many more cuts are made).

The comparison of the one-dimensional model with the two-dimensional model, for much fewer partitions, yielding only say 30*30 rectangles as compared with only say 900 linear pipelets, does not show any significant differentiation in the rate of convergence, as both models appear to yield the same degree of digital closeness to the LOG(1 + 1/d) configuration.

[29] Order of Magnitude Considerations for the Convergence of the Model

As a different perspective on the various partition models, the reason one-dimensional pipe partition model does not lead to a full convergence to the Benford configuration is due to the insufficient order of magnitude (OOM) of the Exponential Distribution itself which hovers around the 2.7 level, while within the field of Benford's Law it is well known that order of magnitude should be approximately at least 3 for full compliance. The intimate connection between Benford's Law and order of magnitude of data is practically universal across all random data, and practically the only exception is the k/x distribution defined over integral powers of ten, such as (1, 10). Even though the support of the Exponential Distribution is the infinite range $(0, \infty)$, in practical terms, real-life Exponential data is never exactly 0, nor infinite. Order of magnitude here is best calculated via the 1% percentile and the 99% percentile, namely as LOG(99% percentile/1% percentile), not via the minimum and the maximum as in LOG(max/min), so as to avoid the potential exaggeration and distortion of order of magnitude due to outliers, and this method of calculation always leads to resultant OOM of approximately 2.7, regardless of the value of parameter Lambda λ.

For the linear two-dimensional partition model on the other hand, resultant order of magnitude of the areas of the random rectangles (i.e. the product of two Exponentials) is theoretically the sum of the individual orders of magnitude of each Exponential, namely 2.7 + 2.7, or 5.4. In reality, it's somewhat less, yet definitely greater than 3. Therefore the linear two-dimensional model fully converges to the Benford digital configuration.

As outlined in Section 2, chapter, 7, resultant theoretical order of magnitude of the product of any two variables X and Y is simply the addition of the individual orders of magnitude, namely $OOM_{X*Y} = OOM_X + OOM_Y$. Yet in reality, for any finite number of computer simulated multiplications of two random variables, resultant OOM of multiplied variables is somewhat lower in actual simulations then the theoretical expectation of $OOM_X + OOM_Y$. In any case, the net effect of the multiplicative process is certainly of increasing OOM and greater variability.

For the actual Monte Carlo simulation mentioned earlier of 499 by 499 grids resulting in 250,000 random rectangles, the OOM of the horizontal set of differences came out as only 2.35, and the OOM of the vertical set of differences came out as only 2.59, while the OOM of their product (i.e. the areas) came out as 3.62, and which is comfortably well above the OOM threshold of approximately 3, and therefore the partition process is practically in perfect agreement with Benford's Law. The above-mentioned OOM values were robustly calculated via the 1% percentile and the 99% percentile, and not via the minimum and the maximum, so as to avoid the potential distortions from outliers, and to prevent them from exaggerating OOM.

[30] Two-Dimensional Linear Random Partition Process in Stages

In the linear partition process, all the marks on the axes are drawn simultaneously in a sense, so that no mark can claim seniority or extra importance as being drawn before any other mark.

Another random partition process is suggested here, one with carefully executed stages, and which also converges to Benford fully, and even more rapidly, as the number of stages grows. As shown in Figure 3.64, the process begins with a single square. Two points are chosen randomly and independently along the horizontal x-axis and the vertical y-axis, leading to two lines drawn within the square as in an asymmetrical cross, and resulting in four rectangles. Figure 3.65 depicts the 2nd stage, which is the continuation of the partition process after the 1st stage of Figure 3.64, and where within each of the four rectangles two random points are chosen in order to draw asymmetrical crosses within them, resulting in sixteen rectangles.

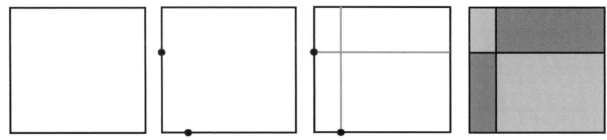

Figure 3.64: Two-Dimensional Partition Process after 1 Stage Yielding 4 Random Rectangles

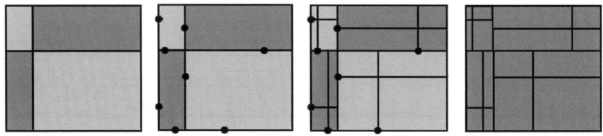

Figure 3.65: Two-Dimensional Partition Process after 2 Stages Yielding 16 Random Rectangles

Figure 3.66 depicts the 3rd stage, which is the continuation of the partition process after the 2nd stage of Figure 3.65, and where asymmetrical crosses are drawn within each of the 16 rectangles, resulting in a total of 64 random rectangles. Surely, Figures 3.64, 3.65, and 3.66 constitute just one unique scenario of the 3-stage partition process, and in principle there are infinitely many other possible scenarios of this 3-stage process.

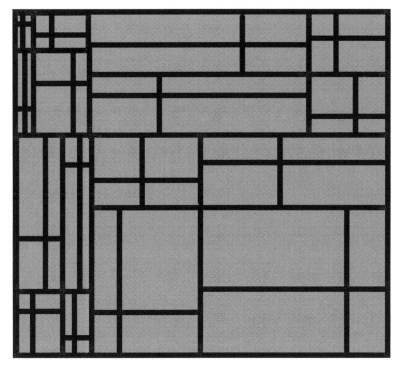

Figure 3.66: Partition Process after 3 Stages Yielding 64 Rectangles

Figure 3.67 depicts 21 distinct Monte Carlo simulations results of the 3-stage random partition process performed on an original 7 by 7 square. Each such process yields 64 random rectangles. Digit configuration recorded for these 21 simulations varies widely as can be seen in Figure 3.67. The last row at the bottom is the calculated average – digit-by-digit – of these 21 simulations. The last column on the right depicts the Sum Squares Deviations (SSD) measure for each simulation.

Digit 1	Digit 2	Digit 3	Digit 4	Digit 5	Digit 6	Digit 7	Digit 8	Digit 9	SSD
26.6%	17.2%	10.9%	9.4%	3.1%	7.8%	10.9%	4.7%	9.4%	89.1
32.8%	12.5%	7.8%	15.6%	3.1%	6.3%	4.7%	9.4%	7.8%	143.6
26.6%	21.9%	10.9%	9.4%	14.1%	9.4%	1.6%	6.3%	0.0%	118.4
39.1%	14.1%	18.8%	4.7%	9.4%	3.1%	4.7%	4.7%	1.6%	182.4
23.4%	10.9%	23.4%	15.6%	6.3%	6.3%	7.8%	3.1%	3.1%	257.0
26.6%	10.9%	18.8%	14.1%	3.1%	15.6%	6.3%	3.1%	1.6%	231.3
26.6%	12.5%	20.3%	6.3%	14.1%	1.6%	4.7%	6.3%	7.8%	188.7
25.0%	26.6%	15.6%	6.3%	7.8%	4.7%	6.3%	1.6%	6.3%	147.5
26.6%	15.6%	12.5%	9.4%	6.3%	6.3%	9.4%	9.4%	4.7%	50.5
43.8%	14.1%	10.9%	6.3%	7.8%	7.8%	3.1%	4.7%	1.6%	230.8
26.6%	10.9%	10.9%	6.3%	7.8%	14.1%	9.4%	9.4%	4.7%	156.5
29.7%	17.2%	15.6%	9.4%	14.1%	6.3%	4.7%	1.6%	1.6%	71.1
37.5%	15.6%	10.9%	6.3%	10.9%	6.3%	4.7%	3.1%	4.7%	87.4
37.5%	18.8%	17.2%	6.3%	4.7%	1.6%	3.1%	6.3%	4.7%	135.1
28.1%	15.6%	12.5%	7.8%	6.3%	7.8%	6.3%	4.7%	10.9%	56.3
28.1%	17.2%	17.2%	4.7%	4.7%	4.7%	7.8%	12.5%	3.1%	126.3
28.1%	21.9%	14.1%	10.9%	10.9%	3.1%	4.7%	1.6%	4.7%	61.9
29.7%	15.6%	15.6%	6.3%	7.8%	10.9%	4.7%	6.3%	3.1%	48.4
21.9%	23.4%	12.5%	12.5%	9.4%	9.4%	6.3%	0.0%	4.7%	145.2
32.8%	12.5%	15.6%	10.9%	7.8%	7.8%	3.1%	4.7%	4.7%	53.4
21.9%	21.9%	18.8%	12.5%	6.3%	6.3%	6.3%	3.1%	3.1%	142.2
29.5%	**16.5%**	**14.8%**	**9.1%**	**7.9%**	**7.0%**	**5.7%**	**5.1%**	**4.5%**	**7.4**

Figure 3.67: 21 Distinct Simulations of 3-Stage Partition - Digit Configuration Varies Widely

On the face of it, nothing of the exact or nearly-exact Benford configuration can be seen in Figure 3.67. Indeed, digit 1 fluctuates from 21.9% to 43.8%, while digit 9 fluctuates even more wildly, from 0% to 10.9%, and SSD values are as high as 50 to 150 approximately. And surely it can easily be argued that although the staged-partition process ultimately leads to Benford as the number of stages (and the number of rectangles) increases, yet, expecting to obtain the Benford configuration so early after merely 3 stages with only 64 generated rectangles is nothing but an illusion and wishful thinking on the part of Benford aficionados and digit enthusiasts (such as this author). Staring at the random structure of the assorted rectangles in Figure 3.66 representing one particular simulation trajectory of the 3-stage random partition process with only 64 resultant rectangles, and employing whatever intuition or personal instinct, one would never expect to get anywhere near the LOG(1 + 1/d) configuration!

Yet, even at this early stage, Benford does indeed manifest itself almost exactly! Albeit in a hidden and indirect way, namely in the calculation of the average – digit-by-digit – of these 21 simulations! Calculated SSD for this average is the fairly low value of 7.4!

Average Digit-By-Digit   - {29.5%, 16.5%, 14.8%, 9.1%, 7.9%, 7.0%, 5.7%, 5.1%, 4.5%}
Benford's Law 1st Digits - {30.1%, 17.6%, 12.5%, 9.7%, 7.9%, 6.7%, 5.8%, 5.1%, 4.6%}

Figure 3.68 depicts the good fit to Benford of the average of the 21 simulations.

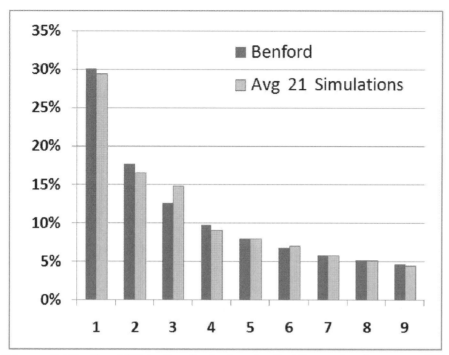

Figure 3.68: Good Fit to Benford after the 3rd Stage for Average of 21 Simulations

Yet, this result actually should not be surprising. Indeed, these 21 distinct simulations could be imagined or thought of as a single complex partitioning of a long 147 by 7 original rectangle made of cardboard or tablecloth, containing within itself 21 squares (of dimension 7 by 7 each), and where each square undergoes three-stage random partition independently of its adjacent squares. For such a vista, the satisfactory result with the low value of 7.4 for SSD is very much expected due to the thorough and very refine partition process of this long rectangle, yielding numerous mini rectangular parts, namely yielding 21*64 or 1344 mini rectangles, all of which naturally should be very close to Benford.

Figure 3.69 follows the long trajectory of one scenario of the staged partition process, from the original 7 by 7 square to the huge set of 65,536 rectangles after the 8th stage. Figure 3.69 starts recording from the 3rd stage the snapshots of the number of the rectangles, the digit configuration, sum of squares deviations (SSD), and order of magnitude (OOM), for each stage. Order of magnitude is calculated after the elimination of the top 1% of values and the bottom 1% of values, so as not to let extreme computer simulation values influence or distort order of magnitude. This is so since even a single outlier here as a minimum or as a maximum can exaggerate order of magnitude a great deal.

Clearly, the 8th stage yields digital results that are practically in perfect agreement with Benford's Law, resulting in having the exceedingly low SSD value of 0.1, and nothing really is going to be gained by continuing the process to the 9th stage. We have managed to obtain practically perfect Benford configuration with only 65,536 rectangles, as compared with the linear model of 500 by 500 partitions with its numerous 250,000 rectangles.

Stage	# Rec.	Digit 1	Digit 2	Digit 3	Digit 4	Digit 5	Digit 6	Digit 7	Digit 8	Digit 9	SSD	OOM
3	64	35.9%	15.6%	12.5%	14.1%	7.8%	7.8%	1.6%	1.6%	3.1%	91.0	4.1
4	256	28.9%	20.7%	11.3%	9.4%	7.4%	7.4%	3.1%	5.1%	6.6%	24.7	4.8
5	1024	29.6%	16.9%	11.6%	10.2%	9.9%	6.2%	5.6%	4.9%	5.3%	6.4	5.8
6	4096	29.4%	17.7%	12.3%	9.7%	8.4%	6.3%	6.3%	5.2%	4.7%	1.3	6.7
7	16384	29.7%	18.1%	12.8%	9.5%	7.5%	6.7%	6.0%	5.0%	4.8%	0.8	7.2
8	65536	30.2%	17.6%	12.4%	9.6%	7.9%	6.8%	5.7%	5.1%	4.7%	0.1	7.8
Benford's Law:		30.1%	17.6%	12.5%	9.7%	7.9%	6.7%	5.8%	5.1%	4.6%	0.0	----

Figure 3.69: Snapshots of # Rectangles, Digit Configuration, SSD, OOM - Stage 3 to Stage 8

Figure 3.70 depicts the histogram of the areas of resultant rectangles after the 8th stage of that particular random trajectory of Figure 3.69, confirming the general principle in random partition processes that predicts decisive quantitative skewness, where the relatively small is much more numerous than the relatively big. A logarithmic scale is used for the vertical frequency axis.

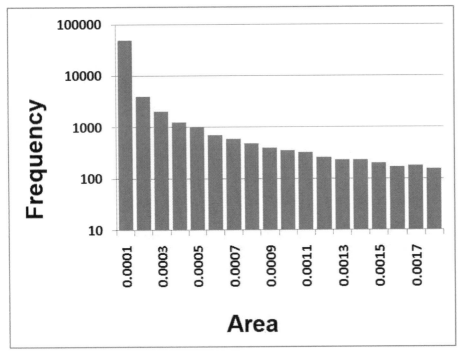

Figure 3.70: Skewed Histogram with a Tail to the Right - 8 Stages – 65,536 Rectangles

[31] Three-Dimensional Linear Random Partition Process

An extension of the linear two-dimensional partition model is a linear partition performed on three-dimensional cubes or rectangular prisms, leading to the multiplication of three Exponential Distributions. Here the focus is on the volume of solid objects, which is the product of three dimensions, the length, the width, and the height. Figure 3.71 depicts one example of such three-dimensional linear random partitions where the x-axis is randomly partitioned into 4 compartments, the y-axis is randomly partitioned into 6 compartments, and the z-axis is randomly partitioned into 4 compartments, resulting in 4*6*4 or 96 random rectangular prisms.

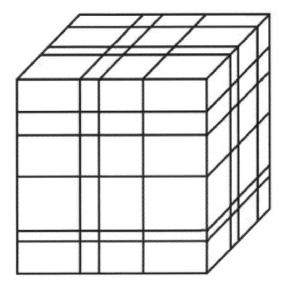

Figure 3.71: Three-Dimensional Random Partition - 4 by 6 by 4

Since three dimensions are multiplied by each other, namely three Exponential Distributions, resultant volume earns an even stronger multiplicative push towards the Benford digital configuration as compared with the two-dimensional random area, although this also requires sufficient number of partitions for each dimension. Order of magnitude here is very large, namely the sum of each order of magnitude of the three Exponentials which theoretically is around 2.7 + 2.7 + 2.7, or 8.1, but in reality only around 4.5 to 5.0 or so, and which easily surpasses the OOM threshold of 3, thus the Benford configuration can be readily obtained here even with a lower number of partitions for each dimension.

Figure 3.72 depicts 9 distinct Monte Carlo simulations results of three-dimensional 9 by 9 by 9 random partition processes performed on an original 7 by 7 by 7 cube. Each such process yields 9*9*9 or 729 random rectangular prisms. The last row at the bottom is the calculated average – digit-by-digit – of these 9 simulations, and which is extremely closed to Benford with the very low SSD value of 1.0. Thus we have managed to obtain in a sense quickly and easily truly excellent Benford configuration with merely nine such processes, each involving 9 + 9 + 9 or 27 actual cutting actions across the cube! Figure 3.73 depicts the excellent fit to Benford of the average of the 9 simulations.

Digit 1	Digit 2	Digit 3	Digit 4	Digit 5	Digit 6	Digit 7	Digit 8	Digit 9	SSD
32.2%	19.3%	11.8%	8.6%	7.7%	5.6%	6.0%	4.8%	3.8%	11.0
29.9%	18.1%	13.3%	9.3%	8.4%	6.0%	6.0%	4.1%	4.8%	2.8
29.9%	18.2%	12.9%	9.3%	7.1%	7.4%	6.4%	4.7%	4.0%	2.8
31.0%	16.5%	12.8%	10.4%	7.7%	5.2%	7.5%	4.5%	4.4%	8.4
32.8%	17.8%	10.6%	8.8%	8.1%	7.3%	5.1%	6.3%	3.3%	15.8
30.2%	18.1%	12.9%	8.1%	9.2%	6.4%	4.8%	4.5%	5.8%	7.4
30.5%	18.5%	12.5%	8.6%	8.0%	6.9%	6.4%	5.3%	3.3%	4.2
30.9%	17.1%	12.9%	9.7%	7.3%	7.3%	5.9%	4.3%	4.7%	2.5
29.4%	18.0%	12.8%	8.9%	8.4%	7.4%	4.9%	5.8%	4.5%	3.2
30.7%	**18.0%**	**12.5%**	**9.1%**	**8.0%**	**6.6%**	**5.9%**	**4.9%**	**4.3%**	**1.0**

Figure 3.72: Nine Distinct Simulations of 9 by 9 by 9 Partition – Average is Practically Benford

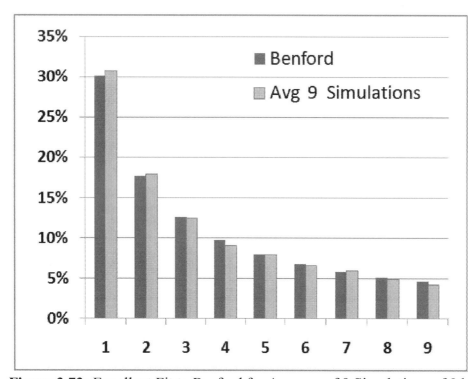

Figure 3.73: Excellent Fit to Benford for Average of 9 Simulations of 9 by 9 by 9

Acknowledgement:

In part, this section also analyzes and expands on the works of Lemons, Kafri, and Miller.

A groundbreaking article titled "On the Number of Things and the Distribution of First Digits" is published by **Don Lemons** in 1986. It explores an averaging scheme on the set of all possible partitions of a real quantity X into smaller real parts $\{\Delta x, 2\Delta x, 3\Delta x, \ldots\}$ without explicitly applying anything from Physics. Lemon arrives at a distribution proportional to $1/x$ which is known to be Benford when defined over ranges with an integral exponent difference between the minimum and the maximum [*namely that LOG(maximum) – LOG(minimum) = Integer*].

Oded Kafri publishes an innovative article in 2009 regarding a particular 'balls and boxes' scheme, after exploring during the previous years possible connections between the principles of Information Theory, Entropy in Thermodynamics, and Benford's Law. The distribution of moveable or flexible balls inside fixed and rigid boxes can be interpreted as the partitioning of a large pile of balls into much smaller piles about to reside inside the boxes.

Steven Miller publishes an article in 2015 presenting a mathematically rigorous partition model termed 'equipartition', successfully leading to the Benford configuration. His model is a variation on Lemons' original insight, substituting integral quantities for real ones; applying results from Integer Partition and Number Theory; as well as 'borrowing' concepts from Physics.

The author would like to thank Don Lemons, Steven Miller, and Oded Kafri for their helpful comments and correspondence regarding the details of their models.

SECTION 4

RANDOM CONSOLIDATIONS AND FRAGMENTATIONS PROCESSES

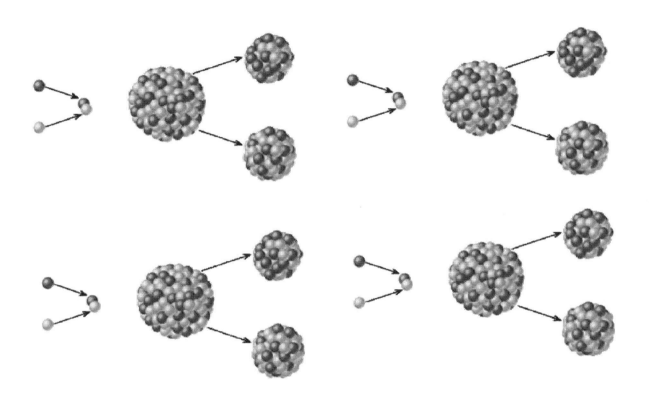

[1] Random Consolidations and Fragmentations Cycles Lead to Benford

In this section it is shown that a process where a large set of identical quantities constantly alternates between minuscule random consolidations (summing two randomly chosen values into a singular value) and tiny random fragmentations (division of one randomly chosen value into two new values) converges eventually to the Benford configuration after sufficiently many cycles. Randomness in selecting the particular quantity to be fragmented, as well as randomness in selecting the two particular quantities to be consolidated, is essential for convergence. Not surprisingly then, fragmentation itself could be performed either randomly say via a realization from the continuous Uniform on (0, 1), or deterministically via any fixed ratio of breakup such as say 25% - 75% or 10% - 90%, and the Benford configuration emerges in either case. This is so since randomness already exists in the system in the determination of which quantities are to be fragmented and consolidated.

The random model discussed in this section is a cyclical process which alternates constantly between (1) the selection of one quantity chosen at random and its breakdown [fragmentation] into two new smaller quantities, (2) followed by the selection of two randomly chosen quantities and their summation [consolidation] into a singular added quantity. Monte Carlo computer simulations results decisively show that such a process leads to a rapid Benford digital convergence. Exploration of the resultant (random) algebraic expressions coming out of the corresponding mathematical model, nicely explains this Benford behavior simply in terms of a random multiplication process, and this is so in spite of the addition terms involved here. In other words, the arithmetical model of the process points to a tug of war between addition and multiplication, and where multiplication ultimately and decisively triumphs over addition.

This consolidations and fragmentations model, abbreviated as **C&F**, is based on **L** balls, all with an identical initial real value **V**, such as for example the initial weight of each ball, and assuming uniformity of mass density, implying equivalency between volume and weight proportions. Each C&F model consists of **C** cycles. Within one full cycle, two opposing processes are performed, one of fragmentation, followed by one of consolidation. The first process is the fragmentation of a single ball chosen at random utilizing the discrete Uniform on $\{1, 2, 3, \ldots, L\}$, as well as the continuous Uniform(0, 1) to decide on the proportions of the two fragments. The second process is the consolidation of two balls chosen at random and fused into a singular and larger ball, utilizing at first the discrete Uniform distribution on $\{1, 2, 3, \ldots, L, L + 1\}$, followed by the utilization of the discrete Uniform on $\{1, 2, 3, \ldots, L\}$, for ball selections. Obviously, after each full cycle, the number of existing balls is unchanged, and it is still L, being the same as in the beginning and as in the end of the entire process. Also, the total quantity of the entire system - namely the overall sum or overall weight of all the balls - is conserved throughout the entire process. The specific description of physical balls of uniform mass density being broken and then fused, and the focus on the weight variable, is an arbitrary one of course, and the generic model is of pure quantities and abstract numbers.

Schematically the process is described as follows:

1) Initial Set = {V, V, V, … L times …V, V, V}
2) Repeat C times:
   (i) Choose one ball at random and split it as in {Uniform(0, 1), 1 – Uniform(0, 1)}.
   (ii) Choose two balls randomly and merge them.
3) Final Set is Benford.

For example, for 7 balls with an initial value of 35 each, that is L = 7, V = 35, we record the initial few cycles as the process develops randomly step-by-step:

{35, 35, 35, 35, 35, 35, 35}
{35, 35, **35**, 35, 35, 35, 35} to be split
{35, 35, 31, 4, 35, 35, 35, 35}
{35, 35, 31, 4, 35, **35**, 35, **35**} to be merged
{35, 35, 31, 4, 35, 70, 35}
{35, 35, 31, **4**, 35, 70, 35} to be split
{35, 35, 31, 2, 2, 35, 70, 35}
{35, 35, 31, 2, **2**, 35, **70**, 35} to be merged
{35, 35, 31, 2, 72, 35, 35}
{**35**, 35, 31, 2, 72, 35, 35} to be split
{25, 10, 35, 31, 2, 72, 35, 35}
{25, 10, 35, 31, 2, 72, **35, 35**} to be merged
{25, 10, 35, 31, 2, 72, 70}

Clearly for such tiny set of balls where L = 7 there could never be any convergence to the logarithmic, even if one goes much further than C = 3 cycles. In fact, at least two digits out of all possible 9 first digits would obtain the embarrassing very low 0% proportion here no matter. It should be noted that the simulation above deliberately avoided fractional values, for pedagogical purposes, keeping the values presented as simple as possible for easy demonstrations.

Monte Carlo computer simulations show that after C full cycles, the weight of the balls is very nearly Benford, given that C > 2*L approximately, although C > 3*L or C > 4*L usually give slightly better results. It is necessary to have a sufficient number of these C cycles so that all or almost all of the balls experience either fragmentation or consolidation (preferably both, and hopefully not merely once, but rather twice or three times). By cycling at least twice as many balls that exist in the system, we ensure that (almost) all the balls undergo transformation of some sort, and that the initial value V is (almost) nowhere to be found among the balls at the end of the entire process. Continuing beyond the required 2*L or 3*L cycles does not ruin the logarithmic convergence thus obtained, and Benford is steadily preserved (or rather further perfected) as more cycles are added. Surely, the other essential prerequisite for logarithmic convergence here is to have a sufficiently large number of balls in the system so that the logarithmic can be properly manifested. Hence the requirement is that L > 200, or for better convergence that L > 300. Falling below 300 or 200 balls for example yields only crude or approximate logarithmic-like results, as in all small data sets aspiring to obey the law.

Core Physical Order of Magnitude (CPOM) shall be in use here to ensure robustness. All C&F processes start out with V as the unique and repeated value, having no variability whatsoever, hence CPOM initially is the lowest possible value of $P_{90\%}/P_{10\%} = V/V = 1$. But at the end of the entire C&F process, CPOM is sufficiently high and Benford behavior is found.

[2] Empirical Examinations of Consolidations and Fragmentations Models

We now run Monte Carlo computer simulations for three different Consolidations and Fragmentations schemes and explore results.

Scheme A:

In Monte Carlo computer simulation of one particular C&F process, 2000 balls, with an initial identical value of 1, undergo 8000 binary cycles, utilizing the random Uniform(0, 1) for fragmentations. The computerized results obtained here were as follows:

Initial Set - {1, 1, 1, ...   2000 times   ... , 1, 1, 1}
Final Set - {0.0000000012, 0.0000000039, 0.0000000227,   ...   15.5, 16.5, 16.9}

Only the first 3 and the last 3 elements from the ordered final set above are shown, namely the extreme values, the 3 biggest and the 3 smallest values, surely to be considered as outliers.

First Digits of Final Set - {29.1, 19.1, 11.9, 9.8, 8.3, 6.8, 6.4, 4.7, 4.2}
Benford's Law 1st Digits - {30.1, 17.6, 12.5, 9.7, 7.9, 6.7, 5.8, 5.1, 4.6}

Figure 4.1 depicts the digital comparison between this C&F process and Benford's Law. The low 4.7 SSD value calculated here indicates that results are very close to the Benford configuration.

Quantitatively, the process takes as input a set of identical numbers with CPOM value of 1, and transforms it into a highly skewed set of numbers, where the small is much more numerous than the big, and where CPOM = $P_{90\%}/P_{10\%}$ = 3.05/0.0014 = 2131. Here 80% of resultant data falls within (0.0014, 3.05).

Figure 4.2 depicts the histogram from 0 to 10.2 of the final resultant set of numbers after the 8000th binary cycle. For better visual clarity a logarithmic vertical scale is used, although this masks the dramatic quantitative fall occurring here.

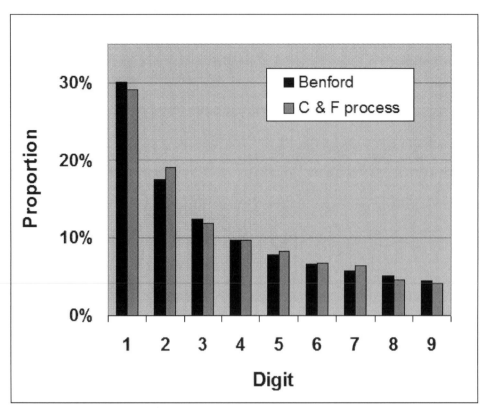

Figure 4.1: C&F with Fragmentation Ratio Uniform(0, 1) – Scheme A

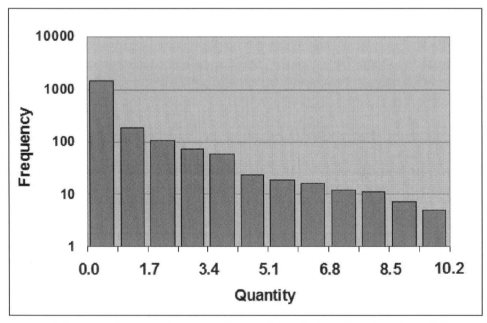

Figure 4.2: Quantitative Configuration with Ratio Uniform(0, 1) – Scheme A

Significantly, the strong tendency towards the logarithmic in consolidations and fragmentations processes is such that randomness regarding the proportions of the fragments of any split ball is not even required. Instead of utilizing the random continuous Uniform(0, 1) to split a ball into two pieces, one may utilize a deterministic fixed ratio called p, and the process converges to Benford just as rapidly! The necessary factor driving convergence here is only the random manner by which the ball which is to be broken is chosen, and the random manner by which the two balls which are to be consolidated are chosen. Such randomness is generated here via the discrete uniform distributions [although other more complex versions and models could be devised and imagined which randomly select balls by taking their sizes into consideration]. No obvious constraint whatsoever seems to exist here for the value of p, and rapid convergence to the logarithmic is found for all the many simulation trials performed by the author with varying p values. Moreover, there seems to be no distinction whatsoever in the rate of convergence whether one utilizes a random model or a deterministic model in splitting a ball. This is so since there exists already 'plenty of randomness' in the system in how balls are selected for fragmentations and for consolidations.

Scheme B:

In another Monte Carlo simulation example utilizing a deterministic fixed 50% - 50% ratio, with 1500 balls all having the identical initial value of 1, being cycled 10,000 times, first digits came out nearly Benford:

Initial Set - {1, 1, 1, …   1500 times   … , 1, 1, 1}
Final Set - {0.0000124, 0.0000153, 0.0000153,   …   13.6, 18.1, 18.6}

Only the first 3 and the last 3 elements from the ordered final set above are shown, namely the extreme values, the 3 biggest and the 3 smallest values, surely to be considered as outliers.

First Digits of Final Set   -   {31.3, 17.0, 12.9, 9.2, 6.0, 7.5, 5.7, 5.6, 4.8}
Benford's Law 1st Digits - {30.1, 17.6, 12.5, 9.7, 7.9, 6.7, 5.8, 5.1, 4.6}

Figure 4.3 depicts the digital comparison between this C&F process and Benford's Law. The low 6.7 SSD value calculated here indicates that results are very close to the Benford configuration.

Quantitatively, the process takes as input a set of identical numbers with CPOM value of 1, and transforms it into a highly skewed set of numbers where the small is much more numerous than the big, and where CPOM is $P_{90\%}/P_{10\%} = 2.78/0.0115 = 241$. Here 80% of resultant data falls within (0.0115, 2.78). Figure 4.4 depicts the histogram from 0 to 10.5 of the final resultant set of numbers after the 10,000th binary cycle.

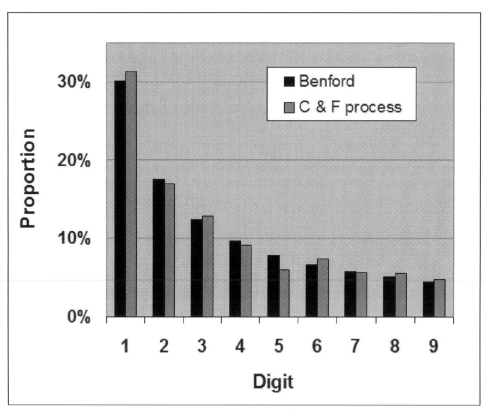

Figure 4.3: C&F of Deterministic Fixed Even Ratio 50% - 50% − Scheme B

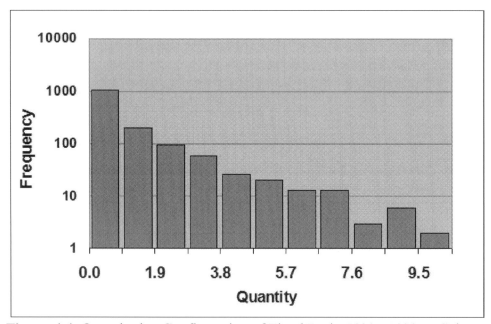

Figure 4.4: Quantitative Configuration of Fixed Ratio 50% - 50% − Scheme B

Scheme C:

In another Monte Carlo simulation example utilizing a deterministic fixed 15% - 85% ratio, with 1000 balls all having the identical initial value 1, being cycled 3000 times, first digits came even closer to Benford:

Initial Set - {1, 1, 1, ...   1000 times   ... , 1, 1, 1}
Final Set - {0.0000000278, 0.0000000278, 0.0000001574,   ...   , 13.1, 13.5, 15.4}

Only the first 3 and the last 3 elements from the ordered final set above are shown, namely the extreme values, the 3 biggest and the 3 smallest values, surely to be considered as outliers.

First Digits of Final Set   -   {30.0, 17.6, 12.7, 10.1, 8.5, 7.4, 5.7, 4.3, 3.7}
Benford's Law 1st Digits - {30.1, 17.6, 12.5,   9.7, 7.9, 6.7, 5.8, 5.1, 4.6}

Figure 4.5 depicts the digital comparison between this C&F process and Benford's Law. The very low 2.5 SSD value calculated for this result indicates an excellent agreement with the logarithmic, having slightly better fit to Benford as compared with the two previous C&F schemes.

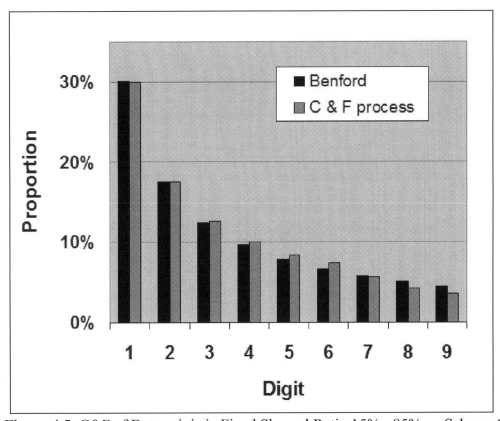

Figure 4.5: C&F of Deterministic Fixed Skewed Ratio 15% - 85% − Scheme C

Of all the three C&F schemes above, scheme C yields the closest digital configuration to the Benford proportion. It is superior to scheme B due to its highly skewed split of 15% - 85% which promotes more intense differentiations in resultant relative quantities, and which drives the digital phenomenon. Scheme A also yields at times highly skewed splits near say the 15% - 85% ratio, but very often it's near the more even splits of around 50% - 50%, or other mild splits around say 45% - 55% and 40% - 60%.

Quantitatively, the process takes as input a set of identical numbers with CPOM value of 1, and transforms it into a highly skewed set of numbers where CPOM is $P_{90\%}/P_{10\%} = 2.92/0.0018 = 1658$. Here 80% of resultant data falls within (0.0018, 2.92). Figure 4.6 depicts the histogram from 0 to 8.8 of the final resultant set of numbers after the 3000th binary cycle.

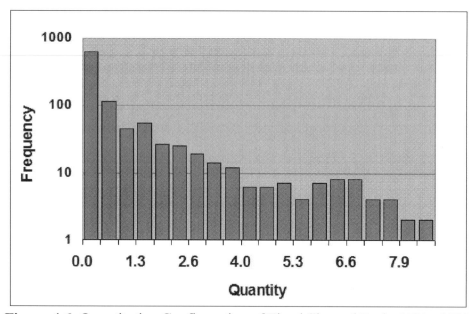

Figure 4.6: Quantitative Configuration of Fixed Skewed Ratio 15% - 85% — Scheme C

As opposed to the case of Random Dependent Partition with sharp differentiation in the rate of convergence to Benford depending on whether the breakup is random or deterministic, here for the Fragmentation and Consolidation model, a deterministic fixed ratio p model yields rapid convergence to Benford just as in the random model. In fact, deterministic models here with highly skewed p and (1 − p) values, such as 15% - 85% for example, seem to converge even faster than random models utilizing the Uniform(0, 1) which is often centered around the more even 50% - 50% or 40% - 60% ratios. The necessary factor driving convergence in C&F models is only the random manner by which the ball which is to be broken is chosen, and the random manner by which the two balls which are to be consolidated are chosen. Since there exists already 'plenty of randomness' in the system in how balls are selected for fragmentations and for consolidations, not much can be gained by adding more randomness with the use of the Uniform(0, 1) – and this explanation applies as well for Chaotic Rock Braking model with a deterministic fixed ratio. In contrast, Random Dependent Partition is able to derive its randomness <u>only</u> from the random manner in how quantities are partitioned, thus Deterministic Dependent Partition models with fixed p ratio converge to Benford extremely slowly after an enormous number of cycles.

[3] Examinations of the Random Emergence of the Algebraic Expressions

It should be noted that the first significant digits configuration of the initial set
{V, V, V, … L times …V, V, V} is as far removed from the Benford configuration as one
could imagine, endowing 0% proportion for 8 digits, and the entire 100% proportion for one
privileged digit, namely for that fortunate digit leading V. This fact dramatizes the decisive and
very rapid digital transformation taking place under repeated consolidations and fragmentations
cycles. Surely, a similar dramatic transformation is occurring at the quantitative level, and
actually it is this quantitative transformation which drives the digital one - not the other way
around. We start out with a single repeated value V, and a histogram showing a tall and very thin
line or a single high bin at V with height of L; then we end up with many diverse and skewed
values, represented by a histogram having many bins such as in Figures 4.2, 4.4, and 4.6, and
with a tail to the right, where the small is numerous and the big is rare.

Let us examine the mathematical expressions being randomly formed as we record the initial six binary cycles of a C&F process of five balls progressing step-by-step - for one particular scenario of chanced development, with 'authentic' randomness in ball selections intended or imagined by the author. Here the balls split randomly into two parts utilizing the Uniform(0, 1), calling each pair of realization as U and (1 – U), or rather as UJ and (1 – UJ) with J as an index :

$\{V, V, V, V, V\}$

$\{\mathbf{V}, V, V, V, V\}$ *to be split*

$\{V*U1, V*(1-U1), V, V, V, V\}$

$\{V*U1, \mathbf{V*(1-U1)}, V, V, V, \mathbf{V}\}$ *to be merged*

$\{V*U1, V*(1-U1) + V, V, V, V\}$

$\{\mathbf{V*U1}, V*(1-U1) + V, V, V, V\}$ to be split

$\{V*U1*U2, V*U1*(1-U2), V*(1-U1) + V, V, V, V\}$

$\{V*U1*U2, V*U1*(1-U2), V*(1-U1) + V, \mathbf{V}, \mathbf{V}, V\}$ *to be merged*

$\{V*U1*U2, V*U1*(1-U2), V*(1-U1) + V, 2V, V\}$

$\{V*U1*U2, \mathbf{V*U1*(1-U2)}, V*(1-U1) + V, 2V, V\}$ *to be split*

$\{V*U1*U2, V*U1*(1-U2)*U3, V*U1*(1-U2)*(1-U3), V*(1-U1) + V, 2V, V\}$

$\{\mathbf{V*U1*U2}, V*U1*(1-U2)*U3, V*U1*(1-U2)*(1-U3), V*(1-U1) + V, \mathbf{2V}, V\}$ *to be merged*

$\{V*U1*U2+2V, V*U1*(1-U2)*U3, V*U1*(1-U2)*(1-U3), V*(1-U1) + V, V\}$

$\{V*U1*U2+2V, V*U1*(1-U2)*U3, V*U1*(1-U2)*(1-U3), \mathbf{V*(1-U1) + V}, V\}$ *to be split*

$\{V*U1*U2+2V, V*U1*(1-U2)*U3, V*U1*(1-U2)*(1-U3), (V*(1-U1) + V)*U4,$
$(V*(1-U1)+V)*(1-U4), V\}$

$\{V*U1*U2+2V, V*U1*(1-U2)*U3, V*U1*(1-U2)*(1-U3), \mathbf{(V*(1-U1) + V)*U4},$
$(V*(1-U1)+V)*(1-U4), \mathbf{V}\}$ *to be merged*

$\{V*U1*U2+2V, V*U1*(1-U2)*U3, V*U1*(1-U2)*(1-U3), V + (V*(1-U1) + V)*U4,$
$(V*(1-U1) + V)*(1-U4)\}$

$\{V*U1*U2+2V, \mathbf{V*U1*(1-U2)*U3}, V*U1*(1-U2)*(1-U3), V + (V*(1-U1) + V)*U4,$
$(V*(1-U1) + V)*(1-U4)\}$ *to be split*

$\{V*U1*U2+2V, V*U1*(1-U2)*U3*U5, V*U1*(1-U2)*U3*(1-U5), V*U1*(1-U2)*(1-U3),$
$V + (V*(1-U1) + V)*U4, (V*(1-U1) + V)*(1-U4)\}$

$\{V*U1*U2+2V, V*U1*(1-U2)*U3*U5, \mathbf{V*U1*(1-U2)*U3*(1-U5)}, V*U1*(1-U2)*(1-U3),$
$\mathbf{V+(V*(1-U1) + V)*U4}, (V*(1-U1) + V)*(1-U4)\}$ *to be merged*

$\{V*U1*U2+2V, V*U1*(1-U2)*U3*U5, V*U1*(1-U2)*U3*(1-U5) + V+(V*(1-U1) + V)*U4,$
$V*U1*(1-U2)*(1-U3), (V*(1-U1) + V)*(1-U4)\}$

$\{\mathbf{V*U1*U2+2V}, V*U1*(1-U2)*U3*U5, V*U1*(1-U2)*U3*(1-U5) + V+(V*(1-U1) + V)*U4,$
$V*U1*(1-U2)*(1-U3), (V*(1-U1) + V)*(1-U4)\}$ *to be split*

$\{(V*U1*U2+2V)*U6, (V*U1*U2+2V)*(1-U6), V*U1*(1-U2)*U3*U5, V*U1*(1-U2)*U3*$
$(1-U5) + V+(V*(1-U1) + V)*U4, V*U1*(1-U2)*(1-U3), (V*(1-U1) + V)*(1-U4)\}$

$\{(V*U1*U2+2V)*U6, (V*U1*U2+2V)*(1-U6), \mathbf{V*U1*(1-U2)*U3*U5}, V*U1*(1-U2)*U3*$
$(1-U5) + V+(V*(1-U1) + V)*U4, V*U1*(1-U2)*(1-U3), \mathbf{(V*(1-U1) + V)*(1-U4)}\}$ *to merge*

$\{(V*U1*U2+2V)*U6, (V*U1*U2+2V)*(1-U6), V*U1*(1-U2)*U3*U5 + (V*(1-U1) + V)*$
$(1-U4), V*U1*(1-U2)*U3*(1-U5) + V+(V*(1-U1) + V)*U4, V*U1*(1-U2)*(1-U3)\}$

Let's examine the five terms after the last cycle:

(V*U1*U2 + 2V)*U6
(V*U1*U2 + 2V)*(1 - U6)
V*U1*(1 - U2)*U3*U5 + (V*(1 - U1) + V)*(1 - U4)
V*U1*(1 - U2)*U3*(1 - U5) + V + (V*(1 - U1) + V)*U4
V*U1*(1 - U2)*(1 - U3)

The initial V weight of each ball appears as a singular factor only once, since in essence it represents the scale of the entire system. Dividing by V (or equivalently setting V = 1) we obtain the 'pure' or dimensionless set of algebraic expressions:

(U1*U2 + 2)*U6
(U1*U2 + 2)*(1 - U6)
U1*(1 - U2)*U3*U5 + ((1 - U1) + 1)*(1 - U4)
U1*(1 - U2)*U3*(1 - U5) + 1+ ((1 - U1) + 1)*U4
U1*(1 - U2)*(1 - U3)

Simplifying a bit, we get:

(U1*U2 + 2)*U6
(U1*U2 + 2)*(1 - U6)
U1*U3*U5*(1 - U2) + (2 - U1)*(1 - U4)
U1*U3*(1 - U2)*(1 - U5) + 1+ (2 - U1)*U4
U1*(1 - U2)*(1 - U3)

It is necessary to keep in mind the randomness involved in building up and writing these five algebraic expressions above. In any case, for this particular random trajectory of events above, out of five expressions, two are sums (minority), and three are products (majority). In fact, since each full cycle yields **3** newly created balls, **2** of which are multiplicative, namely X(p) and X*(1 – p), and **1** of which is an additive, therefore the statistical tendency of the system after numerous cycles is to have approximately 2/3 multiplicative expressions and 1/3 additive expressions. The reason this is only approximately so is that a term such as X(p) converts X into a product if X itself was additive, but it leaves it as a product if X itself was already a product. In the same vein, a term such as X + Y converts X and Y into a sum if X and Y themselves were products, but it leaves them as a sum if X and Y themselves were already sums.

On the face of it, the existence of additive expressions in about one third of all expressions, does not bode very well for Benford digital configuration; the glass is two-third full and one-third empty, and one doesn't know whether he or she should be happy or should be sad. Clearly, the manifestation of tugs of war that occur between addition and multiplication in all Fragmentation and Consolidation models needs to be investigated further.

An abstract Monte Carlo computer simulation program of C&F scheme utilizing a deterministic fixed p and (1 – p) ratios was run, with 43 identical initial values of V, experiencing 52 full cycles. The program was designed to obtain the random algebraic expressions of the process, but not to perform any numerical calculation, leaving V and p as variables. Additive terms are shown as bold types for emphasis. The results after the 52nd cycle are as follows:

1) $((((V)p)p+(((V)p)(1-p))p)p)p$

2) $((V)(1-p))p$

3) $((V)p)p + (V)$

4) $(V+((V)p)(1-p)+(V)p+(V)p+(V)(1-p))p$

5) $(V+((V)p)p)(1-p)p$

6) $V + V + ((V)p)p + V$

7) $(V+(V)(1-p)+((V)p)(1-p) + ((V+V)p+V)(1-p) + ((((V)p)p+(((V)p)(1-p))p)(1-p))p)p$

8) $(V)p + (V)(p)$

9) $(V)(1-p)$

10) $(V+(V)(1-p)+V)(1-p)$

11) $(V+((V)p)p)p + ((V)(1-p)+V)p$

12) $(V)p$

13) $((V)(1-p))p + V + V + (((V)p)p)(1-p) + V$

14) $(((V)p)p)p + (V)(1-p) + (V)p$

15) $((V+(V)(1-p)+V)p)(1-p)$

16) $((((V)p)(1-p))(1-p))(1-p) + ((((V)p)p+(((V)p)(1-p))p)(1-p))(1-p) + (V)p$

17) $(((V+V)p)(1-p))p$

18) $(V+V)(1-p)$

19) $(V+((V)p)p)(1-p)(1-p)$

20) $(V)(1-p) + V + (V+((V)p)(1-p)+(V)p+(V)p+(V)(1-p))(1-p)$

21) $(V+(V)(1-p)+((V)p)(1-p)+((V+V)p+V)(1-p)+((((V)p)p+(((V)p)(1-p))p)(1-p))p)(1-p)$

22) $(V)p + (V)(1-p)$

23) $(((V)p)(1-p))p$

24) $((((V+V)p)(1-p))(1-p))p$

25) $((V)p)(1-p) + (V)p$

26) $(V)p + ((V)p)p + ((V)(1-p))(1-p)$

27) $((V+(V)(1-p)+V)p)p$

28) $((V)(1-p))p$

29) $((((V)p)p+(((V)p)(1-p))p)p)(1-p)$

30) $((V+V)p+V)p + V + (V)(1-p) + (V)(1-p) + ((V+V)p)p$

31) $(((((V)p)(1-p))(1-p))p)p$

32) $(V)(1-p) + (V)(1-p)$

33) $((V)p)(1-p)$

34) $((V)p+V)p$

35) $(V+V)(1-p) + (V)(1-p)$

36) $(((V)p)(1-p))(1-p)$

37) $((V)(1-p)+V)(1-p) + (V)(1-p) + (V)(1-p) + ((V)p)(1-p) + V + V + ((V)p)p$

38) $((((V+V)p)(1-p))(1-p))(1-p)$

39) $(V)p + ((V)(1-p))(1-p) + ((V)p+V)(1-p)$

40) $((V)(1-p))(1-p)$

41) $((V)(1-p))p$

42) $((V)(1-p))(1-p) + V$

43) $(((((V)p)(1-p))(1-p))p)(1-p)$

Out of 43 final expressions at the end of the 52th cycle, 18 are additive and 25 are multiplicative. Simulation runs with 1000+ balls and with 4000+ full cycles consistency show nearly exact ratios of 2/3 multiplicative expressions and 1/3 additive expressions at the end of the processes.

[4] Tug of War between Addition and Multiplication in C&F Models

Let us reiterate what was discussed in Section 2:

Random multiplication processes induce two essential results:

(A) A dramatic increase in skewness – an essential criterion for Benford behavior.
(B) An increase in the order of magnitude – another essential criterion for Benford behavior.

Hence multiplication processes are highly conducive to Benford behavior.

Random addition processes induce two negative results:

(A) Lacking any increase in skewness, and even actively increasing the symmetry of resultant distribution, with added concentration forming around the center/medium.
(B) Lacking any increase in order of magnitude beyond the existing maximum order of magnitude within the set of added variables.

Hence addition processes are highly detrimental to Benford behavior.

In addition, it is necessary to point out to a significant limitation in the effectiveness of the Central Limit Theorem. The CLT's Achilles' heel - in terms of its rate of convergence to the Normal - is the possibility that added variables are highly skewed and that they come with very high order of magnitude (OOM). This is a bad combination for the CLT. Except for Uniforms, Normals, and other symmetrical distributions which converge to the Normal quite quickly after very few additions regardless of the value of OOM of added variables, all other asymmetrical (skewed) distributions show a distinct rate of convergence depending on their OOM value. For skewed variables, whenever OOM is of very high value, CLT can manifest itself with difficulties, and very slowly, only after a truly large number of additions of these random variables. On the other hand, when skewed variables are of very low OOM value, CLT achieves near Normality quite quickly after only very few additions.

For the Consolidations and Fragmentation model, each fragmentation process contributes to the system two products - each with a minimum of 2 multiplicands and possibly more; similarly each consolidation process contributes to the system an additive expression with a minimum of 2 addends and possibly more; and therefore there exists a tug of war here between additions and multiplications with respect to Benford behavior; a struggle between the Central Limit Theorem and the Multiplicative Central Limit Theorem; between the Normal and the Lognormal. Remarkably, even though Benford frequently loses numerous C&F battles, yet he wins the war

251

in the long run. One existing feature here that is partially saving the system from deviation from the logarithmic is that on average only about one-third of the expressions are additive; and that even within those expressions there are plenty of arithmetical multiplicative elements involved. Surely there are some additive expressions with 3 addends which might be quite detrimental to Benford, but they are far and few between; and there are even more menacing expressions with 4 addends, but luckily these are even rarer.

The general [theoretical] understanding gained in Section 2 regarding multiplication and addition processes enables us to thoroughly explain the [empirical] strong logarithmic behavior in all Consolidations and Fragmentations models; namely the reason addition effects do not manage to significantly retard multiplication effects. In a nutshell, the C&F process is Benford because it uses the high OOM variable of Uniform(0, 1) which contributes to high order of magnitude, skewness, and thus Benfordness. As a consequence, the C&F process encounters the Achilles' heel of the Central Limit Theorem and additions are not very effective. Order of magnitude of the Uniform(0, 1) calculated as $LOG(1/0)$ is infinite. CPOM calculated as $P_{90\%}$ divided by $P_{10\%}$ is 0.9/0.1 or simply 9. Surely, the C&F model cannot use any low OOM variable such as say Uniform(5, 7), because it needs to break a whole quantity into two fractions, and this can only be achieved via the high OOM variable Uniform(0, 1). Such high OOM values, coupled with the fact that the terms within the additive expressions almost always involve also some multiplications (which are always skewed), guarantee that the Central Limit Theorem is very slow to act here and that its retarded rate of convergence does not manage to even begin to ruin the general multiplicative tendencies of the system. Skewness for these multiplicative terms hiding within the additive expressions is guaranteed by virtue of simply being multiplication. All multiplication processes yield highly skewed set of values. Since the vast majority of the additive terms here are with only 2, 3, or 4 addends, the Central Limit Theorem does not even begin to manifests itself.

And what about deterministic models of fixed 50% - 50% ratios or fixed 15% - 85% ratios, etc.?! Even though these deterministic models do not apply the Uniform(0, 1) in any way, the effective results of such deterministic p and $(1 - p)$ ratios are also of high OOM values. Occasional terms such as $V*p*p*(1-p)*(1-p)$, or $V*p*p*p*p$, and so forth, are of very low value since $p < 1$, while a fortunate surviving V without any p factor stays as large as V itself, and all that implies high variability. Even a comparison of $V*p$ with $V*p*p*p*p$ indicates the possibility of sufficiency high OOM in the system since the value of the former is much higher than the value of the latter. For example, even for the fixed balanced 50% - 50% ratios model, variations between possible terms are very high, such as say between $V*0.5$ and $V*0.5*0.5*0.5*0.5 = V*0.0625$, guaranteeing sufficiently high order of magnitude and preventing the Central Limit Theorem from manifesting itself.

For both models, for the random utilizing Uniform(0, 1), and for the deterministic utilizing fixed p and $(1 - p)$ ratios, order of magnitude may not be substantial, but it is sufficient to significantly retard CLT given that CLT has only 2 or 3 addends to work with.

[5] Examination of C&F Processes Via Gradual Run Reveals Saturation

Another run of scheme A is performed in order to demonstrate its rate of convergence to the logarithmic, as well as to demonstrate the digital saturation that occurs after sufficient number of cycles. The scheme starts out with 2000 balls having an initial identical value of 1, and utilizing the Uniform(0, 1) for fragmentations, but instead of running 8000 binary cycles and stopping, here it is extended to 13000 cycles, and the scheme is executed slowly and gradually, in stages, in order to be able to take snapshots occasionally, and to examine how first digits, SSD, and CPOM gradually evolve. Figure 4.7 depicts the computerized results in details. Evidently, it takes the system about 2*L cycles [*namely 2*2000 = 4000 cycles*] to achieve its nearly perfect Benford behavior. Beyond 4000 cycles, nothing new is achieved, although the system vigilantly maintains and guards the logarithmic status that it has thus obtained. CPOM though seems to be growing continuously well beyond these 4000 cycles, although it is possible that it might change course and reverse itself further on perhaps.

# of Cycles	1	2	3	4	5	6	7	8	9	SSD	P_{10}	P_{90}	CPOM
0	100%	0%	0%	0%	0%	0%	0%	0%	0%	5634	1.00000	1.00	1
500	59%	14%	6%	5%	4%	4%	3%	3%	3%	951	0.20131	2.00	10
1000	45%	17%	8%	7%	5%	5%	5%	4%	3%	253	0.09119	2.00	22
1500	38%	18%	11%	8%	6%	5%	5%	5%	3%	79	0.04417	2.29	52
2000	35%	19%	12%	7%	7%	6%	5%	4%	4%	38	0.02965	2.47	83
2500	33%	19%	12%	8%	7%	6%	5%	5%	4%	18	0.02116	2.64	125
3000	33%	19%	12%	9%	7%	6%	5%	5%	4%	14	0.01410	2.69	191
4000	31%	17%	13%	10%	8%	6%	5%	5%	5%	2	0.00915	2.84	311
5000	32%	18%	13%	11%	7%	7%	5%	5%	5%	6	0.00346	2.96	856
6000	31%	18%	12%	11%	7%	6%	5%	5%	4%	5	0.00296	2.97	1003
7000	31%	18%	12%	10%	8%	7%	5%	5%	4%	2	0.00188	2.97	1581
8000	30%	18%	13%	11%	7%	6%	6%	5%	4%	2	0.00121	3.09	2543
10000	29%	19%	13%	10%	8%	5%	6%	5%	5%	5	0.00093	2.99	3209
13000	31%	16%	12%	10%	8%	6%	5%	7%	4%	6	0.00086	2.94	3408

Figure 4.7: The Evolution of First Digits, SSD, and CPOM of a Gradual Run of Scheme A

[6] Related Log Conjecture Justifies Logarithmic Behavior of C&F Models

Examining the histogram of log values of gradual run Scheme A - helps in relating the result of C&F models to Related Log Conjecture discussed in chapter 18 of Section 1. Figure 4.8 depicts this histogram after 8000 cycles. Since the original V value for all the balls was 1, it stands to reason that even after 8000 cycles the histogram is at its highest for values around this point which corresponds to the log value of LOG(1) = 0. Earlier in Figure 4.7 it was shown that after 8000 cycles, P_{10} = 0.00121, and P_{90} = 3.09, corresponding to the log values of LOG(0.00121) = -2.92 and LOG(3.09) = +0.49. Indeed, one can easily ascertain that in Figure 4.8 the core of the data (the central 80% of the data) is approximately between -3 and +0.5. All the requirements of Related Log Conjecture are easily satisfied for the curve in Figure 4.8, especially the requirement of having wide enough span on the log-axis, being at least 5 units here, and which is comfortably more than the usual 3 or 4 units required for strong logarithmic behavior. In conclusion: Benford behavior here is in perfect harmony and consistent with Related Log Conjecture.

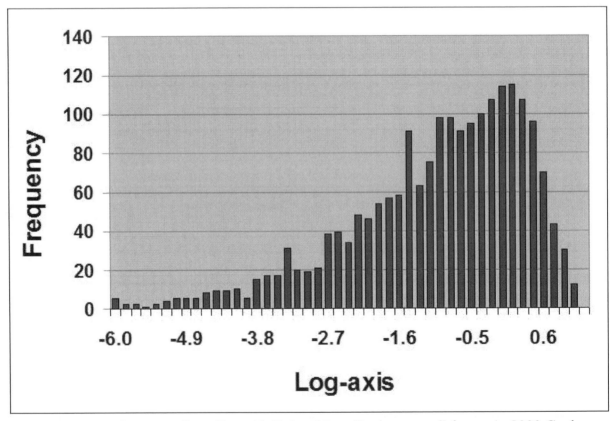

Figure 4.8: Log Histogram Complies with Related Log Conjecture – Scheme A, 8000 Cycles

[7] Partial Convergence to Benford for Models with Few Initial Quantities

Another C&F scheme is run via Monte Carlo computer simulations. The scheme starts out with only 100 balls having an initial identical value of 1, and utilizing the Uniform(0, 1) for fragmentations. It is executed slowly and gradually in stages for a total of 9000 cycles, in order to be able to take snapshots occasionally, and to examine how first digits, SSD, and CPOM gradually evolve. Surely such a meager set of only 100 balls does not have sufficient number of quantities to fully converge to Benford, although the system achieves partial convergence in spite of its small size.

Figure 4.9 depicts the computerized results in details. Evidently, after about 2*L to 3*L cycles (2*100 to 3*100), namely 200 to 300 cycles, the system achieves some kind of digital stability with its convergence to an approximate Benford configuration with SSD stabilizing around the value of about 50. Beyond about 200 or 300 cycles, nothing new is achieved, although the system maintains its approximate logarithmic status it has thus obtained. CPOM on the other hand, grows steadily initially for the first 300 cycles or so; but then it starts reversing itself, fluctuating higher and lower as more cycles are added.

# of Cycles	1	2	3	4	5	6	7	8	9	SSD	P_{10}	P_{90}	CPOM
0	100%	0%	0%	0%	0%	0%	0%	0%	0%	5634	1.00000	1.00	1
50	42%	16%	9%	7%	4%	4%	6%	7%	5%	190	0.07469	2.05	27
100	31%	19%	15%	5%	5%	7%	4%	3%	11%	89	0.04178	2.53	61
150	28%	12%	13%	13%	7%	7%	9%	5%	6%	60	0.03510	2.55	73
200	28%	12%	18%	10%	5%	9%	10%	6%	2%	105	0.01072	3.11	290
250	21%	16%	15%	12%	9%	8%	9%	8%	2%	125	0.00567	3.23	570
300	25%	21%	13%	9%	6%	8%	5%	8%	5%	53	0.00156	2.79	1790
350	27%	22%	9%	14%	4%	8%	4%	7%	5%	84	0.00262	2.77	1059
800	36%	20%	8%	9%	5%	7%	4%	6%	5%	74	0.00034	2.14	6296
2000	28%	16%	17%	7%	9%	7%	8%	7%	1%	57	0.00231	3.18	1377
5000	26%	20%	10%	12%	7%	8%	7%	6%	4%	39	0.00079	3.03	3853
9000	29%	21%	11%	7%	12%	8%	4%	3%	5%	48	0.00087	1.91	2182

Figure 4.9: First Digits, SSD, and CPOM of a Gradual Run Scheme with Only 100 Balls

SECTION 5

PRIME NUMBERS AND DIRICHLET DENSITY

1	2	3	4	5	6	7	8	9	10
11	12	13	14	15	16	17	18	19	20
21	22	23	24	25	26	27	28	29	30
31	32	33	34	35	36	37	38	39	40
41	42	43	44	45	46	47	48	49	50
51	52	53	54	55	56	57	58	59	60
61	62	63	64	65	66	67	68	69	70
71	72	73	74	75	76	77	78	79	80
81	82	83	84	85	86	87	88	89	90
91	92	93	94	95	96	97	98	99	100

[1] The Mysterious Versus the Rational Aspects of Prime Numbers

Prime numbers have intrigued mathematicians for millennia. While the integers are considered to be relatively well-understood, not enough is known about the primes. Indeed there are still many open problems concerning the primes, and this renders them a bit mysterious. Much progress has been made regarding the prime number distribution since the empirical observation that the concentration of primes decreases slightly as bigger integers are considered, starting with Gauss initial conjectures concerning the exact expression of their density among the integers. Although prime numbers are defined deterministically, in the sense that exact rules determine whether an integer is prime or composite, its apparent [distributive] randomness has caused many to speculate about possible stochastic interpretations. On one hand, locally for small ranges of integers, prime numbers seem to be randomly distributed with no other law or pattern than that of chance. On the other hand, globally for large ranges of integers, the distribution of primes tells of a remarkably predictable and steady pattern. This tension between micro local randomness and macro global order renders the distribution of primes a fascinating problem indeed.

[2] Empirical Checks on Distributions of Primes and their First Digits

Prime Numbers are well-known for their paradoxical stand regarding Benford's Law. On one hand they adamantly refuse to obey the law of Benford in the usual sense, namely with regards to normal density measuring the proportion of primes with d as the leading digit. Yet, on the other hand, the Dirichlet density for the subset of all primes with d as the leading digit is indeed LOG(1 +1/d). In this section the superficiality of the Dirichlet density result is demonstrated and explained in terms of well-known and established results in the discipline of Benford's Law, conceptually concluding that prime numbers cannot be considered Benford at all, in spite of the Dirichlet density result. In addition, a detailed examination of the digital behavior of prime numbers shall be outlined, showing a distinct digital development pattern, from a slight preference for low digits at the start for small primes, to a complete digital equality for large primes in the limit as the prime number sequence goes to infinity.

Prime Numbers can be viewed as a dynamic sequence, advancing from 2 forward to higher values, in a conceptually similar sense to exponential growth series or random log walk; as opposed to viewing them as a static collection of unique numbers.

When one compares how the prime numbers sequence advances along the x-axis to exponential growth series, it immediately becomes apparent that it strongly resembles very slow exponential growth rates, rarely passing an Integral Power Of Ten (IPOT) point, and that it keeps slowing down, growing ever more slowly, constantly shifting to lower growth rates. Such state of affairs suggests that any empirical trials regarding primes compliance or non-compliance with Benford's Law should all start and stop at IPOT points. In other words, it is essential that we take the digital pulse of prime numbers on intervals such that an unbiased examination is performed where all possible first digits are equiprobable a priori. A superior approach regarding this boundary issue is not only to pay close attention to IPOT points in all empirical tests of finite subsets of prime numbers, but also to restrict such tests to intervals standing between two adjacent IPOT values, namely intervals on $(10^{INTEGER}, 10^{INTEGER + 1})$, such as (10, 100), (100, 1000), (1000, 10000), $(10^{15}, 10^{16})$, and so forth. To vividly demonstrate the importance of this approach in empirical trials, imagine a digital test performed on all the primes from 2 to 400, or on all the primes in the interval (100, 400). Clearly all such tests are biased, unfairly favoring digits {1, 2, 3}, while strongly discriminating against digits 4 to 9. The results of such misguided tests tell us more about the measuring rod itself (interval chosen) than about the measured object (digital configuration of primes). All such results are worthless as they do not convey any useful information.

Here are some proper empirical results:

Primes between 2 and 100	{16.0, 12.0, 12.0, 12.0, 12.0, 8.0, 16.0, 8.0, 4.0}
Primes between 2 and 1,000	{14.9, 11.3, 11.3, 11.9, 10.1, 10.7, 10.7, 10.1, 8.9}
Primes between 2 and 10,000	{13.0, 11.9, 11.3, 11.3, 10.7, 11.0, 10.2, 10.3, 10.3}
Primes between 2 and 100,000	{12.4, 11.8, 11.4, 11.1, 11.0, 10.6, 10.7, 10.5, 10.5}
Primes between 2 and 1,000,000	{12.2, 11.6, 11.4, 11.1, 11.0, 10.8, 10.7, 10.6, 10.5}

Primes between 10 and 100	{19.0, 9.5, 9.5, 14.3, 9.5, 9.5, 14.3, 9.5, 4.8}
Primes between 100 and 1,000	{14.7, 11.2, 11.2, 11.9, 9.8, 11.2, 9.8, 10.5, 9.8}
Primes between 1,000 and 10,000	{12.7, 12.0, 11.3, 11.2, 10.7, 11.0, 10.1, 10.4, 10.6}
Primes between 10,000 and 100,000	{12.4, 11.8, 11.5, 11.1, 11.0, 10.5, 10.8, 10.5, 10.5}
Primes between 100,000 & 1 million	{12.2, 11.6, 11.4, 11.1, 11.0, 10.8, 10.8, 10.6, 10.5}

[**Note:** the vectors of nine elements refer to proportional percents, starting with digit 1 on the left, and ending with digit 9 on the right; the '%' sign is omitted for brevity.]

A clear picture emerges now regarding digital behavior of the prime numbers. There is a consistent pattern here, slightly favoring low digits, but nothing like the dramatic digital skewness of Benford's Law is found. It also appears as if there is a tendency towards a more equitable digital configuration as higher primes are considered.

For example, the proportions of digit 1 within adjacent IPOT intervals are constantly and monotonically decreasing: 19.0 → 14.7 →12.7 →12.4 →12.2. Does digit 1 aim at the ultimate digital equality of 11.1% (that is 1/9) in the limit as primes go to infinity? Unfortunately, the author has no access to a much bigger list of primes with values beyond one million integers. But this is a blessing in disguise, because in order to learn more about the digital behavior of primes we are forced to employ theoretical reasoning instead of further empirical examinations, and such a path is more durable and reliable, allowing us to probe all primes of whatever sizes, and [virtually] peer deep into infinity.

Figure 5.1 depicts the way the sequence of the 21 prime numbers from 11 to 97 marches along the log-axis. This is done by simply converting these 21 prime numbers into their (decimal) logarithm values and plotting them. The boundary of 11 and 97 is purposely selected so as to stand as close as possible between the IPOT marks of 10 and 100.

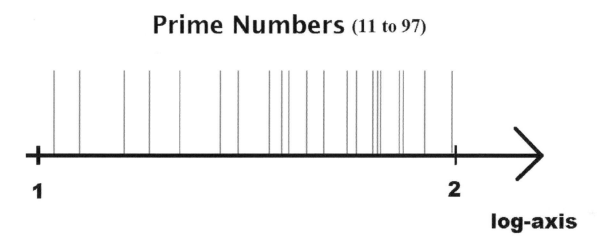

Figure 5.1: Decelerated March Along the log-axis – Primes 11 to 97

If one cannot visually ascertain from Figure 5.1 that this sequence slows down, becoming more and more concentrated towards the right, then the next example of Figure 5.2 would convince anyone of this generic pattern in all plots of logarithms of prime numbers. Figure 5.2 depicts the way the sequence of 143 prime numbers - beginning with 101 and ending with 997 - marches along the log-axis, and where deceleration is quite obvious. A more detailed examination of the log values of these 143 primes is given in Figure 5.3 which focuses separately on the first half and on last half of the entire interval of (2.0, 3.0) for better visualization.

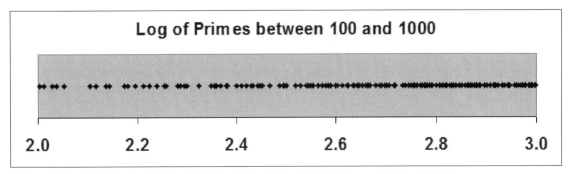

Figure 5.2: Decelerated March Along the log-axis – Primes 100 - 1000

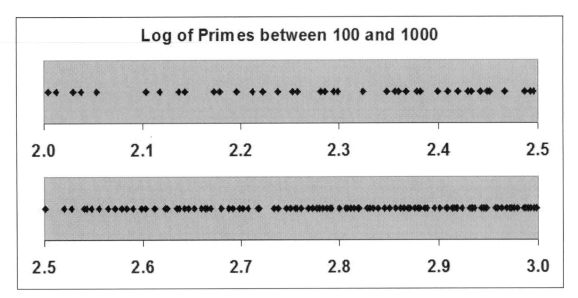

Figure 5.3: Decelerated log-axis March – Primes 100 - 1000 - Detailed

Clearly, distances on the log-axis between consecutive log values of primes generally decrease, and overall points are getting increasing denser. The somewhat 'random' nature of how primes are spread among the integers implies that locally in some less frequent cases distances increase, but globally and most frequently they decrease. The implication of such log pattern is that the density of log of primes is rising on the right. Hence as discussed in Kossovsky (2014) chapter 63, and in particularly as shown in Figure 4.21 there, the expectation here is of an approximate digital equality, or at least of a much milder skewness in favoring low digits – just as was indeed observed and confirmed above empirically. This latest agreement and harmony between the generic theoretical understanding of the entire Benford phenomenon and the particular case of prime numbers (on top of numerous other cases) is highly reassuring.

Figures 5.4 and 5.5 pertain only to the particular interval standing between the adjacent IPOT values 10,000 and 100,000, and one <u>cannot</u> assume that these two figures demonstrate the generic way how all prime numbers behave between any consecutive IPOT points. A probable tendency towards digital equality is a distinct possibility as discussed earlier, and this issue has to be further investigated. At least on this adjacent IPOT interval of (10,000, 100,000), the histogram of the sequence of the primes themselves falls off gently and gradually to the right, while the histogram of the log of primes almost consistently rises.

There are 30 bins being utilized in the construction of the histogram of the prime numbers here, with each bin spanning (100,000-10,000)/30, or 90,000/30, namely the width of 3000 integers. Such global or aggregate perspective on how primes occur is almost smooth, yet it still shows some slight random fluctuations. A much sharper and more refined lens with a narrower bin width of only 50 for example, focusing on how many primes occur every 50 integers, would show many more random and wilder fluctuations in its histogram than the gentle ones seen in Figure 5.4. To mere mortals and humans, the fluctuations in the occurrences of primes do appear somewhat random and chaotic, but to the Goddess of Arithmetic all this appears rational, predictable, and even orderly, and she has proclaimed and determined all this long before the Big Bang noisily erupted up there in the sky some 13.8 billion years ago or so.

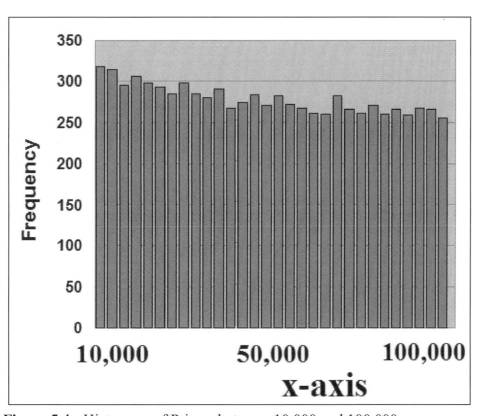

Figure 5.4: Histogram of Primes between 10,000 and 100,000

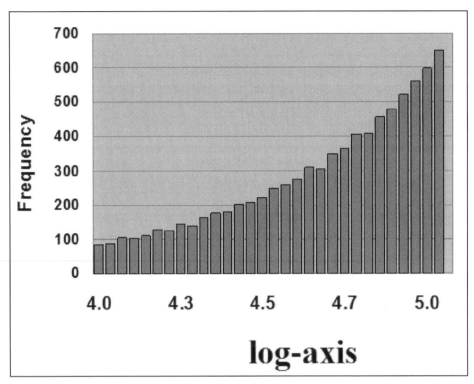

Figure 5.5: Histogram of Log of Primes between 10,000 and 100,000

As discussed in extreme generality in Kossovsky (2014) at the end of chapter 99, sequences that march along the log-axis in a coordinated, organized, and deliberate way with respect to the integer marks, exhibit digital behavior that may not be Benford (unless tiny steps are always taken covering almost all corners and sections of log-axis), while a random and chaotic log march almost always leads to Benford. This statement is merely the grand conceptual guideline, and randomness in the march along the log-axis does not always imply that mantissa is uniform, hence mathematicians are required to provide a rigorous proof for each particular case, showing whether or not it is logarithmic. For the case of the sequence of prime numbers, even though at the micro level the march appears random, at the macro level a clear and decisive pattern is seen of an overall rise along the log-axis between the (global) integer marks, and this pattern guarantees that mantissa is not at all uniform. As shall be demonstrated later, log of primes always rises, in the beginning as well as at infinity; hence Figure 5.5 depicts the true generic shape of all the histograms of log of primes; in contrast to Figure 5.4 of the primes themselves which evolves eventually into flat and uniform histograms at infinity. The huge number of primes residing between each pair of consecutive integer-marks on the log-axis renders those marks definitely global, affecting issues on the macro level. As the sequence of log of the primes passes more and more integer marks on its long march to infinity, mantissa is distributed in a rather skewed manner on its defined range of (0, 1), rising from 0 all the way to 1, depriving low digits of their usual strong Benfordian advantage, and granting high digits (almost and eventually complete) equitable proportions.

[3] Prime Number Theorem Hinting at Digital Configuration

Let us now rigorously prove that the leading digits of prime numbers are equally distributed in the limit as they approach infinity, and also demonstrate that the generic density of log of prime numbers is continuously rising throughout on the macro level, from log(2) all the way to log(infinity). We shall establish both results by deducing them from the asymptotic law of distribution of prime numbers, also widely known as '**The prime Number Theorem**". The function $\pi(x)$ is defined as the number of primes less than or equal to x. Carl Friedrich Gauss is generally credited with first conjecturing that $\pi(x)$ is asymptotically x/ln(x). The notation ln(x) refers to the natural logarithm of x with base e. The notation log(x) refers to the decimal logarithm of x with base 10. Gauss arrived at the conjecture after receiving a book of tables of logarithms as a present at the age of about 15 years in 1792. The book included as an appendix a table of primes, intended just as a mathematical curiosity, and Gauss was able to make this unexpected connection between primes and logarithms. Later in his life Gauss refined his estimate to $\int 1/\ln(t)dt$ with limits of integration running from 2 to x.

$\pi(x)$ = Number of Primes up to x

$$\pi(x) \approx \int_{2}^{x} \frac{1}{\ln(t)} \, dt$$

This integral is strongly suggestive of the notion that the 'density' of primes in the approximate vicinity around t should be 1/ln(t). By 'density' we mean the proportion of consecutive integers centered around t that are primes. This is <u>not</u> the classic definition of density in mathematical statistics where total area sums to 1, and where (dx)*(density) at x indicates the (finite) proportion of all values falling within the tiny interval from x to
x + dx. This is the case for the prime numbers since there are infinitely many of them, and one cannot state that a certain proportion of all the primes falls within any finite
x-axis subinterval. A better perspective about 1/ln(x) is to think of it as the 'Relative Density', comparing counts of primes for a variety of ranges on the x-axis.

Prime Density = Proportion of primes within a range of consecutive integers
Prime Density = (Primes) / (Composites + Primes)
Prime Density = Prime Count / Integer Range
Prime Count = Integer Range * Prime Density

Around the integer 373, the supposed 'density' of primes is 1/ln(373), or 0.169, meaning that about **16.9%** of the integers around 373 are expected to be primes. Indeed, the primes in this vicinity are {349, **353, 359, 367, <u>373</u>, 379, 383, 389**, 397}. Hence from around 351 to around 393, having an interval the size of (393 – 351) = 42, we have 7 primes, and that yields the ratio 7/42, namely **16.7%**. In other words, out of 42 integers with the potential of being primes only 7 turned out to be so, namely only 16.7% of them.

Around the integer 1709, the supposed 'density' of primes is 1/ln(1709), or 0.134, meaning that about **13.4%** of the integers around 1709 are expected to be primes. Indeed, the primes in this vicinity are {1669, **1693, 1697, 1699, <u>1709</u>, 1721, 1723, 1733**, 1741}. Hence from around 1683 to around 1737, having an interval the size of (1737 – 1683) = 54, we have 7 primes, and that yields the ratio 7/54, namely **13.0%**. In other words, out of 54 integers with the potential of being primes only 7 turned out to be so, namely only 13.0% of them.

The two examples above should be further refined by considering wider ranges around the integers, instead of the narrow focus of only about 40 to 50 consecutive integers. The wider the range, the less randomness and fluctuations are observed, and the asymptotic law of distribution of prime numbers is better applied. A good choice perhaps is about 100 consecutive integers, where a remarkable fit is obtained between the theoretical prime density of 1/ln(x), and the empirically observed density. Figure 5.6 depicts the rather good fit between the theoretical 1/ln(x) density and the empirical (actual) prime density for the crude and narrow range of only 27 consecutive integers, examining 548 primes, from the prime 47 to the prime 4079.

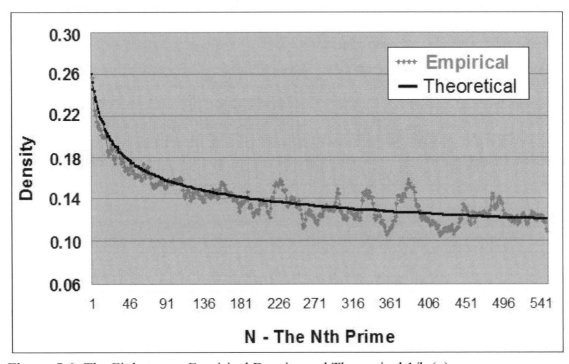

Figure 5.6: The Fit between Empirical Density and Theoretical 1/ln(x)

The above discussion neatly explains why the expression **x/ln(x)** conjectured by young Gauss for the total number of primes up to x consistently underestimates actual counts of primes; always falling short. This is so because the density of primes varies locally, falling as we consider higher integers, being inversely proportional to ln(x). The crude use of x/ln(x) estimates the entire number of integers from 2 to x as simply x, approximating (x - 2) as x, which is actually correct for large x. But it simple-mindedly and lazily estimates the density as a constant 1/ln(x) throughout, from 2 to x, although in reality the density at the last integer x is at its weakest point. Thus (x)*(1/ln(x)) is crudely expressing [Integer Range]*[Prime Density] for the value of [Prime Count]. Since density of primes for all integers less than x is a bit greater than 1/ln(x), such as in 1/ln(x − 568) or 1/ln(x − 37912), and so forth, this crude estimate is always less than the actual count of primes. Gauss the adult, considering the definitive integral, is willing to be diligent and break up the entire interval into many locals ones to account for the variability in the density and the fact that density is higher for lower integers, hence obtaining a much closer fit to the actual count of primes.

From the expression of the density of primes 1/ln(x) one can deduce that ultimately in the limit at infinity first digits are equality distributed within each interval standing between adjacent IPOT values, namely between $(10^N, 10^{(N + 1)})$, where N is an integer.

Value of prime density at the left-most point: $1/ln(10^N)$
Value of prime density at the right-most point: $1/ln(10^{(N + 1)})$

Ratio of left-most point density to right-most point density is:

$1/ln(10^N) / 1/ln(10^{(N + 1)}) = ln(10^{(N + 1)}) / ln(10^N)$

Transforming from base e to base 10 via $LOG_A(X) = LOG_B(X) / LOG_B(A)$:

$ln(10^{(N + 1)}) / ln(10^N) = log(10^{(N + 1)})/log(e) / log(10^N)/log(e) = log(10^{(N + 1)}) / log(10^N) =$
$(N+1)*log(10) / N*log(10) = (N+1)*1 / N*1 = $ **$(N + 1)/N$**

In the limit as primes go to infinity, N also goes to infinity albeit reluctantly and 'much more slowly' (N constitutes approximately the exponent of the largest primes on the way to infinity, not the values of the primes themselves) hence the ratio (N + 1)/N is 1 since the term 1 in the numerator is negligible in the grand scheme of things. In conclusion: leading digit distribution of prime numbers is asymptotically uniform on adjacent IPOT intervals, and therefore leading digit distribution of prime numbers is asymptotically uniform on the entire range to infinity.

Another way of looking at the density of primes is to place its gently falling curve within the series of the following falling curves in order of steepness:

$1/x^3$ $1/x^2$ $1/x$ 1/square-root(x)   1/cube-root(x) $1/\ln(x)$ 1/constant

One can obtain all seven derivatives and make a comparison. Clearly $1/x^3$ is falling the steepest, the fastest, and 1/constant is falling the least; indeed, 1/constant is a uniform flat line, not falling at all. The threshold from which a definite integral from any real point k to infinity diverges is from $1/x$, and all the other curves on its right diverge as well, because in the limit their curves approach a flat line, implying an infinite area to infinity.

The author has access to a list of primes only for the first one million integers (i.e. only up to the integer 10^6), in other words, access to the first 78,498 primes only. In order to obtain digital configurations for higher integers, the expression of the density of primes $1/\ln(x)$ shall be applied, thus enabling us to examine with very high accuracy estimates of the digital proportions within adjacent IPOT intervals of high primes.

Each $(10^N, 10^{(N+1)})$ interval shall be subdivided into nine equally-spaced sub-intervals according to where each digit d leads, namely into the following nine sub-intervals:
$(1*10^N, 2*10^N), (2*10^N, 3*10^N), (3*10^N, 4*10^N), \ldots, (9*10^N, 10*10^N)$.
The prime density for each sub-interval shall be evaluated in its middle, namely at
$\{1.5, 2.5, 3.5, 4.5, 5.5, 6.5, 7.5, 8.5, 9.5\}*10^N$. Using the generic expression discussed earlier
[Number of primes within an interval] = [interval's width]*[density],
the number of primes within say $(1*10^N, 2*10^N)$ for digit 1 is obtained as in:
[interval's width]*[density] = $[2*10^N - 1*10^N]*[1/\ln(1.5*10^N)] = 10^N / \ln(1.5*10^N)$.

In general, for any digit d, the number of primes within the sub-interval is:

$[(d+1)*10^N - (d)*10^N]*[1/ \ln((d + 1/2)*10^N)] =$

$[1*10^N]*[1/ \ln((d + 1/2)*10^N)] =$

$$10^N / \ln((d + 1/2)*10^N)$$

Now we have an almost perfectly precise tool to obtain the digital configurations on any IPOT interval for large primes. Figure 5.7 depicts digit distributions for a variety of IPOT intervals, omitting the first one from 1 to 10. The first 5 rows in Figure 5.7, namely from 10 to 10^6, are the actual digit distributions. From the 6th row on, namely from 10^6 and up, the distributions are the estimates from $10^N/\ln((d + 1/2)*10^N)$. Hence Figure 5.7 is comprised of actual and estimated distributions. The high accuracy of the estimates motivated this mixing. For example, for the interval $(10^5, 10^6)$ results are:

Actual: {12.2%, 11. 6%, 11.4%, 11.1%, 11.0%, 10.8%, 10.8%, 10.6%, 10.5%}
Estimate: {12.2%, 11.7%, 11.4%, 11.1%, 11.0%, 10.8%, 10.7%, 10.6%, 10.5%}

Namely that with one decimal point precision, and considering primes over 10^6, digital proportions are practically identical for actual and estimate values. The higher the IPOT interval considered, the closer is this fantastic agreement between actual and estimate, hence all the values for rest of Figure 5.7 over 10^6 can be taken as actual ones as well. In other words: everything in the entire table of Figure 5.7 is actual! Everything is real!

	1	2	3	4	5	6	7	8	9
[10 , 100]	19.0%	9.5%	9.5%	14.3%	9.5%	9.5%	14.3%	9.5%	4.8%
[100 , 1000]	14.7%	11.2%	11.2%	11.9%	9.8%	11.2%	9.8%	10.5%	9.8%
[1000 , 10000]	12.7%	12.0%	11.3%	11.2%	10.7%	11.0%	10.1%	10.4%	10.6%
[10^4 , 10^5]	12.4%	11.8%	11.5%	11.1%	11.0%	10.5%	10.8%	10.5%	10.5%
[10^5 , 10^6]	12.2%	11.6%	11.4%	11.1%	11.0%	10.8%	10.8%	10.6%	10.5%
[10^6 , 10^7]	12.0%	11.6%	11.3%	11.1%	11.0%	10.9%	10.8%	10.7%	10.6%
[10^7 , 10^8]	11.9%	11.5%	11.3%	11.1%	11.0%	10.9%	10.8%	10.7%	10.7%
[10^8 , 10^9]	11.8%	11.5%	11.3%	11.1%	11.0%	10.9%	10.9%	10.8%	10.7%
[10^9 , 10^{10}]	11.7%	11.4%	11.3%	11.1%	11.0%	11.0%	10.9%	10.8%	10.8%
[10^{10} , 10^{11}]	11.7%	11.4%	11.2%	11.1%	11.0%	11.0%	10.9%	10.8%	10.8%
[10^{11} , 10^{12}]	11.6%	11.4%	11.2%	11.1%	11.0%	11.0%	10.9%	10.9%	10.8%
[10^{12} , 10^{13}]	11.6%	11.4%	11.2%	11.1%	11.1%	11.0%	10.9%	10.9%	10.9%
[10^{13} , 10^{14}]	11.5%	11.3%	11.2%	11.1%	11.1%	11.0%	11.0%	10.9%	10.9%
[10^{14} , 10^{15}]	11.5%	11.3%	11.2%	11.1%	11.1%	11.0%	11.0%	10.9%	10.9%
[10^{15} , 10^{16}]	11.5%	11.3%	11.2%	11.1%	11.1%	11.0%	11.0%	10.9%	10.9%
[10^{16} , 10^{17}]	11.5%	11.3%	11.2%	11.1%	11.1%	11.0%	11.0%	10.9%	10.9%
[10^{17} , 10^{18}]	11.4%	11.3%	11.2%	11.1%	11.1%	11.0%	11.0%	11.0%	10.9%
[10^{18} , 10^{19}]	11.4%	11.3%	11.2%	11.1%	11.1%	11.0%	11.0%	11.0%	10.9%
[10^{19} , 10^{20}]	11.4%	11.3%	11.2%	11.1%	11.1%	11.0%	11.0%	11.0%	10.9%
[10^{20} , 10^{21}]	11.4%	11.3%	11.2%	11.1%	11.1%	11.0%	11.0%	11.0%	11.0%
[10^{21} , 10^{22}]	11.4%	11.3%	11.2%	11.1%	11.1%	11.0%	11.0%	11.0%	11.0%
[10^{143} , 10^{144}]	11.2%	11.1%	11.1%	11.1%	11.1%	11.1%	11.1%	11.1%	11.1%
[10^{144} , 10^{145}]	11.1%	11.1%	11.1%	11.1%	11.1%	11.1%	11.1%	11.1%	11.1%

Figure 5.7: Digital Distributions of Prime Numbers within IPOT Intervals

The table in Figure 5.7 depicts the mild and inverse digital development pattern of sorts for the prime numbers - conceptually akin to the dramatic digital development pattern observed for all random Benford data where digits start out with an approximate digital equality on the left for low values, and end up on the far right for high values with extreme digital inequality. Here for prime numbers, we observe the reversal of the normal positive development of random data, as digits start out with mildly skewed digital configuration, develop negatively, and end up with total digital equality at infinity.

[4] Density of the Logarithms of Prime Numbers

An expression for the 'density' of log of primes shall now be derived applying the Prime Number Theorem. As with $1/\ln(x)$ 'density' for primes which is not a full-fledge density as in mathematical statistics, the density of log of primes to be derived here is also not a full-fledge density in the usual sense, rather it is simply defined as the number of existing log(primes) within 1 unit of log-axis. Surely, if that unity range on the log-axis happened to be bordered exactly by two integers, such as (2, 3) for example, then its mirror-image on the x-axis is an IPOT interval, such as (100, 1000) for the example above.

Let us consider the set of all primes residing within the sequence of consecutive integers from [x] to [x + M - 1] inclusively, where M is a positive whole number.

The estimated number of primes within the integers from [x] to [x + M - 1] is simply the local density of primes times the number of integers in this range (its length).
It should be noted that integer 5 for example occupies one whole unit on the x-axis, namely from 5.0 to 6.0, hence the range of consecutive integers from [x] to [x + M - 1] is from the **left**-most corner of integer x to the **right**-most corner of integer x + M - 1, namely the length of the real x-axis interval (x, x + M) which is $x + M - x$, or simply M.

Prime Count = Integer Range * Prime Density
Prime Count = $[x + M - x] * (1/ \ln(x + M/2))$
Prime Count = $M/\ln(x + M/2)$

The approximate midpoint x + M/2 standing roughly between x and x + M is used to evaluate the prime density $1/\ln(t)$.

The number of log(primes) from log(x) to log(x + M) on the log-axis is the same as in the above expression, namely $M/\ln(x + M/2)$. Hence Log Density of Primes is:

Log Density of Primes = Count of log(Primes) / Range on the log-axis
Log Density of Primes = $M/\ln(x + M/2) / [\log(x + M) - \log(x)]$
Log Density of Primes = $M/\ln(x + M/2) / \log((x + M)/(x))$
Log Density of Primes = $M/\ln(x + M/2) / \log(1 + M/x)$

$$\text{Log Density of Primes} = \frac{M}{\ln(x + M/2) * \log(1 + M/x)}$$

$$\text{Log Density of Primes} = \frac{x}{\ln\left(x + \frac{M}{2}\right) * \left(\frac{x}{M}\right) * \log\left(1 + \frac{M}{x}\right)}$$

$$\text{Log Density of Primes} = \frac{x}{\ln\left(x + \frac{M}{2}\right) * \log\left(\left(1 + \frac{M}{x}\right)^{\frac{x}{M}}\right)}$$

$$\text{Log Density of Primes} = \frac{x}{\ln\left(x + \frac{M}{2}\right) * \log\left((1 + 1/(\frac{x}{M}))^{\frac{x}{M}}\right)}$$

Now we let M go to zero from above to get the log density in the very immediate vicinity of x. Euler's classic result $\lim_{N \to \infty} \left(1 + \frac{1}{N}\right)^N = e$ enables us to determine that in the limit as M goes to zero – and by implication as x/M goes to infinity – log density is:

$$\text{Log Density of Primes} = \frac{x}{\ln(x) * \log(e)}$$

$$\text{Log Density of Primes} = \frac{x}{\left[\frac{\log(x)}{\log(e)}\right] * \log(e)}$$

$$\text{Log Density of Primes} = \frac{x}{\log(x)}$$

Clearly density of log of primes is monotonically increasing as seen from the expression above [being proportional to x, and inversely proportional only to log(x)], hence the observed near/total digital equality of the sequence of the prime numbers themselves.

Admittedly, for a very small range of integers M, the prime density $1/\ln(x)$ is not so meaningful, as the random overtakes the deterministic and chaos rules over order, much less so for M = 1 or for M being a fraction; for how can one concoct a whole prime out of a fraction of an integer?! A fraction of an integer can never be a whole prime! In any case, one must always keep in mind that the so called 'density' here is simply the <u>fraction</u> of integers that are primes. Yet, letting M approach zero in the limit is an appropriate mathematical procedure here since the focus is temporarily being shifted to the nature of the abstract curves, curves or 'densities' from which the seemingly random occurrences of primes can be deduced and predicted.

Interestingly, even though $1/\ln(x)$ is not a proper density probability function, nonetheless if the Monotonic Transformation Technique in mathematical statistics is used here, the results are identical. The notational convention here is as in John Freund book "Mathematical Statistics", 6th Edition, Section 7.3, Theorem 7.1.

$f(x) = 1/\ln(x)$

$y = u(x) = \log(x)$

$x = w(y) = 10^y$

$g(y) = f[w(y)]*|w'(y)|$

$g(y) = [1/\ln(10^y)]*|(10^y)'|$

$g(y) = [1/\ln(10^y)]*|10^y*\ln(10)|$

$g(y) = [1/\ln(x)]*x*\ln(10)$

$g(y) = [1/[\log(x)/\log(e)]]*x*[\log(10)/\log(e)]$

$g(y) = [\log(e)/\log(x)]*x*[1/\log(e)]$

$g(y) = \mathbf{x/\log(x)}$

Figure 5.8 depicts the good fit between the theoretical x/log(x) density of log of primes and empirical density of log of primes for the crude and narrow range of 27 consecutive integers, examining 898 primes, from the prime 47, to the prime 7109.

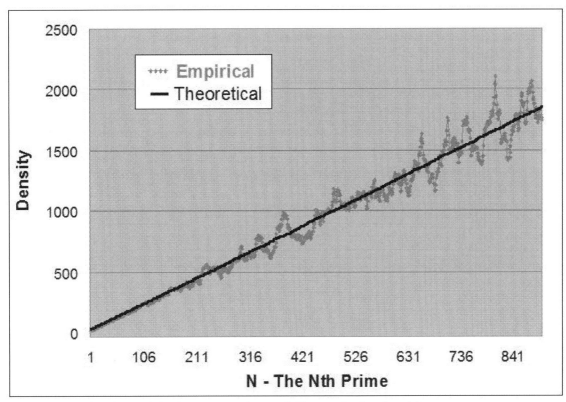

Figure 5.8: Fit of Empirical Log Density and Theoretical x/log(x)

The expression x/log(x) for the density of log of primes is defined as the number of log of primes per 1 unit of log-axis range; hence in essence, it means the number of primes per IPOT interval. Let us evaluate this density expression in the middle of each of the first 6 IPOT intervals from 2 to 10^6 log-wise, namely at 0.5, 1.5, 2.5, 3.5, 4.5, 5.5. The results show a reasonable fit of theoretical with empirical results as follow:

Theoretical count of primes for the six intervals: {6, 21, 126, 904, 7027, 57496}
Empirical count of primes for the six intervals: {4, 21, 143, 1061, 8363, 68906}

For example, for the interval (100, 1000), mid log value is 2.5, corresponding to $10^{2.5}$ or 316.2. Hence log density is evaluated as 316.2/log(316.2), or **126**. This is not too far from the actual number of **143** primes that happened to fall within (100, 1000).

[5] Dirichlet Density for Subsets of Primes

Interestingly, even though prime numbers are not Benford at all, the logarithmic proportions LOG(1 + 1/d) nonetheless pops out here indirectly, as if accidently, not regarding the proportions of numbers with leading digit d, but in another very different sense.

In mathematics, the **Dirichlet density** (also known as the **analytic density**) of a subset of the prime numbers, is a measure of the size of that subset (in comparison to the size of the entire set of all the primes). This measure is easier to use than the natural density.

If A is a subset of the prime numbers, the Dirichlet density of A is the limit (if it exists) of the following ratio as the power s approaches 1 from above:

$$\lim_{s \to 1+} \frac{\sum_{p \in A} \frac{1}{p^s}}{\sum_{p} \frac{1}{p^s}}$$

In other words:

$$\lim_{s \to 1+} \frac{\sum_{only\ the\ primes\ in\ subset\ A} \frac{1}{p^s}}{\sum_{all\ the\ primes} \frac{1}{p^s}}$$

Clearly, the denominator contains the numerator; the numerator is less than or equal the denominator; so that this limit-ratio - if it exists - is less than or equal to 1, and at a minimum it is zero, but it is never negative. Hence Dirichlet density is in [0, 1].

For example, the Mersenne subset of the prime numbers are those primes that are one less than an integral power of two, namely $2^N - 1$, N being an integer. The first seven **Mersenne primes** are $\{2^2 - 1, 2^3 - 1, 2^5 - 1, 2^7 - 1, 2^{13} - 1, 2^{17} - 1, 2^{19} - 1\}$, namely: $\{3, 7, 31, 127, 8191, 131071, 524287\}$. They appear to be quite rare, thinly spread among the entire set of all primes. The Dirichlet density for Mersenne primes is then:

$$\lim_{s \to 1+} \frac{\frac{1}{3^s} + \frac{1}{7^s} + \frac{1}{31^s} + \frac{1}{127^s} + \frac{1}{8191^s} + \cdots to\ \infty}{\frac{1}{2^s} + \frac{1}{3^s} + \frac{1}{5^s} + \frac{1}{7^s} + \frac{1}{11^s} + \frac{1}{13^s} + \frac{1}{17^s} + \frac{1}{19^s} + \cdots to\ \infty}$$

The value of this limit-ratio is zero, indicating conceptually that there are very few Mersenne primes within the entire set of the primes and that they are quite rare indeed. Even though we are dealing with concepts such as hierarchy of infinities and a limit process, a very informal or simple-minded assessment of the rarity of Mersenne primes within all the primes can be given by the extremely low percentage of their occurrences up to the finite integer 524287, namely (7)/(524287), or merely 0.0013%!

[6] Modified Dirichlet Density

As another example, assuming one could obtain the limit as s goes to 1 from above by direct substitution of 1 for s, then for the consideration of the subset of all the primes with 3 as the first leading digit, the Dirichlet density is calculated simply as:

$$\frac{\frac{1}{3} + \frac{1}{31} + \frac{1}{37} + \frac{1}{307} + \frac{1}{311} + \frac{1}{313} + \cdots \; to \; infinity}{\frac{1}{2} + \frac{1}{3} + \frac{1}{5} + \frac{1}{7} + \frac{1}{11} + \frac{1}{13} + \frac{1}{17} + \frac{1}{19} + \cdots \; to \; infinity}$$

The informal and simplistic expectation here is of a non-zero ratio given the substantial 11.1% proportion approximately that primes between integral powers of ten occur having 3 as their first digit [i.e. eventual digital equality within IPOT intervals.]

But direct substitution of 1 for s in the above estimation of the limit is a suspect, since it has not been shown mathematically that plugging in 1 for s is actually the limit itself. Let us demonstrate the challenges and issues involved in attempting to prove this:

The division law of limits:

If the limits of $\lim_{x \to c} f(x)$ and $\lim_{x \to c} g(x)$ both exist, and limit of $\lim_{x \to c} g(x) \neq 0$, then:

$$\lim_{x \to c} \frac{f(x)}{g(x)} = \frac{\lim_{x \to c} f(x)}{\lim_{x \to c} g(x)}$$

The addition law of limits:

If the limits of $\lim_{x \to c} f(x)$ and $\lim_{x \to c} g(x)$ both exist, then:

$$\lim_{x \to c} [f(x) + g(x)] = [\lim_{x \to c} f(x)] + [\lim_{x \to c} g(x)]$$

The Exponential Function $q(x) = B^x$ is continuous at a minimum for all $x \geq 0$ and for all $B \geq 1$ (a condition that all prime numbers satisfy within the repeated p^s expressions of the Dirichlet density, since $p_j \geq 2$). Hence the limit of B^x as x approaches 1 (from above or from below and in general) certainly exists, and it can be obtained by direct substitution. B^x is also always a positive non-zero quantity for all those B and x values. From the division law of limits it follows that for the function $h(s) = \dfrac{1}{p^s}$ where $p \geq 1$ and $s \geq 0$, $\lim_{s \to 1+} \dfrac{1}{p^s} = \dfrac{1}{p^1}$; namely that direct substitution is allowed; and that this limit certainly exists. Consequently, by the addition law of limits, each part of the Dirichlet density separately, namely the numerator as well as the denominator, can be evaluated stepwise, via infinitely many direct substitutions of 1 for s. Yet, unfortunately, the limit of the numerator as well as the limit of the denominator do <u>not</u> exist, since each one diverges (adding infinitely many 1/p terms which never approaches zero close enough yields an ever increasing sum, namely it's infinite). Surely the denominator is decisively a positive non-zero quantity since all prime numbers are positive and there are only addition terms within the Dirichlet expression, not involving any subtractions. Yet, the division law of limits cannot be utilized here for the limit of main division in the Dirichlet density since it's in an indeterminate form where both the numerator and the denominator diverge. Unfortunately, L'Hôpital's rule does not help here either.

In any case, the above simplification – from a limiting process to direct substitution of 1 for s – whether it is justified mathematically in the infinite Dirichlet case or not, shall be utilized in the definition (*hence correctly*) of a modified finite form of Dirichlet in order to gain better theoretical understanding of the forces at play here leading to the logarithmic proportions.

In addition, the quest to perform an approximate computerized test for the Dirichlet density is frustrated since the expression involves an infinite number of primes in the denominator as well as an infinite number of primes with d as the leading digit in the numerator. The suggested next step then in modifying Dirichlet density is to perform partial Dirichlet densities with s = 1 in a piecemeal and finite manner, one integral-power-of-ten-interval at a time, considering a ratio pertaining to two subsets, that of the denominator being the set of all the primes in $(10^{\text{INTEGER}}, 10^{\text{INTEGER}+1})$, and that of the numerator being the smaller subset of all the primes with d as the leading digit in $(10^{\text{INTEGER}}, 10^{\text{INTEGER}+1})$. Such an approach would enable us to examine how Dirichlet density works locally at each IPOT interval level. This ratio shall be termed 'The Modified Dirichlet Density'.

MDD(N, d) ≡ Modified Dirichlet Density(N, d) ≡

$$\frac{\sum_{primes\ with\ leading\ digit\ d\ in\ (10^N,\ 10^{N+1})} \frac{1}{p}}{\sum_{all\ the\ primes\ in\ (10^N,\ 10^{N+1})} \frac{1}{p}}$$

It is hoped that a steady and clear pattern would emerge in these proportions across all IPOT intervals, a pattern which could serve as an indication of the overall proportions for all the prime numbers between 2 and infinity, namely regarding the original all inclusive definition of Dirichlet Density.

Having access to the actual set of all the primes from 2 to 1,000,000 enables the author to perform such calculations for five IPOT intervals. The initial interval of (2, 10) is omitted as being an outlier of sorts, not having enough primes to show any stable pattern.

Empirical results for Modified Dirichlet Density for N values {1, 2, 3, 4, 5} are as follows (one detailed example of the calculations of **44.6%** is given in next page):

Primes 10 to 100   {44.6, 12.4, 9.5, 11.0, 5.7, 5.0, 6.5, 3.7, 1.6}   SSD = 265.6
Primes 100 to 1000   {37.0, 16.1, 11.6, 9.6, 6.4, 6.3, 4.7, 4.5, 3.7}   SSD = 55.1
Primes 1000 to 10,000   {33.1, 18.1, 12.1, 9.3, 7.3, 6.3, 5.0, 4.5, 4.1}   SSD = 11.2
Primes 10,000 to 100,000   {32.3, 18.0, 12.4, 9.3, 7.6, 6.1, 5.4, 4.6, 4.2}   SSD = 6.3
Primes 100,000 to 1,000,000 {32.0, 17.9, 12.4, 9.4, 7.6, 6.3, 5.4, 4.7, 4.2}   SSD = 4.6
Benford's Law first digits: **{30.1, 17.6, 12.5, 9.7, 7.9, 6.7, 5.8, 5.1, 4.6}   SSD = 0**

[Note: The vector of nine elements refers to the values of Modified Dirichlet Density starting with digit 1 on the left, and ending with digit 9 on the right. These values are actually ratios, but thought of as percent, while the '%' sign is omitted for brevity.]

Eureka! An apparent pattern converging to Benford is found <u>locally</u> on each finite interval between adjacent IPOT.

It is reasonable to conjecture that this pattern should persist for higher primes, consistently and gradually approaching the logarithmic proportions in the limit
as primes go to infinity.

Remarkably, this peculiar calculation for each of the 9 digits leads directly to the Benford proportions of LOG(1 + 1/d). It must be emphasized that we are not calculating proportions of primes with d as the leading digit, but rather the sum of the reciprocals of primes with digit d leading, divided by the sum of the reciprocals of all those primes regardless of which d is leading, and all within one IPOT interval. Clearly, this is not just another manifestation of Benford's Law, but rather just a curious coincidence perhaps.

It would be a fallacy to state that: "**The prime numbers are Benford Dirichlet-wise**". They are not! It is only this particular expression concocted out of the primes that somehow happened to be LOG(1 + 1/d).

One detailed example of the empirical calculations above:

The entire set of all the primes between 10 and 100 is:
{11, 13, 17, 19, 23, 29, 31, 37, 41, 43, 47, 53, 59, 61, 67, 71, 73, 79, 83, 89, 97}.

Our modified piecemeal Dirichlet density for digit 1 is calculated as follows:

$$\text{MDD}(1, 1) = \frac{\Sigma \ primes \ in \ (10, \ 100) \ with \ 1 \ as \ the \ leading \ digit \ \frac{1}{p}}{\Sigma \ all \ the \ primes \ in \ (10, \ 100) \ \frac{1}{p}}$$

$$\frac{\dfrac{1}{11} + \dfrac{1}{13} + \dfrac{1}{17} + \dfrac{1}{19}}{\dfrac{1}{11} + \dfrac{1}{13} + \dfrac{1}{17} + \dfrac{1}{19} + \dfrac{1}{23} + \ ... \ + \dfrac{1}{83} + \dfrac{1}{89} + \dfrac{1}{97}}$$

$$\frac{0.0909 + 0.0769 + 0.0588 + 0.0526}{0.0909 + 0.0769 + 0.0588 + 0.0526 + 0.0435 + \ ... \ + 0.0120 + 0.0112 + 0.0103}$$

$$\frac{0.2793}{0.6266} = 0.446 = 44.6\%$$

[7] Explaining Modified Dirichlet Connection to Benford

In order to take the mystery out of this Modified Dirichlet surprising result, let us examine the actual quantity being calculated here, part of which is simply the sum of reciprocals of primes, first digit by first digit. Little reflection is needed to realize that this should yield a decisive advantage for low digits. This follows from the fact that within any particular interval bounded by adjacent IPOT points, a number with a low leading digit implies low quantity, and a number with a high leading digit implies high quantity, namely that leading digits and actual quantities positively correlate (Kossovsky (2014), chapter 143 titled "Digits Serving as Quantities in Benford's Law").

Hence on the IPOT interval (10, 100): the reciprocals of the primes **11** or **23** (with low first digits 1 and 2) yield decisively higher values than the reciprocals of the primes **83** or **97** (with high first digits 8 and 9).

And on the IPOT interval (100, 1000): the reciprocals of the primes **101** or **211** (with low first digits 1 and 2) yield decisively higher values than the reciprocals of the primes **887** or **997** (with high first digits 8 and 9).

Primes within IPOT intervals start out with slightly skewed digital proportions in favor of low digits, but are not too far from equality. In the limit going to infinity, primes on such intervals end up with digital equality, namely that the local histogram or density of primes at infinity is uniform with a flat curve. Hence with no defense given to high digits say in the form of occurring more often than low digits, merely summing reciprocals along digital lines implies a clear advantage for low digits. The only question is by how much exactly do low digits gain over high digits in this reciprocal game?

Between 100 and 1000, the reciprocal of the prime 151 for example is 1/151 or 0.0066, while the reciprocal of the prime 953 for example is 1/953 or 0.0010. Hence the reciprocal of 151 is **6.31** times as large as the reciprocal of 953! Adding slightly to this dichotomy is the fact that early on (before we hit infinity) lower digits occur a bit more frequently (14.7% for digit 1 on the interval [100, 1000]) than high digits (9.8% for digit 9 on the interval [100, 1000]), hence in this example, low digit 1 occurs approximately (14.7)/(9.8) or **1.50** times more frequently than high digit 9. The confluence of these two factors [higher reciprocals and higher frequencies] equals roughly (6.31)*(1.50), or 9.46, being the factor by which reciprocals of digit 1 have an advantage over reciprocals of digit 9, and which is quite close to the actual digit 1 to digit 9 ratio (37.0)/(3.7) or 10.0 of the Modified Dirichlet density empirically calculated above. The classic Benfordian advantage is LOG(1 + 1/1)/LOG(1 + 1/9), or (30.1)/(4.6), namely that digit 1 occurs **6.58** times more frequently than digit 9.

The above discussion justifies the expectation that the pattern of the empirical results obtained earlier for the Modified Dirichlet Density for N values {1, 2, 3, 4, 5} should converge to Benford in the limit as primes go to infinity. This is so since ultimately at infinity all nine digits attain an equal proportion of 1/9 or 11.1% within any single IPOT interval, and thus Modified Dirichlet Density approaches the classic Benfordian digit-1-over-digit-9 advantage of the 6.58 factor (as shall be seen further here in details). In other words, that the driving force behind the gradual digital development of the Modified Dirichlet, from the beginning with results that are slightly skewer than the logarithmic, giving digit 1 a bit more than 30.1%, and into the eventual Benford proportion, is all about digits themselves developing from slight skewness in the beginning favoring low digits, and into equality in the limit. Once digits 'settle into' their equal proportions, Modified Dirichlet Density attains its ultimate logarithmic configuration.

Let us now demonstrate in more details the mechanism at play here, and then prove that Modified Dirichlet Density should approach LOG(1 + 1/d) in the limit as primes go to infinity. We first start with the particular example of all the primes on (10, 100), in order to demonstrate visually how Modified Dirichlet relates indirectly to Benford, and then prove the result in general for any IPOT interval as primes go to infinity.

Figure 5.9 depicts the reciprocals of all the primes on (10, 100), namely of the 21 primes {11, 13, 17, 19, 23, 29, 31, 37, 41, 43, 47, 53, 59, 61, 67, 71, 73, 79, 83, 89, 97}.

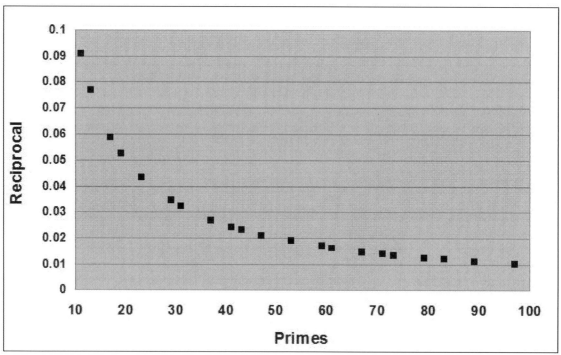

Figure 5.9: The Reciprocals of the 21 Primes on (10, 100)

$$MDD(1, 1) = \frac{\sum primes\ with\ 1\ as\ leading\ digit\ in\ (10,\ 100)\ \frac{1}{p}}{\sum all\ the\ primes\ in\ (10,\ 100)\ \frac{1}{p}}$$

Let us multiply the numerator and the denominator by the average distance between primes for this particular interval between 10 and 100 IPOT points. Calculating the average distance here yields the value 4.3. We get:

$$MDD(1, 1) = \frac{Avg\ Dist * \sum primes\ in\ (10,\ 100)\ 1\ leads\ \frac{1}{p}}{Avg\ Dist * \sum all\ the\ primes\ in\ (10,\ 100)\ \frac{1}{p}}$$

$$MDD(1, 1) \approx \frac{Area\ of\ Reciprocals\ of\ Digit\ 1}{Entire\ Area\ of\ Reciprocals\ of\ all\ Digits}$$

$$MDD(1, 1) \approx \frac{Shaded\ Area\ in\ Figure\ 13}{Entire\ Area\ in\ Figure\ 14}$$

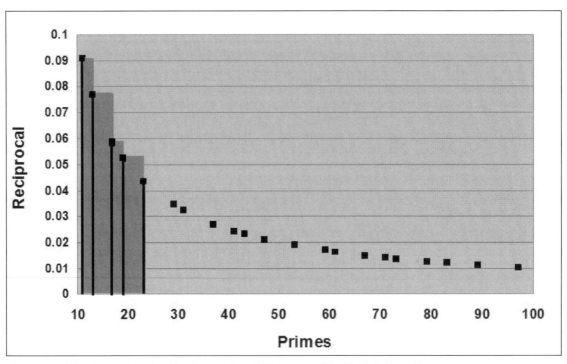

Figure 5.10: Area for the Reciprocals of Primes with digit 1 Leading

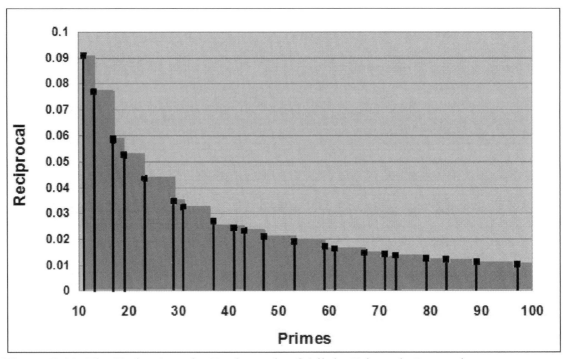

Figure 5.11: The Entire Area for Reciprocals of All the Primes in Interval

Without loss of generality, the drawing in Figures 5.10 and 5.11 utilize the actual distances between each pair of adjacent primes, instead of using the average distance 4.3.
Does it matter much? Certainly there are offsetting and compensating errors here; some rectangles are too short [in relation to the average distance]; others are too long; but overall the errors should cancel each other out; especially for much higher IPOT intervals where the primes are much more numerous, and are more smoothly distributed, as shall be shown shortly.

In order to visually drive the point that low digits always win the reciprocal game within IPOT intervals, Figure 5.12 is added showing the area of reciprocals for high digit 7, which is much more modest in comparison to area of reciprocals for low digit 1 of Figure 5.10.

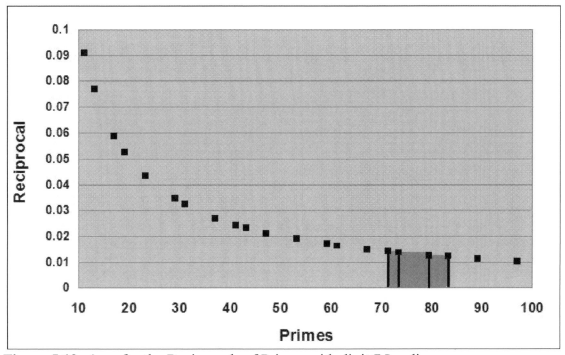

Figure 5.12: Area for the Reciprocals of Primes with digit 7 Leading

Surely, a differential in concentration of primes, not being approximately uniformly spread out at the macro level, would invalidate the above model which converts Modified Dirichlet Density into ratio of two areas (via the singular substitution of the average distance for all actual distances between primes). In reality, at the macro level, distances between primes do increase as higher integers are considered, and this in essence is the Prime Number Theorem; that primes become sparse and rarer as larger integers are considered; but within any given IPOT interval this increase in distance is quite marginal. Moreover, in the limit as primes go to infinity, distances between primes are approximately equal within any particular IPOT interval, therefore the use of the average distance between primes throughout as a singular quantity, for the left part of the interval around digit 1, for the center around digit 5, and for the right part of the interval around digit 9, is certainly justified.

Here are (five empirical and one estimated) results regarding the typical distances between primes (in units of an integer of course).

Two results are given for each IPOT interval:

(1) <u>at the beginning</u> of the interval where digit 1 leads;
(2) <u>at the end</u> of the interval where digit 9 leads:

Average distance between primes on (10, 20) is **2.6**
Average distance between primes on (90, 99) is **6.0**

Average distance between primes on (100, 200) is **4.9**
Average distance between primes on (900, 999) is **7.9**

Average distance between primes on (1,000, 2,000) is **7.4**
Average distance between primes on (9,000, 9,999) is **8.7**

Average distance between primes on (10,000, 20,000) is **9.7**
Average distance between primes on (90,000, 99,999) is **11.4**

Average distance between primes on (100,000, 200,000) is **11.9**
Average distance between primes on (900,000, 999,999) is **13.8**

Average distance between primes farther on $(1*10^{37}, 2*10^{37})$ is **85.6**
Average distance between primes farther on $(9*10^{37}, 10*10^{37})$ is **87.4**

The last pair is the highly accurate estimates utilizing the expression derived earlier $10^N/\ln((d + 1/2)*10^N)$ for the number of primes with digit d leading within IPOT interval, namely: Average Distance = $10^N/[10^N/\ln((d + 1/2)*10^N)] = \ln((d + 1/2)*10^N)$.
The dramatic/gentle dichotomy above within each early/late interval is simply the mirror-image of the fact that low digits are significantly/slightly more frequent than high digits in the beginning before we hit infinity.

Students of Benford's Law are immediately struck by the strong conceptual similarity between the expression of the Modified Dirichlet Density and the most perfectly logarithmic distribution of them all, namely **PDF(x) = k/x** defined over the range between two adjacent IPOT points. Isn't reciprocity the most essential quality of the k/x distribution?! Could the well-known results about k/x distribution explain MDD? (Kossovsky (2014), chapter 60 titled "The Case of k/x Distribution", and chapter 62 titled "Uniqueness of k/x Distribution".)

In order to complete the proof, it is necessary to formalize the comparison between: (1) discrete sums of reciprocals of primes with digit d leading, and (2) continuous area under k/x curve over the interval $((d)10^N, (d+1)10^N)$ with d running from 1 to 9 signifying first leading digits. Let us now demonstrate this comparison more rigorously.

The occurrences of primes on (10, 100) in Figures 5.9, 5.10, 5.11, and 5.12, are very particular, as if each prime has its own unique personality; as if distances between them signify something important. As higher IPOT intervals are considered, occurrences of primes appear more smooth, a bit more predictable, and primes lose their identities, becoming more ordinary and repetitive. Figure 5.13 depicts the reciprocals of all the primes on (100, 1000); visually confirming this generic observation. Figure 5.13 is visually also strongly suggestive of the k/x density curve, much more so than Figure 5.9 is. Other prime-reciprocal curves constructed for much larger IPOT intervals should resemble k/x density even better.

Figure 5.14 depicts the perfectly Benford distribution of k/x on (100, 1000). In order to equate entire area to one, we set $\int k/x \, dx = 1$ with [100, 1000] as the limits of integration, hence $k*[\ln(1000) - \ln(100)] = 1$, so that $k*\ln(1000/100) = 1$, and finally $k = 1/\ln10 = \log(e) \approx 0.434294482$. Clearly $\int \log(e)/x \, dx$ with [(d), (d + 1)] as the limits of integration yields $\log(e)*[\ln(d + 1) - \ln(d)] = \log(1 + 1/d)$, namely Benford's Law.

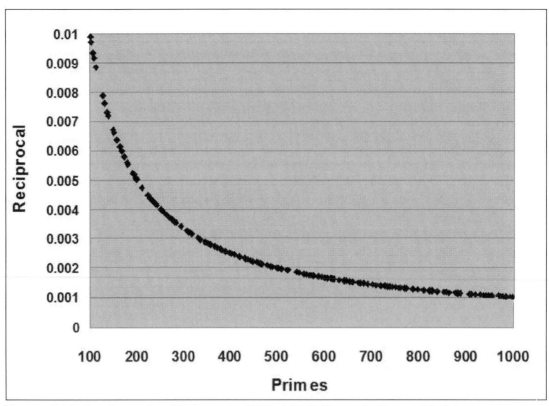

Figure 5.13: Reciprocals of Primes on (100, 1000) Strongly Resemble k/x

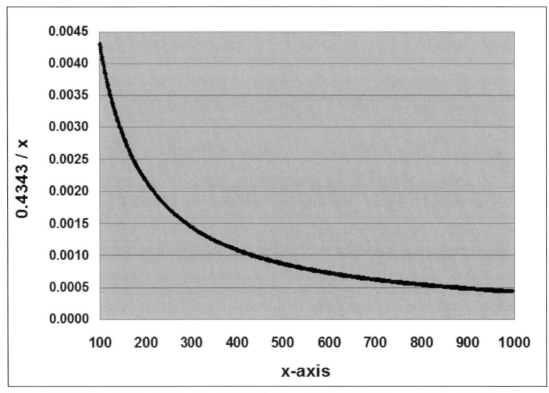

Figure 5.14: The Perfectly Benford Distribution [0.4343]*1/x on (100, 1000)

An additional factor which drives each consecutive prime-reciprocal curve between IPOT points closer towards k/x distribution is that within each new cycle between IPOT there are more primes; they keep growing!

There are   **4**   primes between 1 and 10
There are   **21**   primes between 10 and 100
There are   **143**   primes between 100 and 1,000
There are   **1061**   primes between 1,000 and 10,000
There are   **8363**   primes between 10,000 and 100,000
There are   **68906**   primes between 100,000 and 1,000,000

The expression $10^N/\ln((d + 1/2)*10^N)$ for the 9 digital sections within each IPOT interval obtained earlier from the density of primes $1/\ln(x)$ can be applied in estimating with extremely high accuracy the number of primes within each IPOT interval. An excellent close fit of theoretical with empirical/actual results is obtained as follow:

Empirical count of primes for the six intervals: **{4, 21, 143, 1061, 8363, 68906}**
Theoretical count of primes for the six intervals: **{ 8, 24, 147, 1068, 8379, 68970}**

The empirical growth factors are {21/4, 143/21, 1061/143, 8363/1061, 68906/8363}, namely {5.3, 6.8, 7.4, 7.9, 8.2}. Computer calculations of the estimates of the next prime growth factors between IPOT intervals beyond 1,000,000 applying the expression $10^N/\ln((d + 1/2)*10^N)$ for the 9 digital sections within each IPOT interval yields:
{8.5, 8.7, 8.8, 9.0, 9.1, 9.1, 9.2, 9.3, 9.3, 9.4, 9.4, 9.4, 9.5, 9.5, 9.5, 9.5, 9.6, etc.}.
In the limit, prime growth factor between consecutive IPOT intervals approaches **10-fold**. This is easily explained under the assumption that density $1/\ln(x)$ for primes barely changes for very large x, attainting approximately the same level on $(10^N, 10^{N+1})$ as on $(10^{N+1}, 10^{N+2})$, hence since range of the latter is 10 times as big as the former, the expression [Prime Count] = [Integer Range]*[Prime Density] implies that there are 10 times as many primes in the latter as compared with the former.

Therefore, as much larger primes are considered, each IPOT interval contains huge number of primes, and overall spread is highly smooth, strongly resembling k/x curve. For that reason the model converting the Modified Dirichlet Density into the ratio of two areas is certainly justified for higher primes on the way to infinity when particular distances between primes can be replaced with confidence by the average distance. Let us complete the proof for any MDD(N, d) with sufficiently high N:

$$
MDD(N, d) = \frac{\sum_{(d)10^N}^{(d+1)10^N} \frac{1}{p}}{\sum_{10^N}^{10^{(N+1)}} \frac{1}{p}}
$$

Where only primes are considered within the indices of summation, omitting the composites. Multiplying the numerator and the denominator by the average distance between primes within $(10^N, 10^{(N+1)})$ we get:

$$MDD(N, d) = \frac{(\text{Avg Di st}) * \sum_{(d)10^N}^{(d+1)10^N} \frac{1}{p}}{(\text{Avg Di st}) * \sum_{10^N}^{10^{(N+1)}} \frac{1}{p}}$$

Interpreting the sums as areas under 1/x curve, we write:

$$MDD(N, d) = \frac{\int_{(d)10^N}^{(d+1)10^N} \frac{1}{x} dx}{\int_{10^N}^{10^{(N+1)}} \frac{1}{x} dx}$$

Evaluating the definite integrals, we get:

$$MDD(N, d) = \frac{\ln\left((d+1)10^N\right) - \ln\left((d)10^N\right)}{\ln\left(10^{(N+1)}\right) - \ln\left(10^N\right)}$$

$$MDD(N, d) = \frac{\ln(d+1) + \ln(10^N) - \ln(d) - \ln(10^N)}{(N+1)*\ln(10) - N*\ln(10)}$$

$$MDD(N, d) = \frac{\ln(d+1) - \ln(d)}{N*\ln(10) + \ln(10) - N*\ln(10)}$$

$$MDD(N, d) = \frac{\ln((d+1)/d)}{\ln(10)} = \frac{\ln(1+1/d)}{\ln(10)} = \frac{\log(1+1/d)}{\log(10)}$$

$$MDD(N, d) = \frac{\log(1+1/d)}{1} = \text{The Benford Proportions}$$

[8]   Modified Dirichlet Leads Original Dirichlet to Benford

Could the above result about the Modified Dirichlet Density tell us something about the original all inclusive Dirichlet Density which considers all the primes in one fell swoop? The answer is in the affirmative; in fact the above result implies logarithmic proportions also for the original Dirichlet Density. The challenging issue here is the profound dichotomy between the two densities; namely that while the modified one had already used direct substitution of 1 for s and now comprises of reciprocals; the original one is still a limiting process as variable s approaches 1 from above.

It was shown earlier that from 10^{145} digital proportions of primes are equal considering one decimal precision (see Figure 5.7). Hence beyond 10^{145} one can confidently assume that the Modified Dirichlet Density is nearly Benford. Define k >145 as a sufficiently large power of ten such that from 10^k onward MDD(k, d) is Benford and digital equality prevails for a fixed m-decimal-precision criteria.

Let LMDD(k + 1, d), LMDD(k + 1, d), LMDD(k + 2, d), LMDD(k + 3, d), etc. expressed as **LMDD(k + j, d)** be the Modified Dirichlet Density on the intervals $(10^{k+j}, 10^{k+j+1})$ as a limit process with s approaching 1 from above, and with j running as 0, 1, 2, 3, and so forth, namely MDD as the limits of the ratios, where each 1/p term is written as $1/p^s$, and considering all IPOT intervals beyond 10^k.

Incorporating a limit process in the above definition yields the same result for all MDD ratios as shown earlier via the applications of the addition and division laws of limits, enabling us to evaluate the limits of the individual terms:

$$\lim_{s \to 1+} \frac{1}{p^s} = \frac{1}{p^1}$$

The limit of each LMDD exists since it incorporates only finite terms, namely that the numerator and denominator both converge. Surely LMDD = MDD.

Define $N(j)$ as the numerator and $D(j)$ as the denominator for each LMDD(k + j, d) involving variable s, namely as:

$$LMDD(k + j, d) = \lim_{s \to 1+} \frac{N(j)}{D(j)}$$

for some fixed d value, although d is not shown for brevity.

The sequence of LMDD(k + j, d) with j = 0, 1, 2, 3, and so on, is then the limit as s goes to 1 from above of:

$$\frac{N(0)}{D(0)}, \ \frac{N(1)}{D(1)}, \ \frac{N(2)}{D(2)}, \ \frac{N(3)}{D(3)}, \ \ldots \ , \text{LOG}(1 + 1/d), \ \text{LOG}(1 + 1/d), \ \ldots$$

The notation B is used for the Benford proportions of LOG(1 + 1/d) for the fixed d value in the definition of N(j) and D(j). Hence the following infinitely many relationships hold in the limit as s approaches 1 from above:

$$B = \frac{N(0)}{D(0)} \qquad B = \frac{N(1)}{D(1)} \qquad B = \frac{N(2)}{D(2)} \qquad B = \frac{N(3)}{D(3)} \quad \text{etc. to infinity}$$

Hence:

D(0)*B = N(0)
D(1)*B = N(1)
D(2)*B = N(2)
D(3)*B = N(3)

and so forth to infinity.

The original all inclusive (infinite) Dirichlet Density in essence is built from the various (finite) Limit Modified Dirichlet expressions. For example, for digit 3 it is defined as:

$$\lim_{s \to 1+} \frac{\frac{1}{3^s} + \frac{1}{31^s} + \cdots \frac{1}{307^s} + \cdots \frac{1}{3001^s} + \cdots \frac{1}{30011^s} + \cdots to \ \infty}{\frac{1}{2^s} + \frac{1}{3^s} + \frac{1}{5^s} + \frac{1}{7^s} + \frac{1}{11^s} + \frac{1}{13^s} + \frac{1}{17^s} + \frac{1}{19^s} + \frac{1}{23^s} + \cdots to \ \infty}$$

And one can find within the above expression all the Limit Modified Dirichlet ones for leading digit 3, without overlapping, without repeating any term. In other words, the above expression is comprised solely and exactly of the various Limit Modified Dirichlet ones for leading digit 3. Ignoring the contributions from earlier intervals below 10^k which are negligible as primes go to infinity; as s approaches 1 from above for the original Dirichlet Density, it can be written analytically (finite) term by (finite) term as:

$$\lim_{s \to 1+} \frac{N(0) + N(1) + N(2) + N(3) + \cdots \text{ to infinity}}{D(0) + D(1) + D(2) + D(3) + \cdots \text{ to infinity}}$$

Yet, it is still a limit, not a ratio, since both the numerator and denominator diverge, implying that the limit is in an indeterminate form. Substituting B*D(j) for all the N(j) terms (*true though only when s goes to 1 – but s does indeed go to 1 here!*) we get:

$$\lim_{s \to 1+} \frac{B * D(0) + B * D(1) + B * D(2) + B * D(3) + \cdots \text{ to infinity}}{D(0) + D(1) + D(2) + D(3) + \cdots \text{ to infinity}}$$

$$\lim_{s \to 1+} \frac{B * [\, D(0) + D(1) + D(2) + D(3) + \cdots \text{ to infinity}\,]}{D(0) + D(1) + D(2) + D(3) + \cdots \text{ to infinity}}$$

$$\lim_{s \to 1+} \frac{B}{1} = B$$

Hence the limit of the original Dirichlet density is LOG(1 +1/d) and this completes the proof.

[9]   Conceptual Conclusions

Prime numbers are decisively non-Benford as observed in the normal density of the proportion of primes with digit d leading, in spite of the Dirichlet results obtained in this section. Primes are innately uniformly distributed with a flat density in the limit having digital equality at infinity. Mr. Johann Dirichlet artificially, arbitrarily, and forcefully, dresses the primes with that reciprocal coat, making them appear as if they are Benford. The primes suffer in silence, pretending to go along, but deep down they continue to believe in uniform, equitable, and fair digital distribution, hoping that one day they would be able to tear down that Dirichlet's unnatural dress and act naturally. In the same vein, Dirichlet might as well employ the same reciprocal coat and dress up various other sequences that have nothing to do with the logarithmic distribution, and then proclaim them all to be Benford!?

Any argument in support of the false statement that "The prime numbers are Benford Dirichlet-wise" pointing to the facts that MDD and DD are LOG(1 + 1/d), should also be countered by pointing to the sequence $X_{N+1} = X_N + 1$ beginning with $X_1 = 1$ (namely the sequence of all the integers Z), and constructing a similar limit-ratio (involving not only the primes but also the composites) such as:

$$\lim_{s \to 1+} \frac{\sum \text{integers with leading digit } d \ \frac{1}{z^s}}{\sum \text{all the integers} \ \frac{1}{z^s}}$$

Surely, the arguments given in this section regarding the connection to k/x distribution should apply here as well, and even more perfectly and smoothly so, hence this points to the same logarithmic proportions of LOG(1 + 1/d); yet no one would dare claiming that the integers are Benford! They are not!

In reality prime numbers do not care much about integral-powers-of-ten intervals; they merrily march forward along the integers at their own pace, totally disregarding them; nor do they have any special esteem for number 10 – an annoying member of their archrival and competitive group - the composites. They live and behave as if the number 10 does not matter much. Surely they also do not pay any attention whatsoever to the particular number systems invented by the various civilizations scattered about randomly across the universe, as they float eternally up there well above all such lowly and arbitrary local inventions. Conclusions about the primes in digital Benford's Law, either in base 10 or even employing a generic base for all positional number systems, do not interest the primes much. Yet the primes are quite enthusiastic about what was found in connection with the General Law of Relative Quantities (GLORQ) which assigns the primes the bin-equality proportion of 1/D ultimately at infinity [Kossovsky (2014), Section 7, Chapter 132] and which is number system invariant.

Acknowledgement:
The author wishes to thank the distinguished mathematician **George Andrews** for his helpful comments and suggestions regarding this section on prime numbers.

SECTION 6

EXPONENTIAL GROWTH SERIES

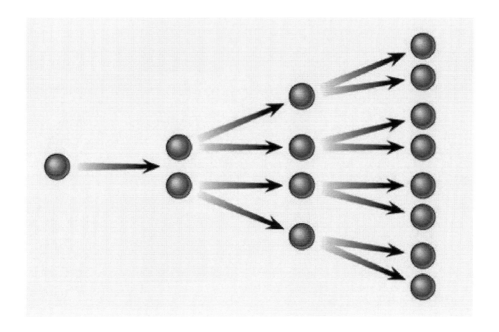

[1] Normal Versus Anomalous Exponential Growth Series

Exponential growth occurs when the growth of a given quantity [the added value due to growth] is proportional to the quantity's current value, so that 5% exponential growth for example yields only 5 when the quantity is 100, but it yields 50 when the quantity is 1000.

Discrete Exponential growth series are expressed as:

$$\{B, BF, BF^2, BF^3, \ldots , BF^N \}$$

where B is the <u>base</u> (initial) value, P is the constant <u>percent</u> growth rate, and F is the constant multiplicative <u>factor</u> relating to growth rate as in F = (1 + P/100). Throughout this section we shall operate under the assumption of our decimal base 10 number system, although of course everything can be generalized to other bases.

For example, for initial base value B = 10, and 7% growth rate per period (or year), the implied F factor is F = (1 + 7/100) = 1.07. We let quantity 10 grow for 34 years, hence the series is:

$$\{10, \; 10*1.07, \; 10*1.07^2, \; 10*1.07^3, \; 10*1.07^4, \; \ldots , \; 10*1.07^{34}\} =$$

{10.0	10.7	11.4	12.3	13.1	14.0	15.0	16.1	17.2	18.4	19.7	21.0
22.5	24.1	25.8	27.6	29.5	31.6	33.8	36.2	38.7	41.4	44.3	47.4
50.7	54.3	58.1	62.1	66.5	71.1	76.1	81.5	87.2	93.3	99.8}	

First digits are emphasized in bold font:

{**1**0.0	**1**0.7	**1**1.4	**1**2.3	**1**3.1	**1**4.0	**1**5.0	**1**6.1	**1**7.2	**1**8.4	**1**9.7	**2**1.0
22.5	**2**4.1	**2**5.8	**2**7.6	**2**9.5	**3**1.6	**3**3.8	**3**6.2	**3**8.7	**4**1.4	**4**4.3	**4**7.4
50.7	**5**4.3	**5**8.1	**6**2.1	**6**6.5	**7**1.1	**7**6.1	**8**1.5	**8**7.2	**9**3.3	**9**9.8}	

The choice of 34 years was deliberate, so as to get as close to Benford as possible by ensuring that the last term is approximately 10-fold the initial base value of 10. In other words, that the exponent difference between maximum 99.8 and minimum 10.0 is LOG(99.8) – LOG(10.0) = 1.9991 – 1.0000 = 0.9991, and which is very close to an integral value (namely integer 1). The phrase 'exponent difference' refers to 'log difference', namely the difference between the log values of the minimum (first element) and the maximum (last element). For example, the exponent of 100 is 2, because LOG(100) is 2, and because 10^2 = 100, meaning that the exponent necessary to raise 10 to resulting in 100 is exactly 2. For example, for the series {8, 16, 32, 64, 128, 256, 512, 1024}, exponent difference is LOG(1024) – LOG(8) = 3.0103 – 0.9031 = 2.1072. The term <u>Exponent Difference</u> is better known as <u>Order of Magnitude</u> in the literature.

Digit distribution regarding the 7% exponential growth series above for the vector of nine digits {1, 2, 3, 4, 5, 6, 7, 8, 9} with '%' sign omitted is:

7% Growth from 10 for 34 Periods: {31.4, 17.1, 11.4, 8.6, 8.6, 5.7, 5.7, 5.7, 5.7}
Benford's Law 1st Digits Order: {30.1, 17.6, 12.5, 9.7, 7.9, 6.7, 5.8, 5.1, 4.6}

Quoting from Kossovsky (2014) chapter 98 titled "Singularities in Exponential Growth Series": "Due to their discrete and finite nature, real-life exponential growth series can never be formally considered as exactly logarithmic by pure mathematicians, only in the limit possibly". Discreteness in general causes deviations from the logarithmic, and could possibly be remedied by considering infinitely many elements and the limiting case. For mere mortals such as statisticians and scientists, the pressing issue is not some highly theoretical and mathematically exact digital behavior, but rather the practical consideration of an approximate Benford behavior. Does it matter much that digit 1 has earned 29.8% or 30.5% while digit 9 has earned only about say 4.2%? No! Because such is the typical behavior of all real-life data sets obeying Benord's Law, and no one should be disturbed at all if the exact theoretical 30.1% proportion of Benford is not found. Hence, in extreme generality, what is [practically] necessary for exponential growth series to behave approximately or nearly logarithmically is having plenty of elements, although for low growth rates there is an additional requirement, namely that exponent difference between minimum and maximum should be as close as possible to an integer. One straightforward way of achieving this last requirement for low growth rates is to make sure that the series starts and ends approximately at integral powers of ten, such as 10 & 100, 1 & 1000, or 100 & 1000000, and so forth. Additional discussions about this issue are given in later chapters of this section.

Let us further explore exponential growth series in general by examining how logarithm values of the series progress forward at each new period. As a consequence, this exploration would indirectly lead to better understanding of what happens to mantissa and its resultant distribution after numerous periods. The decimal log series of any exponential growth series is simply:

{ LOG(B),
  LOG(B) + 1*LOG(F),
  LOG(B) + 2*LOG(F),
  LOG(B) + 3*LOG(F),

        , ... ,

  LOG(B) + N*LOG(F) }

It is certainly proper to think of LOG(F) namely LOG(1 + P/100) as being a fraction, as it is so in any case up to 900% growth of P, else the integral part can be ignored as far as digital configuration is concerned. For example, for 2% growth, LOG(1 + 2/100) = 0.009. For 50% growth, LOG(1 + 50/100) = 0.176. Even for 180% growth, LOG(1 + 180/100) = 0.447 which is a fraction. Therefore, related log series of the exponential growth series is simply a series with constant additions (accumulation) of LOG(F) from an initial base of LOG(B). While this related log series grows ever larger, mantissa on the other hand is constantly oscillating between 0 and 1, as it typically (for low growth) takes many small steps forward, then suddenly one large leap backwards whenever log overflows an integer, and so forth. This is so since mantissa is obtained by constantly removing the whole part of the log *(whenever growth series ≥ 1, that is whenever LOG(series) is positive or zero, an assumption which can be taken for granted here)*. Therefore, mantissa is clearly seen as being uniform on (0, 1) as more and more points of newly minted mantissa keep falling there on a variety of points, covering an ever increasing 'portion' of the entire (0, 1) range. Even though the process of exponential growth series is truly deterministic, if one were to visually follow these 'rapid' mantissa additions onto (0, 1) space without getting severely dizzy, it would all seem quite random, disorganized, and highly chaotic, and it is precisely this nature of mantissa creation that yields uniformity to its final overall distribution! As an example, we examine an exponential 30% growth series, starting from an initial base 3, having factor F = 1.30, and with the implied LOG(F) = 0.1139433. Figure 6.1 depicts part of that series in details. What should be carefully noted here is that mantissa always re-enters into (0, 1) interval at different (newer) locations, namely at 0.047, 0.072, 0.098, and so on. This is a necessary condition for logarithmic behavior, as it guarantees that mantissa is well-spread over the entire (0, 1) range in an even and uniform manner, covering all corners and segments. In a sense it guarantees that mantissa creation is random and unorganized with respect to (0, 1) space, and thus that (almost) always new mantissa values are being created.

Series	Log	Mantissa
3.0	0.477	0.477
3.9	0.591	0.591
5.1	0.705	0.705
6.6	0.819	0.819
8.6	0.933	0.933
11.1	1.047	0.047
14.5	1.161	0.161
18.8	1.275	0.275
24.5	1.389	0.389
31.8	1.503	0.503
41.4	1.617	0.617
53.8	1.730	0.730
69.9	1.844	0.844
90.9	1.958	0.958
118.1	2.072	0.072
153.6	2.186	0.186
199.6	2.300	0.300
259.5	2.414	0.414
337.4	2.528	0.528
438.6	2.642	0.642
570.1	2.756	0.756
741.2	2.870	0.870
963.6	2.984	0.984
1252.6	3.098	0.098
1628.4	3.212	0.212
2116.9	3.326	0.326

Figure 6.1: Normal Logarithmic Exponential Series with 30% Growth

Figure 6.2 depicts the way the 30% growth series almost always re-enter differently after each integer mark on the log-axis.

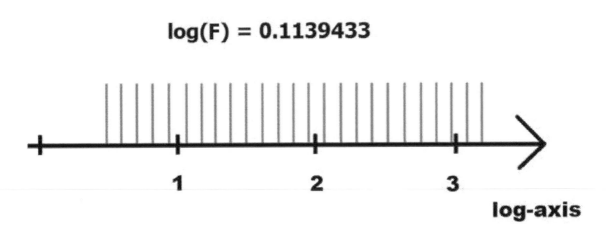

Figure 6.2: The March on the Log-Axis of Normal Logarithmic Series with 30% Growth

Such logarithm and mantissa vista leads to the detection of some peculiar digital singularities. The argument above falls apart whenever the fraction LOG(F) happens to be such that exactly T whole multiples of it add up to unity, as in the fractions 0.50, 0.25, 0.10, 0.125, 0.05, and so forth, in which case constant LOG (F) additions lead to re-entering (0, 1) always at the same point, and taking the same type of steps over and over again focusing only on a few selected fortunate points having some very strong concentration (density), all the while ignoring all the other points or sections on the interval (0, 1). Such growth rate results in some quite uneven mantissa distribution on (0, 1) and yields decisively non-logarithmic digital distribution for the exponential series itself. In other words, beyond the creation of a few initial mantissa points, the series does not create any new mantissa, but just repeats those few old ones over and over again. Algebraically, the series is non-logarithmic and **rebellious (anomalous)** whenever there exists an integer T such that LOG(F)*T = 1. One should always be reminded that only uniformity of mantissa yields logarithmic behavior.

As an example, we examine one such anomalous exponential growth series starting from the initial base value of 10, growing at 77.8279% per period, and thus having factor F = 1.778279 and the implied problematic LOG(F) = LOG(1.778279) = 0.25 = 1/4, where exactly 4 (called T) multiples of it add up to 1. Figure 6.3 depicts part of that series in details. Mantissa always re-enters into (0, 1) interval at the same location, namely at 0.000, and then always takes the same subsequent 'long' steps landing at 0.250, 0.500, and 0.750, intentionally skipping 'numerous' points in between them. Obviously such state of affairs cannot result in any logarithmic behavior because mantissa is not uniform, and the series is anomalous. This conclusion is so regardless of the initial value (base) of the series, and the fact that here the series starts at 10 is irrelevant. For example, for the 77.8279% anomalous growth series above, had the base value been 8 instead of 10, then mantissa would always re-enter into (0, 1) interval at 0.153, and then land exclusively on the points 0.403, 0.653, 0.903, over and over again. These points are all 0.25 units apart.

Series	Log	Mantissa
10.0	1.000	0.000
17.8	1.250	0.250
31.6	1.500	0.500
56.2	1.750	0.750
100.0	2.000	0.000
177.8	2.250	0.250
316.2	2.500	0.500
562.3	2.750	0.750
1000.0	3.000	0.000
1778.3	3.250	0.250
3162.3	3.500	0.500
5623.4	3.750	0.750
10000.0	4.000	0.000
17782.8	4.250	0.250
31622.8	4.500	0.500
56234.1	4.750	0.750
100000.0	5.000	0.000
177827.9	5.250	0.250
316227.8	5.500	0.500
562341.3	5.750	0.750

Figure 6.3: Anomalous Non-Logarithmic Series with 77.8279% Growth

Figure 6.4 depicts the 77.8279% anomalous growth series with base value of 10 as it re-enters the log-axis always at the same location after each integer mark.

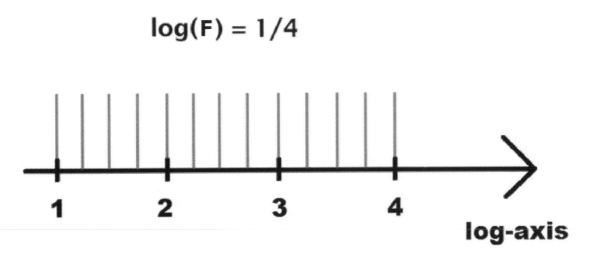

Figure 6.4: The March on the Log-Axis of Anomalous 77.8279% Growth Series - Basic Type

Yet, even for those rebellious non-logarithmic rates, some comfort and relief can be found whenever LOG(F) [the width of the steps by which log of the series advances along the log-axis] is sufficiently small as compared with unit length, because then no matter how repetitive, peculiar, and picky mantissa chooses the points upon which to stamp on (0, 1), each step is still so tiny that it has no choice but to walk almost all over the interval covering most corners and locations of (0, 1). Simply put: the creature is such that its legs are too short to be truly picky, so it cannot jump and skip much, and it is reduced to walking almost all over the interval, willingly or unwillingly. Only a walking creature with long legs can be effectively picky and successfully avoid certain segments lying on the ground. A giraffe can successfully avoid a 50-centimeter hole or gap on the ground, but a tiny ladybug cannot, no matter how carefully it walks.

Therefore, as the value of LOG(F) decreases, becoming very small, say less than 0.01 (a rational 1/100, designated as 1/T), its spread over (0, 1) is fairly good and quite even, so that its digit distribution is very nearly logarithmic, even though upon closer examination it still stamps cautiously and discretely on the log/mantissa-axis in even but tiny steps of 0.01-width each.

Figure 6.5 depicts such a scenario with LOG(F) = 1/100 where the growth rate is 2.329%. The percent growth rate is calculated via 1/100 = LOG(1 + Percent/100), then taking 10 to the power of both sides of the equation we get: $10^{1/100} = 10^{LOG(1 + Percent/100)}$, therefore $10^{1/100} = (1 + Percent/100)$, and $10^{1/100} - 1 = Percent/100$, so that $100[10^{1/100} - 1] = 2.329\%$. Digit distribution is {30.0, 17.9, 12.1, 10.0, 8.0, 6.0, 6.0, 5.0, 5.0} for such a series starting at an initial base 3 and growing for 1000 periods; where SSD comes out as 1.1, and such very low SSD value indicates nearly perfect Benford behavior.

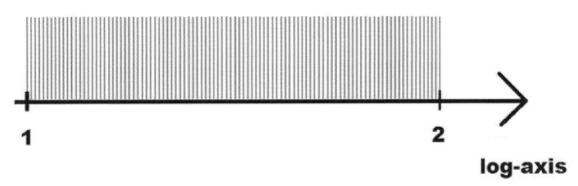

log(F) = 1/100

1   **2**

log-axis

Figure 6.5: Uniformity of Mantissa is Achieved via Small Rational Steps

Clearly, the series with LOG(F) = 1/N in the limit as N goes to infinity is perfectly logarithmic **by definition**, since this is indeed the very meaning of uniformity of mantissa!

For example, let us consider the rational series with LOG(F) = 1/1,000,000. Here for each tiny sub-interval on the log-axis with width of 0.0005 say, there are always (that is: uniformly) 500 points. Hence the interval on (0.0005, 0.0010) has 500 points; the interval on (0.0010, 0.0015) has 500 points; the interval on (0.8005, 0.8010) has 500 points as well, and so forth for any interval of 0.0005 width, and which is the very meaning of the density being uniform! No excuses, explanations, or reference to exceptions, are needed in stating this, except that formally the definition is stated in a continuous – infinitely refined – sense. Hence surely, for the above series with LOG(F) = 1/1,000,000, if one insists on scrutinizing density fanatically on even tinier sub-intervals such as (0.0000000000001, 0.0000000000002) then mantissa is not uniform, although it is uniform indeed for all practical purposes.

Figure 6.6 depicts various anomalous growth rates by varying integer T from 1 to 20, as well as evaluating T at 25, 35, 40, 50, 100, and 200. For each anomalous growth rate in this table, an actual computer simulation (calculation rather) of the relevant exponential series is run, using the first 1000 elements, all starting from the initial base value of 3. Digital results from such computer runs are displayed in the table, along with their associated SSD values. Figure 6.6 demonstrates that eventually when T gets to be over 100 or so, deviations from the logarithmic become very small and insignificant.

T	LOG(F)	% Growth	1	2	3	4	5	6	7	8	9	SSD
1	1	900.000%	0.0	0.0	100.0	0.0	0.0	0.0	0.0	0.0	0.0	9155.8
2	0.5	216.228%	0.0	0.0	50.0	0.0	0.0	0.0	0.0	0.0	50.0	4947.6
3	0.333	115.443%	33.3	0.0	33.3	0.0	0.0	33.3	0.0	0.0	0.0	1701.8
4	0.25	77.828%	25.0	22.7	2.3	0.0	25.0	0.0	0.0	0.0	25.0	1063.3
5	0.2	58.489%	40.0	0.0	20.0	20.0	0.0	0.0	20.0	0.0	0.0	926.9
6	0.167	46.780%	16.6	16.6	16.7	16.7	0.0	16.7	0.0	0.0	16.7	619.8
7	0.143	38.950%	28.6	14.2	14.3	14.3	14.3	0.0	0.0	14.3	0.0	262.9
8	0.125	33.352%	25.0	24.5	0.5	12.5	12.5	0.0	12.5	0.0	12.5	424.9
9	0.111	29.155%	33.3	11.1	22.3	0.0	11.1	11.1	0.0	11.1	0.0	362.6
10	0.1	25.893%	30.0	10.0	20.0	10.0	10.0	0.0	10.0	0.0	10.0	236.7
11	0.091	23.285%	36.4	14.1	13.1	9.1	9.1	9.1	0.0	9.1	0.0	130.3
12	0.083	21.153%	24.9	16.6	16.8	8.4	8.4	8.3	8.3	0.0	8.3	97.4
13	0.077	19.378%	30.8	22.8	7.9	7.7	7.7	7.7	7.7	7.7	0.0	84.8
14	0.071	17.877%	28.4	20.9	7.7	14.4	7.2	7.2	0.0	7.1	7.1	103.6
15	0.067	16.591%	33.2	19.4	7.2	13.4	6.7	6.7	6.7	6.7	0.0	80.3
16	0.063	15.478%	31.0	12.4	12.6	12.6	6.3	6.3	6.3	6.3	6.2	43.5
17	0.059	14.505%	35.3	11.6	17.7	5.9	11.8	5.9	5.9	5.9	0.0	141.9
18	0.056	13.646%	27.5	16.5	16.8	5.6	11.2	5.6	5.6	5.6	5.6	56.6
19	0.053	12.884%	31.4	15.6	15.9	10.6	5.3	5.3	10.6	5.3	0.0	71.0
20	0.05	12.202%	30.0	15.0	15.0	10.0	10.0	5.0	5.0	5.0	5.0	21.2
25	0.04	9.648%	28.0	16.0	16.0	8.0	8.0	8.0	4.0	4.0	8.0	40.1
35	0.029	6.800%	28.1	16.8	14.5	8.7	8.7	5.8	5.8	5.8	5.8	13.1
40	0.025	5.925%	30.0	17.5	12.5	10.0	10.0	5.0	7.5	5.0	2.5	14.5
50	0.02	4.713%	30.0	16.0	14.0	10.0	8.0	6.0	6.0	4.0	6.0	8.8
100	0.01	2.329%	30.0	17.9	12.1	10.0	8.0	6.0	6.0	5.0	5.0	1.1
200	0.005	1.158%	30.0	17.5	12.5	10.0	8.0	6.5	6.0	5.0	4.5	0.2
Ben	=====	======	30.1	17.6	12.5	9.7	7.9	6.7	5.8	5.1	4.6	0.0

Figure 6.6: Benford is Eventually Found in Small Steps of Anomalous Rational Series

Anomalous series of the form described above, where the fraction LOG(F) happens to be such that exactly whole multiples of it add up to unity, are actually just one particular type. More generally, whenever whole multiples of LOG(F) add up any integer, be it 2, 3, or any other integral number, we encounter the same dilemma of having uneven mantissa distribution on (0, 1), resulting in non-logarithmic digital distribution for the exponential series itself. To recap, **General Types** of anomalous exponential growth rates are found when exactly **T** whole multiples of LOG(F) add up exactly to any integer **L**, and not just to unity.

For example, 5 whole multiples of 0.4 yield exactly the value of 2 units of distance spanning log/mantissa interval, hence its related 151.1886% growth series has a non-logarithmic digit distribution. The percent growth rate is calculated via $0.4 = 2/5 = LOG(1 + Percent/100)$, which leads to $2/5 = LOG(1 + (151.1886)/100) = LOG(2.511886)$. Figure 6.7 depicts in details part of that series starting from the initial base value of 10. Figure 6.8 depicts the selective and repetitive way the series marches along the log-axis, in coordination and synchronization with integer marks. Clearly this rebellious growth series is not Benford at all.

Series	Log	Mantissa
10.0	1.000	0.000
25.1	1.400	0.400
63.1	1.800	0.800
158.5	2.200	0.200
398.1	2.600	0.600
1000.0	3.000	0.000
2511.9	3.400	0.400
6309.6	3.800	0.800
15848.9	4.200	0.200
39810.7	4.600	0.600
100000.0	5.000	0.000
251188.6	5.400	0.400
630957.3	5.800	0.800
1584893.2	6.200	0.200
3981071.7	6.600	0.600
10000000.0	7.000	0.000
25118864.3	7.400	0.400
63095734.4	7.800	0.800
158489319.2	8.200	0.200
398107170.6	8.600	0.600

Figure 6.7: Anomalous 151.1886% Growth Series

$$\log(F) = 2/5$$

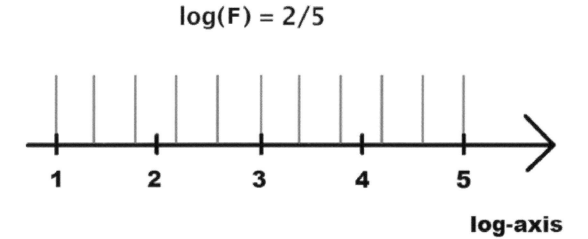

Figure 6.8: The March on the Log-Axis of Anomalous 151.1886% Series - General Type

The general rule regarding Benford behavior for any exponential growth series is as follows: whenever the fraction LOG(1 + P/100) equals the rational number L/T, non-logarithmic behavior for the series with P% growth rate is found. In other words, for both L & T being positive integers, the exponential series is rebellious whenever:

LOG(F) = L/T

The **Basic Types** anomalous exponential series discussed earlier, such as the one of Figures 6.3 and 6.4, are those where L = 1 and where the rationality rule there is that the series is not Benford whenever LOG(F) = 1/T with T being a positive integer.

Conceptually it helps keeping in mind that for rebellious rates, either of the general type or of the basic type, the interpretation of the rationality condition can be stated as follows:

LOG(F) = [in L whole log units] / [we cycle exactly T periods]

The set of first digits obtained in one cycle of T periods, will repeat itself in the next cycle of T periods, and also in the next one, and so forth. This is so since the same series values are repeated in each cycle of T periods, except for the decimal point which moves once to the right (namely that mantissa values are perfectly repeated in each cycle of T periods). This is why T plays such a decisive role in determining the magnitude of the deviation from Benford, while L is almost totally irrelevant to the magnitude of deviation.

For large value of T, say over 100, the series has already a fairly large number (namely T) of distinct mantissa values and a nearly logarithmically decent set of first digits, and repeating it on each new cycle of T periods does not diminish its approximate Benfordness.

For a very small value of T, say 5, the series repeats these 5 first digits over and over again on each cycle of T periods, and there is no hope of ever breaking out of this vicious digital cycle.

It should be noted that L > T cases are distinct real possibilities, although they all correspond to extremely high growth rates over 900%. Cases where L = T correspond to LOG(F) = 1, which implies that F = 10, so that (1 + P/100) = 10, and therefore P = 900%.

In addition, care should be exercised always to avoid non-reduced forms such as L/T = 2/214 and L/T = 500/11200, which are really of 1/107 and 5/112 reduced forms respectively.

The non-reduced form 4/10 for example, with L = 4 and T = 10, is associated with the interpretation that L/T = (in 4 whole log units)/(we cycle 10 periods), but upon careful examination of how such a series marches along the log-axis one could note that the more basic interpretation is L/T = (in 2 whole log units)/(we cycle 5 periods), and that the 4/10 interpretation is simply a pair of two distinct 2/5 cycles mistakenly considered as one. Non-reduced forms should be avoided as they do not represent any new anomalous possibilities distinct from the sufficient and complete set of all reduced formed possibilities.

[2] The Simplistic Explanation of the Rarity of Anomalous Series

On the face of it, we can expect not merely most, but rather practically all exponential growth series to behave logarithmically. This appears to be so simply because that for growth rate to be rebellious the fraction LOG(1 + P/100) must equal exactly some rational number L/T, and this is quite rare. In fact, Mathematics teaches us that when a number is picked at random from any continuous set of real numbers, there is zero probability of obtaining a rational number and 100% probability of obtaining an irrational number. Figure 6.9 humorously dramatizes the argument.

Admittedly, quite often, financial rates, economics-related rates, and others, are quoted rationally as fractions, such as say 5½ % yearly interest rate on a 30-year bond investment, but one should not confuse the set of growth rates with the set of LOG operations performed on these rates. Such rational quotes in finance do not usually lead to anomalous series, because it's the equality LOG(1 + Rational-Rate/100) = (Rational L/T) which should hold. A possible set of rates on say the real continuous interval (0%, 100%) is mapped into a different set of LOG(1 + Rate/100) on the real continuous interval (0, LOG(2)), and with the same resultant rarity of the rationals.

Figure 6.9: The Simplistic Explanation of the Rarity of Anomalous Series

307

The above argument attempts to utilize the fact that the irrationals vastly outnumber the rationals. Back in the time of Pythagoras around 500 BC, mathematicians refused to believe that irrational numbers even existed. A major principle or dogma of the Pythagorean Sect was the claim that whole numbers ruled the world, and that everything could be explained in terms of integers or fractions of integers. When the Pythagoreans were able to prove that the diagonal of the square is irrational it caused a major intellectual crisis. Many centuries later, at the end of the 1800's, the mathematician Georg Cantor made the startling discovery that the irrational numbers are actually more numerous than the rational numbers! The rational fractions are countably infinite whereas the irrationals are uncountably infinite. The Pythagoreans would have been quite shocked and further dismayed had they learnt of Georg Cantor's discoveries!

[3] The Indirect Damage Done to Normal Series by Anomalous Series

Yet, the above argument of why exponential growth series are rarely anomalous is too simplistic, as it indirectly relies on the assumption of highly abstract and idealized infinitely-long exponential growth series having infinite number of growth periods, and which is irrelevant to finite real-life exponential growth series. This is so because for an irrational normal growth series that is finite, merely being in the neighborhood of a rational rebellious series [namely having nearly the same growth rate as shown in Figures 6.10 and 6.11] is problematic and logarithmic behavior is severely disrupted, unless a large number of growth periods is considered to overcome the closeness to the series (and that number depends on the intensity of the rebelliousness of the anomalous series and on the closeness to it). The only remedy for an irrational normal series of finite length residing near a rational rebellious series is increasing its length, i.e. considering many more growth periods. Such a cure may be too demanding though in terms of the number of growth periods necessary for good behavior, and sometimes even standard personal computers are not sufficiently powerful to perform this task. Hence, for any irrational normal growth series of finite number of growth periods, one can theoretically find a nearby rational rebellious series in the neighborhood - extremely close to it! Would this supposedly render many or most finite exponential growth series rebellious!?

For example, for the rational rebellious series with a growth rate of **93.069773%** and where LOG(1 + 93.069773/100) = 0.285714286 = 2/7, starting at the initial base value of 3, and growing for 300 periods, first digits are {28.6, 14.3, 14.3, 14.3, 14.3, 0.0, 0.0, 14.3, 0.0} resulting in high SSD value of 261.7.

Just being in the neighborhood of this 93.069773% rebellious series causes problems. For the irrational normal series **93.15%**, starting at the initial base 3, and growing for 300 periods, first digits are with SSD = 166.5, hence this short and finite series is not Benford. But let us increase the number of periods in order to 'remedy' this deviation and manifest its true Benfordian nature:

300 periods: {28.3, 14.3, 14.3, 14.3, 4.0, 10.3, 0.0, 12.7, 1.7} SSD = 166.5
500 periods: {28.6, 14.2, 14.4, 12.4, 4.2, 10.6, 1.4, 7.6, 6.6} SSD = 83.6
600 periods: {30.3, 14.3, 14.3, 10.3, 6.0, 8.8, 3.5, 6.3, 6.0} SSD = 31.6
700 periods: {30.3, 16.3, 14.1, 9.0, 7.1, 7.6, 5.0, 5.4, 5.1} SSD = 7.4
900 periods: {29.9 , 17.2, 12.8, 10.2, 8.3, 6.2, 5.1, 6.2, 4.0} SSD = 3.0

Hence with 900 periods we were able to cure and bring the 93.15% series back to Benford. Yet, getting even closer to the 93.069773% rebellious series, say for the series **93.09%**, starting at base 3, would spell serious trouble even with 900 periods. This is so since this series 'mimics' even better the rebellious series. Here digits are {28.6, 14.3, 14.3, 14.2, 5.3, 9.0, 0.0, 14.2, 0.0} with SSD = 186.5, even after 900 periods. Since the 93.09% series is even closer to the 93.069773% rebellious series as compared with the 93.15% series above, many more periods are needed to remedy it, and merely 900 periods are simply not sufficient for it.

Let us explain why an irrational series in close proximity to a rational one should suffer a similar non-Benfordian fate, at least in the beginning, for the first tens of periods, or perhaps for the first hundreds or so periods.

In order to give a general explanation, let us consider the log values of the above rebellious series 93.069773% having the rationality condition of LOG(F) = 2/7, as well as log values of the above irrational normal series 93.15%, both starting at an initial base 3. The goal is to follow how these two series develop their log values, and by extension to examine how they develop their mantissa values, in the beginning, for the initial 22 periods. The following two sets of data refer to their mantissa values as well as to their first digits (the 93.069773% series is shown on top and the 93.15% series is shown at the bottom below it):

1	2	3	4	5	6	7	8	9	10	11	12	13
0.477	0.763	0.049	0.334	0.620	0.906	0.191	0.477	0.763	0.049	0.334	0.620	0.906
0.477	0.763	0.049	0.335	0.621	0.907	0.192	0.478	0.764	0.050	0.336	0.622	0.908

14	15	16	17	18	19	20	21	22
0.191	0.477	0.763	0.049	0.334	0.620	0.906	0.191	0.477
0.194	0.480	0.766	0.051	0.337	0.623	0.909	0.195	0.481

1	2	3	4	5	6	7	8	9	10	11	12	13
3	5	1	2	4	8	1	3	5	1	2	4	8
3	5	1	2	4	8	1	3	5	1	2	4	8

14	15	16	17	18	19	20	21	22
1	3	5	1	2	4	8	1	3
1	3	5	1	2	4	8	1	3

The 93.15% series grows a tiny bit faster than the 93.069773% series, having an extra ≈0.08% growth rate, hence after 22 periods it reaches the mantissa value of 0.481 which is just slightly higher than the mantissa value of 0.477 of the slightly slower 93.069773% series. Clearly, these two series differ only by very little in how they generate their mantissa values early on.

And what is the difference here on leading digits for these initial 22 periods? None whatsoever! They both generate the same (non-Benfordian) first digits! This is so because the mapping from mantissa to first digits is quite flexible and slack, so that tiny differences in mantissa values usually do not affect first digits. Chapter 16 Section 1 titled 'Benford's Law as Uniformity of Mantissa' discusses the nine compartments within the [0, 1) mantissa space, where each compartment points to a unique first digit. These compartments are: [0, 0.301), [0.301, 0.477), [0.477, 0.602), [0.602, 0.699), [0.699, 0.778), [0.778, 0.845), [0.845, 0.903), [0.903, 0.954), and [0.954, 1.000).

For example, for 14th period, the two series generate the distinct mantissa values of 0.191 and 0.194, yet both values are firmly inside the [0, 0.301) compartment belonging to digit 1. Figure 1.15 of chapter 16 illustrates clearly the slack nature of the mapping of mantissa values into first digits. Yet, since these tiny differences in mantissa values are cumulative, eventually, when many more periods are considered, the 93.15% series begins to generate quite different mantissa values as compared with the 93.069773% series, and consequently its overall digit distribution becomes closer and closer to the Benford configuration after many more additional periods.

For any irrational normal series such as say 40.0% growth rate where LOG(1 + 40/100) = 0.146128035678238…etc., supposedly one could find a pair of L and T integers such that approximately but very nearly 0.146128035678238 ≈ L/T. A finite irrational growth series in the neighborhood of another rational series having strong deviation from Benford experiences a great deal of deviation itself, and it requires a large number of periods in order to manifest its true and innate Benfordian character. In practical terms, such a cure is not available or not realistic when real-life growth series are concerned.

[4] Empirical Tests of Digital Configurations of Exponential Growth Series

Do rational series in close proximity to irrational series really adversely affect logarithmic behavior for many or even most finite series? Fortunately they do not! They hardly ever disrupt digital configuration! And this fact can be verified empirically with the aid of the computer. The line of reasoning above, worrying about neighboring rational series adversely affecting many or most of the irrationals series turned out to be mere 'theoretical paranoia'. Let us empirically test a large variety of finite exponential growth series [with length from 822 growth periods to 3000 growth periods] for their compliance or non-compliance with Benford's Law.

Three distinct random checks shall be performed: (1) for rates randomly chosen from 1% to 5%, (2) for rates randomly chosen from 1% to 50%, (3) for rates randomly chosen from 1% to 890%. The rates are chosen with equal probability as in the continuous Uniform(1, 5), Uniform(1, 50), and Uniform(1, 890), respectively. Each randomly chosen growth rate is then utilized to create several exponential growth series with that rate applying a variety of growth periods, from the shortest one of only 822 periods, to the longest one of 3000 periods; all beginning at the same initial base value of 3. The need to create for each random rate several series with distinct growth periods emanates from the general requirement for an approximate integral exponent difference between the first and the last elements in obtaining logarithmic behavior, namely that LOG(maximum) – LOG(minimum) ≈ an integer, as in the k/x distribution.

The computer program then chooses that series with the lowest SSD, and which is often equivalent to choosing the series having exponent difference closest to an integer. The variety of growth periods that are being employed for each randomly chosen growth rate are: {3000, 2897, 2800, 2697, 2597, 2297, 2284, 2262, 1930, 1759, 1433, 1268, 822}, and the program then registers only the minimum SSD of all these thirteen series, so as to arrive at the most Benford-complying series, and by implication often at that series where LOG(maximum/minimum) is closest to an integer. Such a 'crude' method of avoiding [in the approximate] non-integral exponent difference is not effective for extremely low growth rates between 0% and 1%, and this fact provided the motivation to avoid this challenging range on the percent-axis, and to consider only rates above 1%. Extremely low rates below 1% grow very slowly, and therefore such series often take very many periods (>>3000) to obtain integral exponent difference. Any computer program attempting to determine whether such low rate series are Benford would need to be much more sophisticated and to allow numerous periods.

Let us first outline that part of the empirical results coming out of the computer calculations regarding the supposed damage done to irrational series due to existing in the neighborhood of rational series.

As a dramatic example from the empirical computerized results thus obtained, we focus on the rebellious percent range around the theoretical rationality LOG(F) = 2/3 with its associated growth rate of 364.1588%. This is a highly rebellious growth rate with T = 3, having the exceedingly high SSD value of about 1500, indicating severe deviation from Benford. As can be seen in Figure 6.10, such a highly disruptive rate affects wide range all around it. Surely, affected range is such only due to the fact that the irrational series around it have no more than 3000 periods as programmed in the computer scheme, and surely substantially longer lengths of, say, one million periods would remedy almost all of them except those residing extremely near 364.1588% [and even those could also be remedied via billions or trillions growth periods]. The closer is the irrational to the disruptive rational, the more periods are needed in order to overcome the bad influence of the rebellious neighboring series, and this is the reason for the approximate bell-shape round curve seen in Figure 6.10.

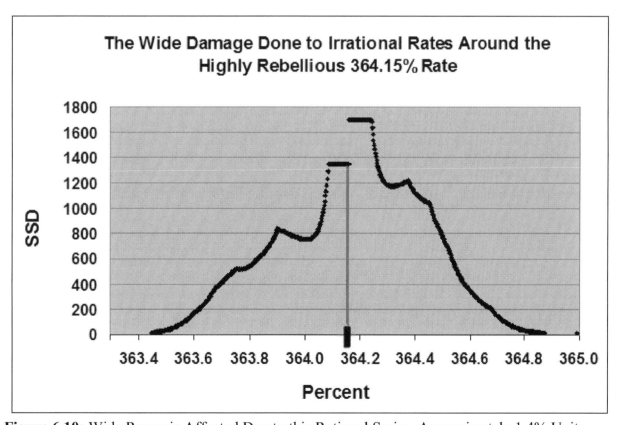

Figure 6.10: Wide Range is Affected Due to this Rational Series, Approximately 1.4% Units

Note: the percent growth rate is calculated via $2/3 = \text{LOG}(1 + \text{Percent}/100)$, then taking 10 to the power of both sides of the equation we get: $10^{2/3} = 10^{\text{LOG}(1 + \text{Percent}/100)}$, therefore $10^{2/3} = (1 + \text{Percent}/100)$, and $10^{2/3} - 1 = \text{Percent}/100$, so that $100[10^{2/3} - 1] = 364.1588\%$.

The lower the T value of the rebellious rate, the more significant is the deviation from Benford, and the wider is affected range. The higher the T value, the milder is the deviation from Benford, and the narrower is affected range. Hence, in sharp contrast to the example of Figure 6.10 with its affected wide range of about 1.4% units, another example shall be explored focusing on the rebellious range round the theoretical rationality $\text{LOG}(F) = 13/19$ and its associated growth rate of 383.2930%. This series has the relatively large T value of 19, thus its deviation from Benford is mild. The empirical result obtained here shows a much narrower affected range for this rate, as well as much reduced intensity of deviation, with SSD consistently below the value of 76. As can be seen in Figure 6.11 which depicts the empirical results around 383.2930%, such mildly disruptive rate affects only a very narrow range around it of about 0.15% units.

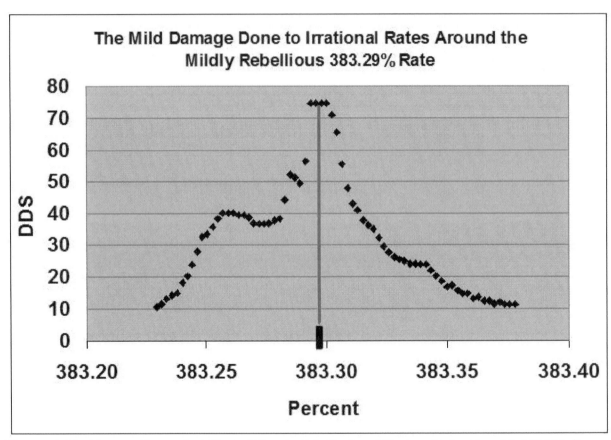

Figure 6.11: Narrow Range is Affected Due to this Rational Series, Only About 0.15% Units

Note: the percent growth rate is calculated via $13/19 = \text{LOG}(1 + \text{Percent}/100)$, then taking 10 to the power of both sides of the equation we get: $10^{13/19} = 10^{\text{LOG}(1 + \text{Percent}/100)}$, therefore $10^{13/19} = (1 + \text{Percent}/100)$, and $10^{13/19} - 1 = \text{Percent}/100$, so that $100[10^{13/19} - 1] = 383.2930\%$.

The [false] alarm is now raised, and the worry that many or most [finite] exponential growth series are disrupted by neighboring rational series now seems all too convincing. The milder case of Figure 6.11 gives some hope for better logarithmic results, while the severe case of Figure 6.10 is a bad omen for logarithmic behavior.

Yet in reality, the following thorough empirical tests encompassing growth rates from 1% to 5%, from 1% to 50%, and from 1% to 890%, decisively show that there is not much to worry about, and that almost all exponential growth series are Benford or very nearly so – assuming that these series are of considerable length, say over 1000 periods.

For 30,000 simulations of such random growth rates between 1% and 5%, only two came out with SSD over the mild value of 10! These two mildly rebellious series were 4.81134164367% with 13.9 SSD, and 4.81139819992% with 10.4 SSD; both located at the upper range near the 5% boundary, and both belong to the L/T = 1/49 theoretical rationality. Other rationality-related growth rates found there with even lower deviations were: 4.71269871352% with only 4.11 SSD; also located at the upper range near the 5% boundary, and belonging to the L/T = 1/50 theoretical rationality, as well as 4.91365456231% with only 4.75 SSD; also located at the upper range near the 5% boundary, and belonging to the L/T = 1/48 theoretical rationality. Average SSD for all of these 30,000 low growth rates is the highly satisfactory and extremely low value of 0.12. The median gives an even more satisfactory and incredibly low value of 0.03.

For 30,000 simulations of random growth rates between 1% and 50%, we get the following results: Out of 30,000 rates, only 278 (namely 0.9%) came out with SSD over the mild value of 10. Out of 30,000 rates, only 101 (namely 0.3%) came out with SSD over the relatively high value of 50. Out of 30,000 rates, only 55 (namely 0.2%) came out with SSD over the high value of 100. Average SSD for all of these 30,000 growth rates is the satisfactory very low value of 0.9. The median (being a much more robust measure of centrality) gives the extremely low value of 0.02. As known from Descriptive Statistics, the average usually exaggerates and overestimates centrality, while the median could at times underestimates centrality.

For 30,000 simulations of random growth rates between 1% and 890%, we get the following results: Out of 30,000 rates, 2,515 (namely 8.4%) came out with SSD over the mild value of 10. Out of 30,000 rates, 1070 (namely 2.5%) came out with SSD over the relatively high value of 50. Out of 30,000 rates, 455 (namely 1.5%) came out with SSD over the high value of 100. Average SSD for all of these 30,000 growth rates is the moderate value of 10.0. The median gives the very low value of 0.6. The median in this case might be underestimating centrality.

Conclusion: For finite series of considerable length with at least 822 periods up to 3000 periods: not only most, but the vast majority of exponential growth series are nearly perfectly Benford! Moreover, LOW rates with growth below approximately 5% are practically all nearly perfectly Benford with the exceptions of only very few and extremely rare mildly rebellious cases. For HIGH rates with growth approximately over 5% there exists a small non-Benford minority, although the vast majority of them do satisfactorily obey the law of Benford. [Note: the choice of 3 as the base value for all of these empirical tests is practically irrelevant to this conclusion.]

[5] The Proper Explanation of the Rarity of Anomalous Series

Let us now give the proper and more profound explanation of why the vast majority of exponential growth series are rarely rebellious and do obey the law of Benford nearly perfectly:

The reason why almost all finite exponential growth series with considerable length [of over 1000 periods say] are Benford is because of a pair of limitations placed on the values of T and L.

Limitation on the value of T:

Deviation due to coordinated and rational walk along the log-axis is limited to series with relatively low T values, while series with high T values are very close to Benford approximately in spite of their rationality. As seen in Figure 6.6, whenever T is roughly over 50, deviation from the logarithmic is fairly small. If one is willing to tolerate SSD values up to say approximately 10 or 15 - considering all series with SSD value lower than say 10 or 15 as sufficiently near Benford - then any rational series with T over 50 should not be considered as truly rebellious, but rather as approximately Benford.

Hence T is limited as in: $\mathbf{T \leq 50}$.

Limitation on the value of L:

For any fixed interval on the percent-axis, as in the exclusive consideration of the range $(0\%, P_{MAX}\%)$, and the exploration of all the rebellious rates that might be laying there, T and L values are restricted as in $L/T \leq LOG(1 + P_{MAX}/100)$.

Hence L is limited as in: $\mathbf{L \leq T{*}LOG(1 + P_{MAX}/100)}$.

For any possible T value, the value of L is restricted by this ceiling on the upper side.

This pair of severe limitations on T and L values places a firm cap on how many rational L/T series with significant deviation from Benford exist on any given percent-axis interval. And it renders the set of all rational series with significant deviation from Benford not merely finite, but also very small in size! Moreover, as shall be demonstrated in the next two numerical examples, this pair of limitations on T and L values implies that the vast majority of T and L rational combinations are of relatively high T values around 30 to 50, relating to mild deviations from Benford, and that only a minority of such rational combinations are of relatively low T values, relating to significant deviations from Benford.

In some limited sense, high T values correspond with low growth, and low T values correspond with high growth – but this is so only for L being fixed at a constant value. Yet, the true nature of T value is not about low growth rate versus high growth rate, but rather about the number of periods the rebellious series go through in order to return to the same mantissa value. In the same vein, and in some limited sense, low L values correspond with low growth, and high L values correspond with high growth – but this is so only for T being fixed at a constant value.

As a consequence of the above discussion, it is noted that the T = 50 and L = 1 case yields the rebellious series with the lowest possible growth rate [under this pair of limitations on T and L values], and that lowest possible rebellious growth rate is 4.71285480509%. For this series, LOG(1 + 4.71285480509/100) = 0.02 = 1/50.

For all low growth rates below 4.71285480509%, deviation is very small, and thus they are considered to be nearly Benford. This fact alone saves all growth rates within the (0%, 4.7128%) percent-axis interval from the fate of being anomalous [assuming that SSD values below 10 or 15 are considered mild and that therefore such series are considered to be practically Benford].

Under this pair of limitations on T and L values, all combinations of T and L values other than the T = 50 and L = 1 case, yield growth rates higher than 4.71285480509%. To prove this assertion one needs only to note that all other cases of T = 50 where L > 1 such as 2/50, 3/50, 4/50, and so forth, represent higher growth rates than that of the 1/50 case. In addition, all other cases of L = 1 where T < 50, growth rates are higher as well. For example, 1/49, 1/48, 1/47 are all of higher growth rate than that of the 1/50 case. The same argument applies for example to 2/49, 3/48, 4/47, 5/46 and so forth, which are all of higher growth rate than that of the 1/50 case.

In a nutshell:

(1)/(50) ≤ (any L ≥ 1)/(any T ≤ 50)

and this is so for any L and T combination.

[6] A Brief Example of the Consequences of L and T Limitations

The first brief example regarding the very limited number of pairs of T and L possibilities which are effectively disruptive to Benord behavior is given for the exclusive consideration of the percent-axis interval on **(0%, 100%)**. Here T and L values are restricted as follows:

$T \leq 50$ and $L/T \leq LOG(1 + 100/100)$
$T \leq 50$ and $L \leq T*LOG(2)$
$T \leq 50$ and $L \leq T*0.30103$

The set of all possible T and L disruptive pairs is then:

$T = 50$ and $L \in \{1, 2, 3, \dots 12, 13, 14, 15\}$
because $L \leq 50*0.30103$, namely $L \leq 15.05$

$T = 49$ and $L \in \{1, 2, 3, \dots 12, 13, 14\}$
because $L \leq 49*0.30103$, namely $L \leq 14.75$

$T = 48$ and $L \in \{1, 2, 3, \dots 12, 13, 14\}$
because $L \leq 48*0.30103$, namely $L \leq 14.45$

$T = 47$ and $L \in \{1, 2, 3, \dots 12, 13, 14\}$
because $L \leq 47*0.30103$, namely $L \leq 14.15$

$T = 46$ and $L \in \{1, 2, 3, \dots 12, 13\}$
because $L \leq 46*0.30103$, namely $L \leq 13.85$

and so forth

$T = 14$ and $L \in \{1, 2, 3, 4\}$
because $L \leq 14*0.30103$, namely $L \leq 4.21$

$T = 13$ and $L \in \{1, 2, 3\}$
because $L \leq 13*0.30103$, namely $L \leq 3.91$

and so forth

$T = 7$ and $L \in \{1, 2\}$
because $L \leq 7*0.30103$, namely $L \leq 2.11$

and so forth

$T = 4$ and $L \in \{1\}$
because $L \leq 4*0.30103$, namely $L \leq 1.20$

T cannot fall below 4. For example, T cannot be 3, because then $L \leq 3*0.30103$, and this implies that $L \leq 0.90$, but this is a contradiction since L must be a positive integer, namely that at a minimum L must be integer 1. Hence the possibilities of T = 3, T = 2, T = 1 are eliminated here.

Therefore, the complete set of all possible {T, L} disruptive pairs is:

{50, 1}, {50, 2}, {50, 3}, ... {50, 13},{50, 14},{50, 15}
{49, 1}, {49, 2}, {49, 3}, ... {49, 13},{49, 14}
{48, 1}, {48, 2}, {48, 3}, ... {48, 13},{48, 14}
{47, 1}, {47, 2}, {47, 3}, ... {47, 13},{47, 14}
{46, 1}, {46, 2}, {46, 3}, ... {46, 13}

    … etc. …

{14, 1}, {14, 2}, {14, 3}, {14, 4}
{13, 1}, {13, 2}, {13, 3}
{12, 1}, {12, 2}, {12, 3}
{11, 1}, {11, 2}, {11, 3}
{10, 1}, {10, 2}, {10, 3}
{9, 1}, {9, 2}
{8, 1}, {8, 2}
{7, 1}, {7, 2}
{6, 1}
{5, 1}
{4, 1}

This yields the very small set of only 360 disruptive pairs. Yet, there are many pairs that are not in reduced form, and they should be eliminated.

For example:

{50, 2}	is really {25, 1}	2/50 = 1/25
{50, 15}	is really {10, 3}	15/50 = 3/10
{12, 2}	is really {6, 1}	2/12 = 1/6

All of which yields an even smaller set of only 232 {T, L} disruptive pairs in purely reduced form on the percent-axis interval (0%, 100%).

[7] A Detailed Example of the Consequences of L and T Limitations

A second example fully analyzed and detailed - regarding the very limited number of T and L pair possibilities which are effectively disruptive to Benford behavior - is given for the exclusive consideration of the percent-axis interval on **(0%, 900%)**. Here T and L values are restricted as follows:

$T \leq 50$ and $L/T \leq LOG(1 + 900/100)$
$T \leq 50$ and $L \leq T*LOG(10)$
$T \leq 50$ and $L \leq T*1$
$T \leq 50$ and $L \leq T$

Therefore, the complete set of all possible {T, L} disruptive pairs is:

{50, 50}, {50, 49}, {50, 48}, ... {50, 3}, {50, 2}, {50, 1}
{49, 49}, {49, 48}, {49, 47}, ... {49, 2}, {49, 1}
{48, 48}, {48, 47}, {48, 46}, ... {48, 1}

  … etc. …

{3, 3}, {3, 2}, {3, 1}
{2, 2}, {2, 1}
{1, 1}

This yields the small set of only $(N^2 + N)/2 = (50^2 + 50)/2 = 1275$ pairs.

It should be noted that pairs with low T values [high intensity deviation] are by far fewer than pairs with high T values [low intensity deviation], therefore most of these pairs cause relatively mild deviation from Benford and only on relatively narrow and limited neighboring ranges. For example, there are only $(N^2 + N)/2 = (10^2 + 10)/2 = 55$ pairs with low T values from 1 to 10 causing intense and wide damage, while the vast majority of them, namely $1275 - 55 = 1220$ pairs, are with T values of over 10, causing relatively milder and narrower damage. In other words, the upside down pyramid-like structure of the above set of all possible {T, L} pairs implies that most of the pairs are around the top near T = 50 case; that fewer are around the center near the T = 25 case; and that only a tiny minority are around the bottom near the T = 1 case. Such a structure for the set of {T, L} pairs bodes well for Benford compliance in general, because low T values where most significant deviations occur are much fewer and rarer than high T values where only insignificant and minor deviations occur.

The above set of 1275 pairs contains numerous redundant cases. This is so since many pairs are not in reduced form and should be eliminated. This purge yields an even smaller set of only 774 reduced form disruptive {T, L} pairs. For example, the first pair {50, 50} is really the last pair {1, 1}; the pair {50, 2} is really {25, 1}, and so forth.

Hence only 774 explosive mines are laid out there quite sparsely on the large minefield of all the real numbers from 0% to 900%. These 774 land mines are literally few and far between, and they are rarely encountered. Most of the mines are of distinct sizes and gunpowder concentration (namely distinct T values), hence usually they cause different types of damage when activated, killing or maiming to a large or small degree. Is it really very dangerous for a normal exponential growth series to walk over this relatively sparse minefield? Well, for finite and relatively short series without many growth periods, the answer depends on the width of the damaged range around each land mine. For infinitely long series with infinite number of growth periods, walking over this minefield is quite safe, because affected range around each land mine approaches zero in the limit as the length of the series goes to infinity, rendering these mines as a non-threatening set of 774 real points on the infinitely uncountable real range (0%, 900%).

Let us contemplate the danger of walking over this minefield for finite series of considerable length with 1000 to 3000 periods approximately.

If we assume that all of these 774 mines are of the low intensity type as seen in Figure 6.11 relating to the 13/19 theoretical rationality where only a narrow range of about 0.15 units was affected, then simple-minded calculations lead to the conclusion that this would supposedly cover only 0.15*774 = 116.1 units on the percent-axis range, out of a total of 900 units, assuming that there is no overlapping of affected ranges. This still represents a significant portion of affected ranges, namely 13% of the total range.

If we assume that all of these 774 mines are of the high intensity type as seen in Figure 6.10 relating to the 2/3 theoretical rationality where a wider range of about 1.4 units was affected, then simple-minded calculations lead to the conclusion that this would supposedly cover a whapping 1.4*774 = 1083.6 units on the percent-axis range, out of a total of 900 units, assuming that there is no overlapping of affected ranges. This represents more than the entire possible range of 0% to 900%! The implication of this scenario would be that all finite exponential growth series with about 1000 to 3000 periods are rebellious!

Yet, the empirical results decisively showed that nearly all finite series of considerable length with 1000 to 3000 periods approximately are Benford! This implies that the fear of the impact due to rational series is exaggerated, and that the damage done on the percent-axis to irrational series is overestimated.

In a nutshell, the reason for the very mild and very limited damage done on the entire range by these 774 land mines is that the vast majority of T values here are of relatively high value of around 25 to 50. These high-T-value series cause very mild and very narrow damage, and progressively so as one nears the 50 value mark. The scenario of Figure 6.10 relates to T = 3, and this is rather a very rare case within the set of 774 land mines. In addition, the scenario of Figure 6.11 relates of T = 19, and even this case is not of the most frequent types within the set of 774 land mines. The shape of the entire set of {T, L} disruptive pairs shown in the previous page is that of an upside down pyramid-like structure, therefore, overall, out of 774 cases, those with relatively low T values causing significant and wide damage constitute only a very small minority, representing truly only very few cases. All this renders the (0%, 900%) range very much Benford-like.

Let us define the theoretical magnitude of deviation from Benford as **500/T**. The motivation behind this definition emanates from the fact that the larger the value of T the less deviation from Benford is found for the rebellious rational series, namely that deviation is inversely proportional to T. Since L practically does not play any role in the magnitude of deviation, it is left out of this definition.

The smallest possible magnitude of theoretical deviation is of the T = 50 case, and it is calculated as 500/50, namely 10. The largest possible magnitude of theoretical deviation is of the T = 1 case, and it is calculated as 500/1, namely 500, yet this case is discarded and excluded, and only deviations from the 2nd largest T = 2 case are to be considered. This top value of theoretical deviation is calculated as 500/2, namely 250.

The reason for this exclusion is that T = 1 case relates to the highest possible growth rate where P = 900%, and perhaps it is wise to avoid the inclusion of the boundary of the percent-axis interval of (0%, 900%). [Since L ≤ T it follows that L ≤ 1, so that L must be 1 as well, therefore L/T = 1/1 = LOG(1 + P/100), so that 10 = (1 + P/100), and finally we obtain P = 900%.] The other reason for this [singular] exclusion is to be able to give a good visual presentation of all the other [773] less dramatic cases where 500/T ≤ 250, as shown clearly in Figure 6.12. This unique case of T = 1 where 500/T is 500 can be considered as an outlier in some sense, and including it would needlessly distort and expand the scale for all the other cases in Figure 6.12.

Figure 6.12 depicts these 773 (namely 774 – 1) theoretical rebellious rates, where the vertical axis expressing 500/T is plotted versus the horizontal axis expressing the percent growth. The count of the points in the entire chart of Figure 6.12 where 500/T values are 50 or higher shows that there are only 31 such points out of a total of 773 points, namely that only 4% of all the points are with truly strong deviation from Benford. Clearly, the vast majority of the points in Figure 6.12 are with low 500/T < 50 deviation values.

In addition, it should be noted that the overall frequency of the points (crude 'density' of sorts) falls off as focus shifts from low percent growth to higher percent growth, getting more diluted on the right. For example, from 0% up to 100% there are 232 points, from 100% up to 200% there are 138 points, and from 800% up to 900% there are only 32 points. The vector of frequencies per 100% interval is {232, 138, 94, 77, 62, 52, 45, 41, 32}, falling off monotonically to the right. In contrast, average magnitude of deviation - namely average of 500/T per 100% interval - is nearly steady, and it's approximately of the constant value 19.5.

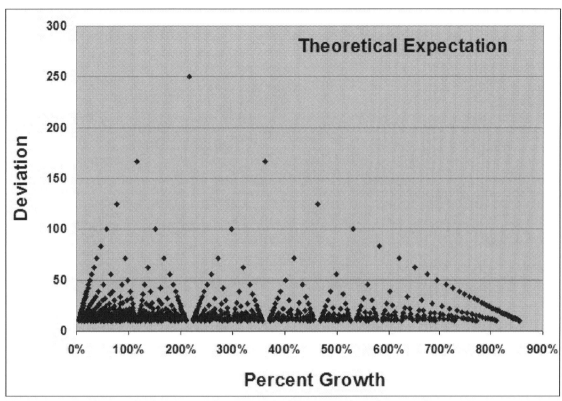

Figure 6.12: Theoretical Expectation of Rebellious Rational Rates from 0% up to 900%

[8] Excellent Fit between Empirical Results and Theoretical Expectation

Consequently, another computer program is run, empirically checking exponential growth series for any possible deviation from Benford. Here for this 2nd empirical check, instead of choosing growth rates randomly as was done previously in the 1st empirical check, this program methodically examines growth rates in order, from 1% all the way to 890%, advancing forward in tiny % increments. This is so in order to cover for sure and without any random glitches all corners and all sub-intervals within the entire interval of (0%, 900%) – except for the problematic part at the end of (890%, 900%). The aim is to show excellent correspondence between what is expected theoretically due to any possible rational march along the log-axis and what is actually found empirically. If such good fit is found, then as a consequence, this could indirectly imply that no other causes or reasons for deviation from Benford are observed for exponential growth series in extreme generality except for the argument discussed in this section [assuming that sufficient number of elements of the series are considered for good compliance, as well as adherence to the constraint of an integral exponent difference for low rates].

The program starts at 1% and ends at 890%, in tiny refine increments of 0.00215714598%. It checks 412,118 rates in total, namely nearly half a million exponential growth series. It uses the quantity 3 as the initial base value for all the series. The program only registers those series with deviation larger than the somewhat arbitrary value of 8.88 SSD. Any obedient series with SSD less than 8.88 is deleted and forgotten. The motivation for the choice of the cutoff SSD value of

321

8.88 is to aim just slightly below the round number of 10, in an attempt to correspond to the theoretical list of rebellious rational rates with a cap of T at 50. That cap at 50 for the value of T restricts SSD values for all the theoretical rational series to over 10 or over 15 approximately.

Just as was done for the random selection of growth rates, here too, in order to allow low rates to comply with the logarithmic requirement that exponent difference between the maximum and the minimum should be approximately close to an integer, a variety of growth periods are used via the set {3000, 2897, 2800, 2697, 2597, 2297, 2284, 2262, 1930, 1759, 1433, 1268, 822}. This is the same set of 13 growth periods that was used in the 1st program. The program then registers only the minimum SSD for all these series, so as to arrive at the most complying series.

The program has registered 37,453 rebellious growth rates with SSD over 8.88 – out of 412,118 in total. In other words, about 9.1% of the series were found to deviate by registering over the 8.88 SSD threshold value. Figure 6.13 depicts those empirical rebellious rates with SSD over 8.88 on the interval (1%, 890%). It should be emphasized that there are various gaps and empty sub-ranges here for the obedient series, in spite of the fact that the chart appears continuous.

Note: The earlier simulations in chapter 4 of random growth rates between 1% and 890% resulted in having 8.4% of all the series with SSD over 10. In comparison, here the SSD threshold is lowered from 10 to 8.88, allowing a few more series between 8.88 and 10 to be considered rebellious, and this is why slightly higher 9.1% ratio is obtained here instead of 8.4%.

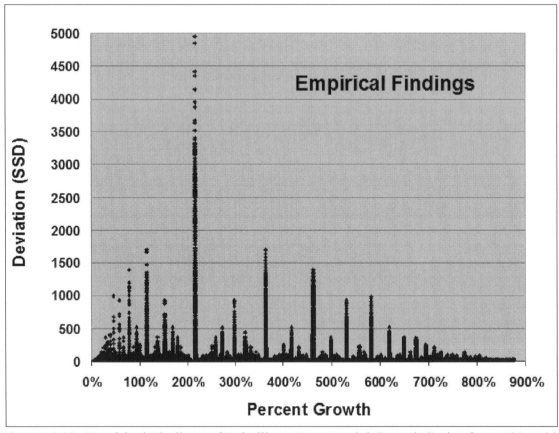

Figure 6.13: Empirical Findings of Rebellious Exponential Growth Series from 1% to 890%

The moment of truth has arrived. Let us now check compatibility between theoretical and empirical results by putting these two charts next to each other as shown in Figure 6.14.

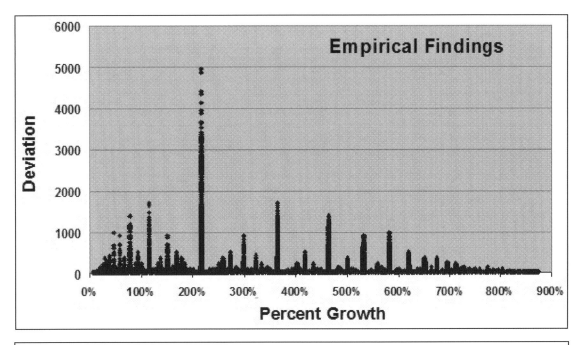

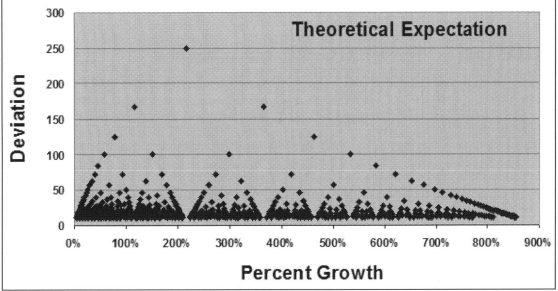

Figure 6.14: Compatibility between Theoretical & Empirical Rebellious Series 1% to 890%

Hooray! The empirical and the theoretical agree!

Furthermore, a closer scrutiny of various smaller sub-intervals within the entire 1% to 890% range reveals a near perfect fit between the theoretical and the empirical. As for but one example of these careful examinations, Figure 6.15 depicts the excellent fit for rates between 1% and 50%. Comparisons of various other sub-intervals show the same good fit as well.

Figure 6.15 clearly reveals the various gaps and empty sub-ranges that exist for the obedient series in the empirical computerized study. This helps to repudiate the false appearance of continuity in Figure 6.13.

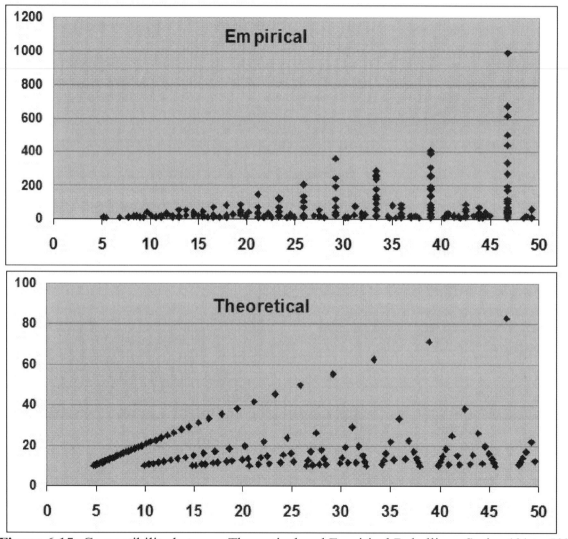

Figure 6.15: Compatibility between Theoretical and Empirical Rebellious Series 1% to 50%

Empirically, for the entire region below 5%, not a single rebellious rate was detected; as all growth rates from 0% to 5% had their SSD values below 8.88. This empirical result is almost in perfect harmony with the theoretical threshold of approximately 5% that was calculated earlier (actually 4.71285480509% to be precise), and which showed that there exist only 3 rational rebellious series just below 5% - under the assumption of the theoretical pair of limitations on T and L values. These 3 theoretical rational rates are: 4.71285480509%, corresponding to the theoretical rationality of 1/50; 4.81131341547%, corresponding to the theoretical rationality of 1/49; and 4.91397291363%, corresponding to the theoretical rationality of 1/48.

These 3 theoretical rational series have been missed out by the empirical computer program due to the relatively low SSD generated exactly at these 3 points, as well as around their neighborhood perhaps. Under the parametrical assumptions of the computer program, namely the base of 3 and with these 822 to 3000 periods, at exactly the theoretical series of 4.71285480509%, SSD is 8.77; at exactly the theoretical series of 4.81131341547%, SSD is 8.47; and at exactly the theoretical series of 4.91397291363%, SSD is 7.85. Since the SSD threshold of the computerized empirical scheme is set at 8.88, these theoretical rationalities must have been missed out by a hair.

Figure 6.16 depicts discrete-like comparisons between all 'large' empirical deviations with SSD values over 100, referring only to the central point of each such rebellious sub-range, together with all those 'large' theoretical deviations with T < 14, namely all those with theoretical magnitude of deviation 500/T > 38.5. These two cutoff points, namely T < 14 and SSD > 100, correspond to each other in the approximate, since SSD of the rationality 1/14 under the assumptions of the empirical parameters is 101.5, thus SSD value of 100 could serve perhaps as a nice round figure for the cutoff point between 'large deviations' and 'non-large deviations' of the empirical results. As can be seen in Figure 6.16, there is an excellent fit between the theoretical and the empirical for all these 'large' deviations from Benford.

The excellent fit found here between the empirical results and the theoretical expectation strongly confirms the conclusion that there exist no other causes or reasons for deviations of exponential growth series from the Benford configuration except for the rationality of their LOG(F). And even though the next concluding statement may sound highly puzzling and paradoxical to the psychologist, psychiatric, or social scientist, yet it should be believed, namely that: "Any **deviant** digital behavior of a growth series differing from the Benford configuration is always exclusively associated and correlated with the **rationality** of its related LOG(F) fraction". There seems to be nothing else adversely affecting digital behavior except for the argument given in this section regarding the rational march along the log-axis.

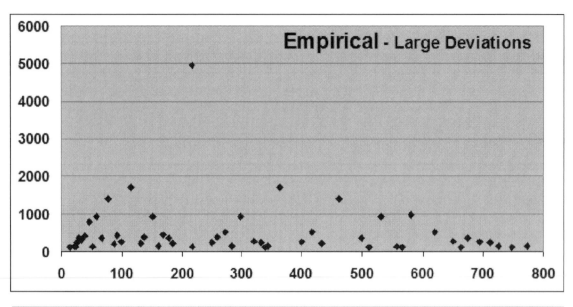

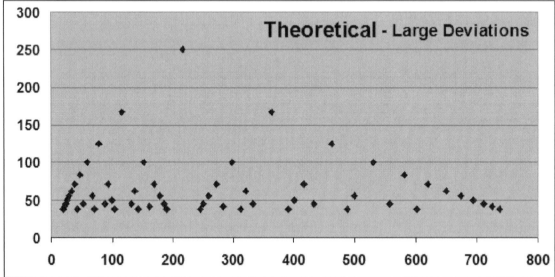

Figure 6.16: Compatibility between Theoretical & Empirical in Large Deviations

[9] Wider Damage Done to Short Normal Series by Anomalous Series

The excellent fit between the empirical and the theoretical seen above; and the very positive conclusion of chapters 4 and 8, namely that the vast majority of exponential growth series are nearly perfectly Benford; both assume that the series are of considerable length, with at least 822 periods up to 3000 periods.

For shorter series with length of only100 periods or so for example, the effects of the rational series are felt far and wide, leaving many more series with considerable deviation from Benford.

Consequently, in order to examine how very short series behave digitally, another computer program is run, empirically checking short series for deviation from Benford. The program starts at 2% and ends at 890%, in tiny refine increments of 0.0098375462%. It checks 90,266 exponential growth series in total. It uses the quantity 3 as the initial base value for all the series.

Just as was done in the previous two empirical programs, utilizing a variety of lengths, here too, in order to allow low rates to comply with the logarithmic requirement that exponent difference between the maximum and the minimum should be approximately close to an integer, a variety of growth periods are used via the set {101, 102, 103, 104, 105, 106, 107, 108, 109, 110}. The program then registers only the minimum SSD for all these series, so as to arrive at the most complying series.

The deliberate avoidance of the range (1%, 2%) is due to the fact that the set of the above choices of 10 lengths is still not sufficient for very low growth rates below 1%, and many more choices for the length are needed there in order to comply with the logarithmic requirement that exponent difference should be close to an integer.

The program has registered 27,554 growth rates with SSD over 10, out of the 90,266 total. In other words, about 30.5% of the series were found to deviate by registering over the 10 SSD threshold value. About a third of very short exponential growth series are not really Benford! And this highly pessimistic conclusion is all due to these 773 land mines of the rational series sparsely laying there on the long minefield from 2% to 890%!

Such high **30.5%** portion of affected [short] series with SSD over 10 is compared with only **8.4%** portion of affected [long] series with SSD over 10 as was seen earlier in chapter 4.

Figure 6.17 depicts the wider damage done to the entire neighborhood of the rational series 364.1588% growth rate - associated with the rationality LOG(F) = 2/3 – affecting many short irrational series all around it due to its presence. This chart is derived from the above computer program utilizing very short series of lengths varying from 100 to 110 periods, and starting from the initial base value of 3. The comparison between Figure 6.17 and Figure 6.10 reveals that short series suffer by far greater damage from the 364.1588% rebellious rational series. Here for short series with only 100 to 110 growth periods, much wider range of about 5.5% is adversely affected, as compared with only about 1.4% affected range of the longer series with 822 to 3000 growth periods.

The same conclusions and results are expected to be seen for the neighborhood of the rational series of 383.2930% growth rate - associated with the rationality LOG(F) = 13/19. As seen in Figure 6.11, only very narrow range of about 0.15% units is affected, but this is so due to the consideration of long finite series with 822 to 3000 growth periods. Much shorter series of about 100 growth periods should show a wider range of affected series than merely 0.15%.

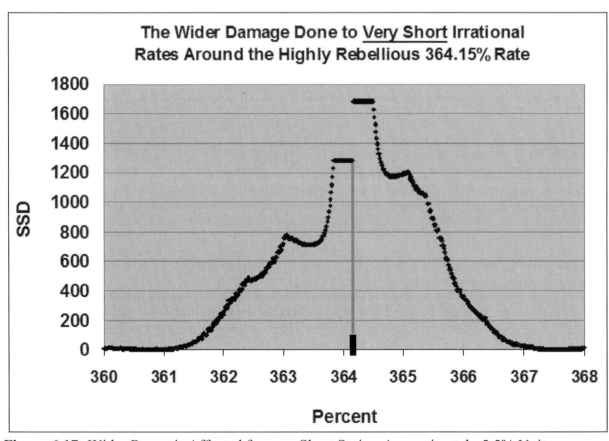

Figure 6.17: Wider Range is Affected for very Short Series, Approximately 5.5% Units

[10] Random Exponential Growth Series

Exponential growth series are constructed with a constant (fixed) multiplicative factor $F_{CONSTANT}$. Another possibility to consider is that of the random multiplicative factor F_{RANDOM}. For example, random factors constantly chosen from the Uniform on (1.23, 1.67) for each growth period could be utilized, leading to an exponential growth series with haphazard and constantly changing growth rate, randomly fluctuating between 23% and 67%. We shall denote U_N as the Nth realization from the Uniform Distribution, standing for the multiplicative factor F_{RANDOM}.

$\{B, BF, BF^2, BF^3, \ldots, BF^N\}$ standard exponential growth series

$\{B, BU_1, BU_1U_2, BU_1U_2U_3, \ldots, BU_1U_2 U_3\ldots U_N\}$ random exponential growth series

Such random selections of growth rates yield an overall uniform log distribution just as the fixed growth rate series usually do, assuming there are plenty of elements in the series. Let us examine how the series advances along the log-axis:

$\{LOG(B),$
$LOG(B) + LOG(U_1),$
$LOG(B) + LOG(U_1) + LOG(U_2),$
$LOG(B) + LOG(U_1) + LOG(U_2) + LOG(U_3),$

… etc. …

$LOG(B) + LOG(U_1) + LOG(U_2) + LOG(U_3) + \ldots + LOG(U_N) \}$

Clearly log series here is an additive random walk on the log-axis. It can be assumed that each new element gives rise to totally new mantissa value, and this in turns implies an overall uniform and flat 'density' on the mantissa space of (0, 1) albeit in a discrete manner. Hence the random growth series is Benford!

In contrast with the digital pitfalls of deterministic anomalous exponential growth series, random exponential growth series on the other hand are always nicely logarithmic. Here we never stumble upon the perils of LOG(F) whose multiples always add up exactly to an integral value on the log-axis in a consistent manner (i.e. anomalous series). Repeated additions of random $LOG(F_{RANDOM})$ values result in covering the entire (0, 1) mantissa space fairly and evenly whenever there are plenty of such accumulations.

As an example of a specific random model, log values march forward along the log-axis as in $LOG_{N+1} = LOG_N + Uniform(0, 1)$, as depicted in Figure 6.18. More generally, log values may be modeled as in: $LOG_{N+1} = LOG_N + [Almost Any Positive Random Variable]$.

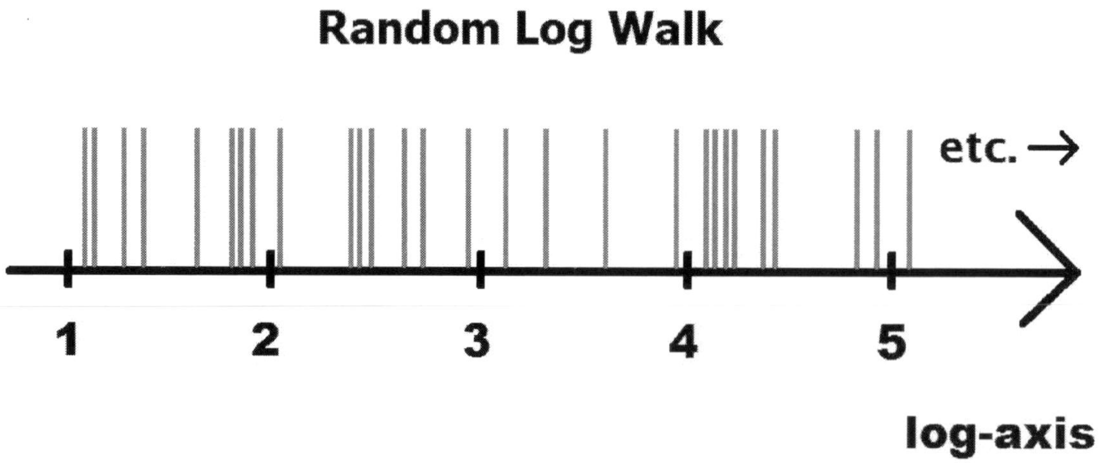

Random Log Walk

Figure 6.18: Uniformity of Mantissa via Random Additions of Uniform(0, 1)

[11] Super Exponential Growth Series

Classic exponential growth series such as $\{B, BF^1, BF^2, BF^3, BF^4, \ldots, BF^N\}$ are characterized by having a constant F multiplicative factor, and where the set of factors in going from one element to the next element is the vector $\{F, F, F, F, F, \ldots, F\}$ with $F > 1$. Super exponential growth series is a case where the factors themselves are growing exponentially, and where logarithmic behavior is found in spite of the rather odd nature of such series!

The set of the growing factors themselves is defined as $\{F^1, F^2, F^3, F^4, F^5, \ldots, F^N\}$ where F stands for the initial factor as well as for the factor by which the factors themselves are growing. The super exponential series itself is then written in terms of its construction, term by term, as:

$$\{B, \quad B(F^1), \quad BF^1(F^2), \quad BF^1F^2(F^3), \quad BF^1F^2F^3(F^4), \quad BF^1F^2F^3F^4(F^5), \quad BF^1F^2F^3F^4F^5(F^6), \quad \ldots, \\ BF^1F^2F^3F^4F^5F^6 F^7F^8 \ldots F^{N-2}F^{N-1}(F^N)\}$$

which is simplified as the super series:

$$\{B, BF^1, BF^3, BF^6, BF^{10}, BF^{15}, BF^{21}, \ldots, BF^{(N*N + N)/2}\}$$

This is essentially of the same format of the classic exponential growth series, but instead of having the exponents of the factors increasing sequentially as in 1, 2, 3, 4, 5, 6, etc., they are expanding more rapidly as in 1, 3, 6, 10, 15, 21, etc., namely as in $(N*N + N)/2$. In one particular computer calculation example, base B is chosen as 100, and the factor of the factors F is chosen as 1.0008. The 1st digits distribution of this super exponential growth series considering the first 1328 elements is {31.4, 16.8, 12.3, 9.9, 7.4, 5.6, 5.9, 5.6, 5.0}, and SSD is a rather low value of 4.3, signifying a great deal of closeness to the logarithmic.

Let us examine how the series advances along the log-axis:

$\{LOG(B),$
$LOG(B) + LOG(F),$
$LOG(B) + LOG(F) + 2*LOG(F),$
$LOG(B) + LOG(F) + 2*LOG(F) + 3*LOG(F),$
$LOG(B) + LOG(F) + 2*LOG(F) + 3*LOG(F) + 4*LOG(F),$
$LOG(B) + LOG(F) + 2*LOG(F) + 3*LOG(F) + 4*LOG(F) + 5*LOG(F),$
$LOG(B) + LOG(F) + 2*LOG(F) + 3*LOG(F) + 4*LOG(F) + 5*LOG(F) + 6*LOG(F),$
$\ldots$ etc. $\ldots \}$

Clearly, **the super growth series speeds up along the log-axis**, namely that distances between consecutive log values of the super series are increasing, assuming the series represents exponential growth so that $F > 1$ and thus LOG(F) is positive, as opposed to exponential decay where $F < 1$ and where LOG(F) is negative.

It is very hard to imagine any scenario here where a particular value of F could lead to a situation where exactly N number of cycles (periods) always yields an integral value thus leading to an anomalous series where only very few selected mantissa values are being generated over and over again by the series. Hence it seems that any possible pitfall of LOG(F) = L/T rational number is totally irrelevant here to logarithmic behavior. Surely, the super exponential growth series does not march in constant steps of LOG(F), but rather in steps of [increasingly] multiple values of LOG(F).

The rationale for the logarithmic behavior of super exponential growth series is that it speeds up along the log-axis in a random and haphazard fashion as far as mantissa is concern. In other words, it is logarithmic because the series tramps upon the log-axis in a disorganized and uneven manner in relation to the log integers, resulting in the accumulation of all sorts of mantissa values. This in turns guarantees that mantissa is approximately uniform since the process gives no preference to any type or any sub-set of mantissa space, but rather keeps picking truly 'random' mantissa values as the series marches along.

Figure 6.19 depicts one such [imaginary] accelerated march of super exponential series along the log-axis, where log values are clearly speeding up and distances between consecutive log values are constantly increasing. See Kossovsky (2014) chapter 99 for more details and discussion about super exponential growth series.

Accelerated Log March - Super Exponential Growth

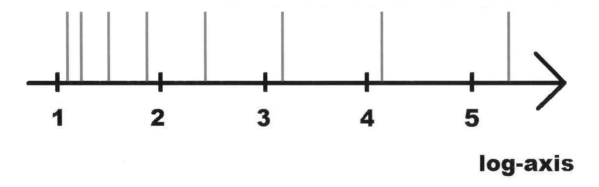

Figure 6.19: Uniformity of Mantissa via Accelerated Log-Axis March

[12] The Factorial Sequence as Super Exponential Series

The Factorial Sequence $\{N!\} = \{1!, 2!, 3!, 4!, \ldots\}$ is logarithmic in the limit as N goes to infinity, or in a practical sense and approximately so as N becomes quite large. The sequence is considered to be a deterministic multiplicative process; there is nothing random about it.

Examining log distances between consecutive elements of the Factorial Sequence we get:

$LOG(X_{N+1}) - LOG(X_N)$
$LOG((N + 1)!) - LOG(N!)$
$LOG(N!(N + 1)) - LOG(N!)$
$LOG(N!) + LOG(N + 1) - LOG(N!)$

$LOG(N + 1)$

Clearly, the Factorial Sequence marches along the log-axis in an accelerated way where distances between consecutive elements are constantly increasing, as in super exponential growth series. This nicely explains the digital logarithmic nature of the sequence! The sequence keeps minting totally new mantissa values in an unorganized way and without any coordination with respect to integral log marks. Figure 6.20 depicts the accelerated march along the log-axis for the first 19 elements of the Factorial Sequence.

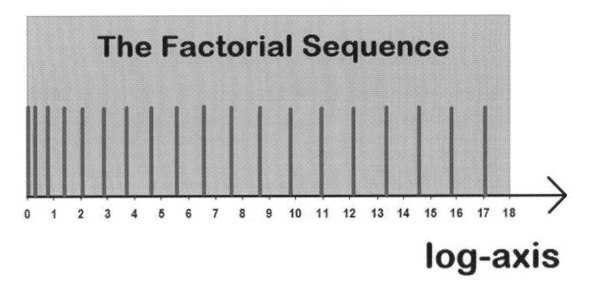

Figure 6.20: Accelerated March Along the Log-Axis – Factorial Sequence

An alternative vista of the Factorial Sequence is to view it simply as an exponential growth with increasing growth factors. This vista is surely justified when written as $\{N!\} = \{1!, 2!, 3!, 4!, \ldots\}$ = $\{1, (1)*2, (1*2)*3, (1*2*3)*4, \ldots\}$. Hence the sequence can be written as in: $X_{N+1} = N*X_N$. In conclusion: The Factorial Sequence is also a super exponential growth series, and therefore its accelerated march along the log-axis is certainly expected, consistent, and in harmony with all that was discussed in the previous chapter.

Empirically testing the finite and short Factorial Sequence $\{1!, 2!, 3!, \ldots, 168!, 169!, 170!\}$ yields the first digits distribution $\{31.8, 17.1, 12.9, 7.1, 7.1, 5.9, 3.5, 8.2, 6.5\}$, with SSD = 30.1.

[13] The Self-Powered Sequence as Super Exponential Series

The Self-Powered Sequence $\{N^N\} = \{1^1, 2^2, 3^3, 4^4, ...\}$ is logarithmic in the limit as N goes to infinity, or approximately so as N becomes large. This sequence is a deterministic one, yet it differs profoundly from the classic exponential series as there exists no obvious connecting link between consecutive terms by the way of a simple multiplicative factor as in $X_{N+1} = Factor*X_N$.

Examining log distances between consecutive elements of the Self-Powered Sequence we get:

$LOG(X_{N+1}) - LOG(X_N)$
$LOG((N+1)^{(N+1)}) - LOG(N^N)$
$(N+1)*LOG(N+1) - N*LOG(N)$
$N*LOG(N+1) + LOG(N+1) - N*LOG(N)$
$N*LOG(N+1) - N*LOG(N) + LOG(N+1)$
$N*LOG((N+1)/N) + LOG(N+1)$
$N*LOG(1+1/N) + LOG(N+1)$
$LOG((1+1/N)^N) + LOG(N+1)$

$\lim_{N\to\infty} \left(1+\frac{1}{N}\right)^N = e$, hence in the limit as N goes to infinity, log distance in between is:

$LOG(e) + LOG(N+1)$

Clearly, the Self-Powered Sequence marches along the log-axis in an accelerated way, and where distances between consecutive elements are constantly increasing, as in super exponential growth series. This nicely explains the digital logarithmic nature of the sequence! The sequence keeps minting totally new mantissa values in an unorganized way and without any coordination with respect to integral log marks. Figure 6.21 depicts the accelerated march along the log-axis for the first 14 elements of the Self-Powered Sequence.

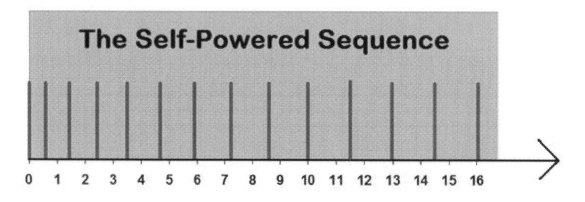

Figure 6.21: Accelerated March Along the Log-Axis – Self-Powered Sequence

Empirically testing the finite and short Self-Powered Seq. $\{1^1, 2^2, 3^3, ... , 141^{141}, 142^{142}, 143^{143}\}$ yields the first digits distribution $\{36.4, 14.7, 16.8, 9.1, 6.3, 2.8, 4.2, 7.0, 2.8\}$, with SSD = 93.6.

[14] Four Distinct Styles of Log-Axis Walk Leading to Benford

Figure 6.22 summarizes the four distinct ways a sequence could march along the log-axis yielding uniformity of mantissa; (1) Normal exponential growth series with LOG(F) being an irrational number; (2) Anomalous exponential growth series with a rational LOG(F) of a very small value; (3) Super exponential growth series, the Factorial Series, the Self-Powered Series, all characterized by accelerated march along the log-axis where distances between consecutive log values are constantly increasing; (4) Random walk along the log-axis.

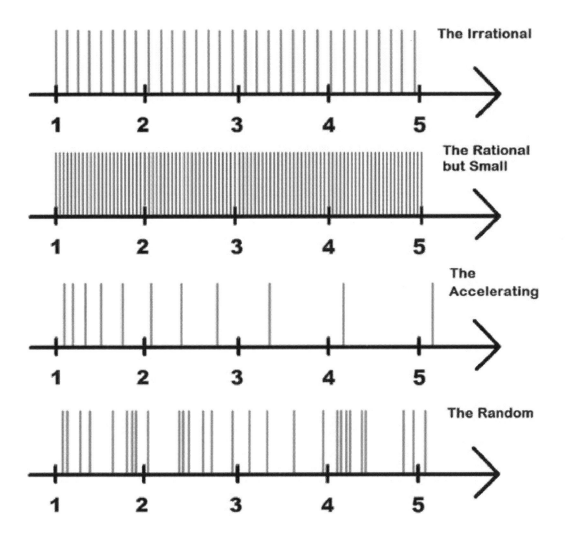

Figure 6.22: Four Distinct Logarithmic Ways for Sequences to March along the Log-Axis

What should we expect a priori - in extreme generality - from the vast majority of mathematical sequences with regards to Benford behavior, without performing the tedious and endless task of examining each and every one of them? The obvious answer is that the vast majority of them are surely Benford, because it's a bit rare for any particular sequence to march along the log-axis in a coordinated and organized way with respect to the integer marks of the log-axis. Such a deliberate march requires a great deal of effort and close attention on the part of the sequence, and very few sequences are willing to work so hard so as to obtain the abstract, intangible, and vague goal of uniformity of mantissa. Most sequences prefer to march merrily, freely, and relaxingly along the log-axis as they see fit, guaranteeing Benford behavior for the corresponding values they generate on the x-axis.

[15] Continuous Exponential Growth Series

Thus far we have dealt only with discrete exponential growth series where quantity count takes place at the end of a particular length of time, say at the end of each month, or at the end of each year. Perhaps the quantity actually jumps at an instant over and over again at the end of each period of time, such as a bank account gaining interest at the last minute of the last day of each month, or earning interest at an instant just before midnight on each December 31 day.

For a continuous exponential growth series, the growing variable X is a function of time, and it is expressed as $X(T) = BF^T$ where B is the initial base value at time 0, F is the fixed multiplicative factor of the growth rate per one integral unit time period, and T is the continuous real time variable, comprising not only of integral values such as 1 year, 2 years, 3 years, and so on, but also of the unaccountably infinite values of all the real points in between the integers. Assuming that the time unit in the definition of factor F is measured in years, then quantity X is being continuously recorded say every minute, or even every second if humanly possible, and not only once a year. Hence the data set in focus here is not merely the small collection of say December 31 readings for several years or few decades, but rather the extremely large data set comprising the detailed evolution of quantity X examined each minute, or even each second if possible, and so on, recording X extremely frequently each tiny interval of time. Of interest is the digital configuration of such large data set, and the quest to demonstrate that the data set - having proper span by standing exactly between integral powers of ten or having an integral exponent difference - is perfectly logarithmic in the limit as time measurement shrinks from hours, to minutes, to seconds, and then to infinitesimal small time intervals, approaching zero from above in the limit. In reality for all practical purposes, recording quantity X each minute is considered to be sufficiently small as a time frame for Benford behavior when the growth period in the definition of F is measured in years.

As an example, let us consider 5% exponential growth from X = 1 to X = 10. Here variable X as a function of time is expressed as $X(T) = 1*(1.05^T)$. The motivation for the start at 1 emanates from the fact that at 1 we start anew the natural cycle of the 1st digits. The motivation for the termination at 10 emanates from the fact that at 10 we end the first complete cycle of the 1st digits. Such a deliberate span from 1 to 10 ensures that all nine digits would cycle exactly once and fully so, so that each digit would be allowed to express its full potential; and that all the digits enjoy equal opportunity. Figure 6.23 depicts this particular continuous exponential growth series from 1 to 10 with F = 1.05, namely 5% growth per period. Clearly, for any X value within the interval [1, 2) the first digit is always 1, such as for X values 1.345, 1.876503, 1.5, 1.0, 1.999, 1.06214, and so on. In general here, all values falling on the sub-interval [d, d + 1) are such that the first leading digit is d (assuming d is an integer from 1 to 9).

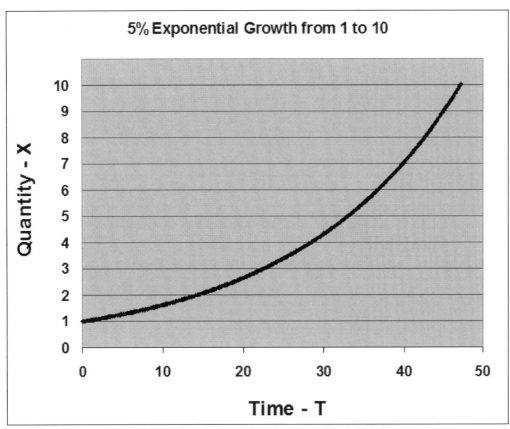

Figure 6.23: Continuous 5% Exponential Growth from 1 to 10

Hence, the amount of time it takes the series to grow from 1 to 2 is equivalent to the amount of time the series spends while digit 1 is leading; and the amount of time it takes the series to grow from 2 to 3 is equivalent to the amount of time the series spends while digit 2 is leading, and so on. What is needed to be calculated here is simply the relative proportions of the time spent on all the nine first digits.

Figure 6.24 depicts the critical moments when the series encounters those crucial integral values of 1, 2, 3, 4, 5, 6, 7, 8, 9, and 10, when the first digit changes.

Figure 6.25 depicts the time intervals the series spends on each digit, as well as the proportion of time each digit earns, and which 'happened' to be exactly Benford!

t - time	X - Quantity
0.0	1.0
14.2	2.0
22.5	3.0
28.4	4.0
33.0	5.0
36.7	6.0
39.9	7.0
42.6	8.0
45.0	9.0
47.2	10.0

Figure 6.24: Integral Quantity Points & Associated Time

Change in X	Time Interval	% Time Interval
1 to 2	14.2	30.1%
2 to 3	8.3	17.6%
3 to 4	5.9	12.5%
4 to 5	4.6	9.7%
5 to 6	3.7	7.8%
6 to 7	3.2	6.8%
7 to 8	2.7	5.7%
8 to 9	2.4	5.1%
9 to 10	2.2	4.7%

Figure 6.25: Relative Time Spent on the Various Nine Digits is Benford

In practical terms and as far as actual data sets are concerned in the context of Benford's Law, what is meant abstractly by the phrase "the relative proportion of the time spent on digit d" is simply equivalent to the proportion of numbers the data analyst has actually written down recording the growth with digit d leading, as compared with all the written numbers recorded while quantity develops from 1 all the way to 10.

For example, let us assume that the period is defined as one year, and that the data analyst measures the growing quantity every minute as it develops from 1 all the way to 10. Assuming a year with 365 days, 24 hours a day, and 60 minutes an hour, then this yields 365*24*60 or 525,600 total minutes per year. The time it takes to reach quantity 10 is 47.2 years, hence the data analyst has recorded (47.2)*(525,600) = 24,808,320 data points (for each minute) in total. The data analyst has observed that from year 33.0 to year 36.7 digit 5 was consistently leading the numbers during these 3.7 years, and which consists of (3.7)*(525,600) = 1,944,720 data points. Hence according to these observations, the proportion of digit 5 leading here is calculated as (1,944,720)/(24,808,320) = 0.0784, and which is in accordance with Benford's Law.

Figures 6.26 and 6.27 depict clearly the time intervals on the T-axis corresponding to first digit occurrences on the X-axis.

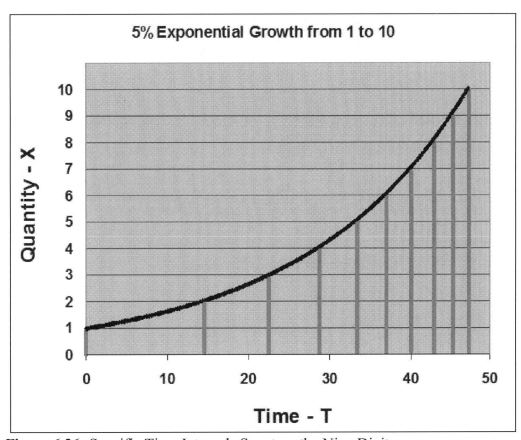

Figure 6.26: Specific Time Intervals Spent on the Nine Digits

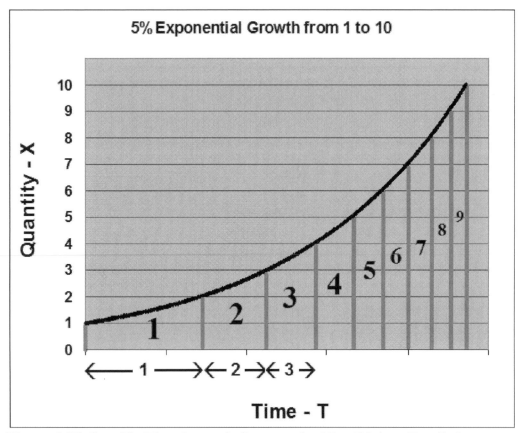

Figure 6.27: Specific Time Intervals Spent on the Nine Digits – Detailed

Let us consider **the general continuous case** of the exponential growth series $\mathbf{X(T) = 1*F^{T}}$ having **any growth rate with factor F**, with an initial quantity 1 at time 0, and limited to the progression from 1 to 10. The time T_2 spent to get to quantity 2 is obtained from the expression itself as follows:

$2 = F^{T2}$

$LOG_F(2) = LOG_F(F^{T2})$

$LOG_F(2) = T_2 \cdot LOG_F(F)$

$LOG_F(2) = T_2 \cdot 1$

$LOG_F(2) = T_2$

The time T_3 spent to get to quantity 3 is obtained from the expression itself as follows:

$$3 = F^{T_3}$$
$$LOG_F(3) = LOG_F(F^{T_3})$$
$$LOG_F(3) = T_3 \cdot LOG_F(F)$$
$$LOG_F(3) = T_3 \cdot 1$$
$$LOG_F(3) = T_3$$

The total time T_{10} spent to get to quantity 10 is obtained from the expression itself as follows:

$$10 = F^{T_{10}}$$
$$LOG_F(10) = LOG_F(F^{T_{10}})$$
$$LOG_F(10) = T_{10} \cdot LOG_F(F)$$
$$LOG_F(10) = T_{10} \cdot 1$$
$$LOG_F(10) = T_{10}$$

The time T_{10} is the time of one full cycle for all digits 1, 2, 3, 4, 5, 6, 7, 8, and 9.

Hence, time spent between 2 and 3 where 1st digit is always 2 =
Time to get to 3 – Time to get to 2 =
$$LOG_F(3) - LOG_F(2) =$$
$$LOG_F(3/2)$$

And as a proportion in comparison to the total time needed for one full cycle for all the digits in the progression from 1 to 10:

Overall time proportion for digit 2 = $LOG_F(3/2)$ / $LOG_F(10)$

Utilizing the logarithmic identity
$$LOG_A(X) = LOG_B(X) / LOG_B(A)$$
for the numerator and for the denominator we get:

$$LOG_F(3/2) / LOG_F(10)$$
$$[LOG_{10}(3/2) / LOG_{10}(F)] / [LOG_{10}(10) / LOG_{10}(F)]$$
$$[LOG_{10}(3/2)] / [LOG_{10}(10)]$$
$$[LOG_{10}(3/2)] / [1]$$
$$LOG_{10}(3/2)$$
$$LOG_{10}((2 + 1)/2)$$
$$LOG_{10}(1 + 1/2)$$

Which is the proportion for digit 2 according Benford's Law!

Let us prove this in general for the time spent between (d) and (d+1), where 1st digit is always d:

The time T_d spent to get to quantity d is obtained from the expression itself as follows:

$d = F^{T_d}$
$LOG_F(d) = LOG_F(F^{T_d})$
$LOG_F(d) = T_d \cdot LOG_F(F)$
$LOG_F(d) = T_d \cdot 1$
$LOG_F(d) = T_d$

Similarly the time spent to get to quantity $(d + 1)$ is $LOG_F(d + 1) = T_{d + 1}$.

Hence, time spent between (d) and $(d + 1)$ where 1st digit is always $d =$
Time to get to $(d + 1) -$ Time to get to $(d) =$
$LOG_F(d + 1) - LOG_F(d) =$
$LOG_F((d + 1)/d)$

And as a proportion in comparison to the total time needed for one full cycle for all the digits in the progression from 1 to 10:

Overall time proportion of digit $d = LOG_F((d + 1)/d) / LOG_F(10)$

Utilizing the logarithmic identity
$LOG_A(X) = LOG_B(X) / LOG_B(A)$
for the numerator and for the denominator we get:

$LOG_F((d + 1)/d) / LOG_F(10)$
$[LOG_{10}((d + 1)/d) / LOG_{10}(F)] / [LOG_{10}(10) / LOG_{10}(F)]$
$[LOG_{10}((d + 1)/d)] / [LOG_{10}(10)]$
$[LOG_{10}((d + 1)/d)] / [1]$
$LOG_{10}((d + 1)/d)$
$LOG_{10}(d/d + 1/d)$
$LOG_{10}(1 + 1/d)$

Which is exactly Benford's Law!

Little reflection is needed to realize that the above proof can be extended to include exponential growth spanning any adjacent integral powers of ten points such as growth from 10 to 100, from 100 to 1000, or from 1000 to 10000, and so on. The transformation of (1, 10) growth into (10, 100) growth is leading-digits-neutral, since digit 3 leads the number 3.987055 say, just as it does lead the number 39.87055. Moreover, even for growth spanning non-adjacent multiple powers of ten points, such as say from 1 to 10000, the same argument as in the above proof holds separately for each sub-interval spanning adjacent integral powers of ten, so that the entire span of (1, 10000) is Benford since it is so on each of its constituent sub-intervals (1, 10), (10, 100), (100, 1000), and (1000, 10000). Furthermore, any continuous exponential growth series spanning integral exponent difference such as those growing say from 3 to 30, from 7.8233 to 78.233, or from 43.129 to 43129, and in general growing from R to $R*10^{INTEGER}$, are all perfectly Benford as well. In a nutshell, for the example of 3 to 30 growth, viewed in terms of the generic 1 to 10 growth, the 'missing' sub-interval of (1, 3) on the left, is being exactly and perfectly substituted by the 'extra' sub-interval of (10, 30) on the right.

The above proof has been constructed in extreme generality, for any continuous growth rate of factor F, and without requiring LOG(F) to be irrational (except that the series is required to start at 1 and to terminate at 10). Hence, on the face of it, this last result seems to contradict the constraint which excludes anomalous exponential growth series from logarithmic behavior. As an example, for the continuous **5.92537251773%** yearly growth, the yearly steps taken along the log-axis are in units of LOG(1 + 5.92537251773/100) = LOG(1.0592537251773) = 0.025 = 1/40 = rational number, yet the above proof guarantees logarithmic behavior here!? Nonetheless, harmony and consistency prevail, and our delicate and complex digital edifice does not fall into ruin and contradiction in the least, since what is actually recorded here is not how quantity grows per year, but rather how quantity grows say per second, and that rate is the extremely low **0.0000001825%** growth, calculated as 100*[(60*60*24*365)th root of 1.0592537251773 - 1]. LOG(F) for this very low growth rate per second is calculated as the extremely tiny value of LOG(1 + 0.0000001825/100) = 0.0000000007927447761.

Admittedly, this extremely low rate of 0.0000001825% is also anomalous since its LOG(F) is also rational, yet the exact logarithmic behavior of continuous exponential growth series with rational LOG(F) comes under the protective umbrella of the exception given for anomalous rates whenever LOG(F) is sufficiently small compared with unit length on the log-axis (i.e. growth rate is very low – relating in general to high T values). The fact that continuous exponential growth series is considered here as opposed to the discrete integral case implies that we take the population pulse each very tiny interval of time, say each second, and therefore the growth rate per each such infinitesimal small time interval is extremely small and approaches 0% growth rate from above in the limit. Not only does the creature lack long legs here for skipping undesirable mantissa sections, but it has such extremely short ones that it is literally forced by default to walk all over the (0, 1) mantissa interval, not omitting any spots or corners at all, hence the series is perfectly and exactly logarithmic.

[16] The Discrete Embedded within the Continuous

Turning the **continuous** 5% exponential growth from 1 to 10 into a **discrete** series by measuring quantity $X(T) = 1*(1.05^T)$ once each period (say once a year), namely utilizing only the integral values of T = {0, 1, 2, 3, … , 47}, yields 48 elements (including the initial base value considered as the 1st element):

1.00	1.05	1.10	1.16	1.22	1.28	1.34	1.41	1.48	1.55	1.63	1.71
1.80	1.89	1.98	2.08	2.18	2.29	2.41	2.53	2.65	2.79	2.93	3.07
3.23	3.39	3.56	3.73	3.92	4.12	4.32	4.54	4.76	5.00	5.25	5.52
5.79	6.08	6.39	6.70	7.04	7.39	7.76	8.15	8.56	8.99	9.43	9.91

In Figure 6.28 the black dots represent the quantities for each integral time value. It should be noted that these 48 points are perfectly embedded within the continuous curve of 5% growth of the previous charts as seen in Figures 6.23 to 6.27.

Here 1st digits distribution is {31.3, 16.7, 12.5, 8.3, 8.3, 6.3, 6.3, 6.3, 4.2}, and SSD = 6.3, having only slight deviation from the perfect Benford behavior of the continuous growth.

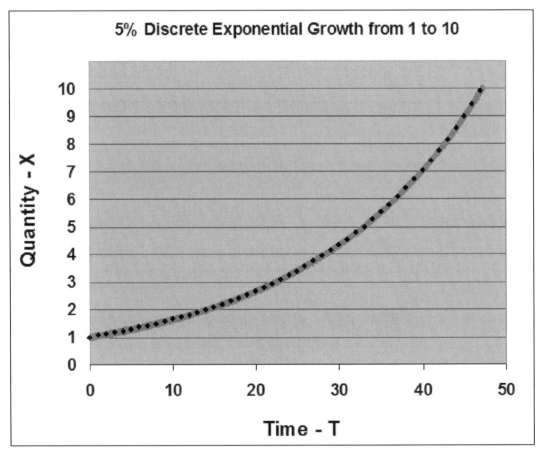

Figure 6.28: The Discrete 5% Growth Embedded within the Continuous 5% Growth

Since this 5% exponential growth series is discrete; and since 5% is not as small a rate as to be considered immune to rebellious behavior, there exists the possibility of it being an anomalous series. As it happened, LOG(1.05) = 0.0211893, and this is not a rational number. Moreover, the nearest rational growth series are the mildly rebellious series of 4.91397291% belonging to the rationality L/T = 1/48 = 0.0208333, and 5.02110796% belonging to the rationality L/T = 1/47 = 0.0212766. Both of these potentially disruptive series are quiet mild in nature, and they are not as near our 5% normal series, all of which guarantees the approximate logarithmic behavior of the 5% discrete series.

Turning the **continuous** 5% exponential growth from 1 to 10 into a **discrete** series by measuring quantity $X(T) = 1*(1.05^T)$ once every 3 periods (say once each 3-year interval), namely utilizing the particular set of values of T = {0, 3, 6, 9, 12, 15, 18, 21, 24, 27, 30, 33, 36, 39, 42, 45}, yields 16 elements:

1.00	1.16	1.34	1.55	1.80	2.08	2.41	2.79	3.23	3.73	4.32	5.00
5.79	6.70	7.76	8.99								

In Figure 6.29 the black dots represent the quantities for each 3-year time interval. These 16 points are perfectly embedded within the continuous curve of 5% growth, although the effective (cumulative) growth factor from one point to the next, namely the growth factor for each 3-year period, is $(1.05)^3 = 1.157625$, and thus the associated growth rate here is actually 15.8%.

Here 1st digits distribution is {31.3, 18.8, 12.5, 6.3, 12.5, 6.3, 6.3, 6.3, 0.0}; having some noticeable deviation from the perfect Benford behavior of the continuous; especially having the unusual 0% proportion for digit 9, and with SSD = 58.1. The reason for the deterioration in digital fit to Benford for the higher 15.8% growth of the 3-year measurements program is due to lack of sufficient number of data points, being too-discrete-like. The yearly measurements program of discrete 5% growth enjoys the advantage of having more points, and thus being more continuous-like. Yet it should be emphasized that both of these discrete series represent (approximately) the same overall growth from 1 to about 10 – and both are embedded within the same continuous 5% growth curve. The 3-year measurements program grows from 1 to 8.99, while the yearly measurements program grows from 1 to 9.91, and such similarity in overall growth should be acknowledged and considered.

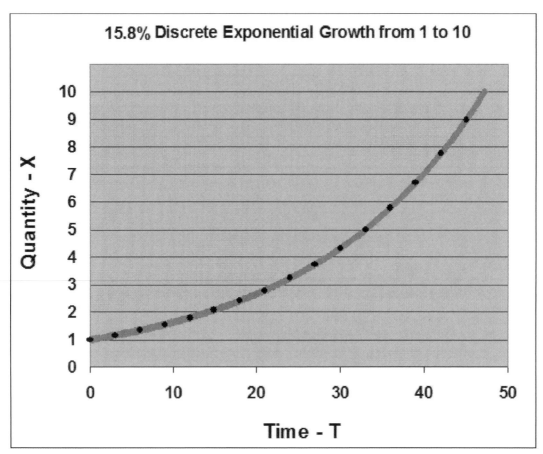

Figure 6.29: The Discrete 15.8% Growth Embedded within the Continuous 5% Growth

Turning the **continuous** 5% exponential growth from 1 to 10 into a **discrete** series by measuring quantity $X(T) = 1*(1.05^T)$ once every 5 periods (once each 5-year interval), namely utilizing the particular set of values of T = {0, 5, 10, 15, 20, 25, 30, 35, 40, 45}, yields the very small set of 10 elements:

1.00   1.28   1.63   2.08   2.65   3.39   4.32   5.52   7.04   8.99

In Figure 6.30 the black dots represent the quantities for each 5-year time interval. These 10 points are perfectly embedded within the continuous curve of 5% growth, although the effective (cumulative) growth factor from one point to the next, namely the growth factor for each 5-year period, is $(1.05)^5 = 1.276281$, and thus the associated growth rate here is actually 27.6%.

Here 1st digits distribution is {30, 20, 10, 10, 10, 0, 10, 10, 0}, with SSD = 123.6. The markedly larger deviation from the Benford digital configuration of the continuous case can be squarely blamed on the severe lack of sufficient number of data points here. The series is being severely punished for its discreteness. Yet, potentially, this 27.6% series can still remedy itself by simply continuing to march forward along the 5% continuous growth curve, well beyond 8.99 and even well beyond 100, accumulating many more points and passing numerous integral power of ten points along the way, and then successfully becoming nearly logarithmic.

Since $LOG((1.05)^5)$ is not a rational number; the series is then not rebellious, and its re-entrance points into each new integral power of ten interval is random-like and different, generating a variety of mantissa values. Since the 5% territory upon which the 27.6% series stamps upon is itself perfectly Benford, then randomly 'picking' points from it would eventually lead to a set of values which are also Benford.

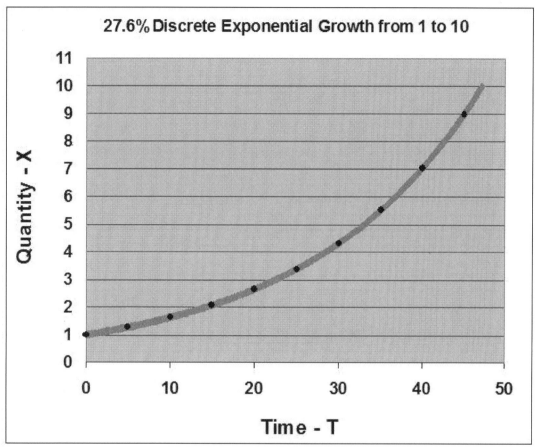

Figure 6.30: The Discrete 27.6% Growth Embedded within the Continuous 5% Growth

It must be emphasized once again that the same physical <u>continuous</u> growth (say a particular bacterial population growing in a lab) yields distinct digital configurations depending on how often <u>discrete</u> readings of population count takes place. As a much more illustrative and dramatic example than the 1-year versus 3-year or 5-year data recording schemes above, let us consider **5% per minute** exponential growth of the laboratory culture bacteria Salmonella Choleraesuis from an approximate initial level of 100,000 as in $X(T) = 100000*(1.05^T)$, and two data analysts named George and Frank having two distinct measuring styles. George is very fastidious and hard-working, thus he constantly takes measurements of the colony every minute. Frank on the other hand is quite laid-back and relaxed, his mind is focused more on personal matters than work, and he makes the effort to check the colony only once every hour. The general manager at the laboratory supervising George and Frank obliges them to come up with 170 carefully measured data points. This would require hard-working George only little less than 3 hours, while for Frank the task would require about 7 days. For George who stops by the laboratory every <u>minute</u>, the growth rate is 5%, while for Frank who stops by every hour, the growth factor F is $(1.05)^{60} = 18.67918$, namely a whopping 1767.9% <u>hourly</u> growth rate.

Frank fast hourly growth series has the huge 214.86 order of magnitude. The first 5 elements and last 3 elements are shown below:

100000	1867919	34891199	651739184	12173957374 ... etc. ...
$2.074*10^{217}$	$3.874*10^{218}$	$7.236*10^{219}$		

George slow minute growth series has the small 3.58 order of magnitude. The first 5 elements and last 3 elements are shown below:

100000	105000	110250	115762	121550 ... etc. ...
345631204	362912765	381058403		

Digital results for George and Frank are as follow:

George Series - Low 5% Growth:	{33.5, 19.4, 13.5, 8.2, 6.5, 5.9, 4.7, 4.7, 3.5}
Frank Series - High 1767.9% Growth:	{30.0, 17.1, 12.9, 10.0, 7.1, 7.1, 6.5, 4.1, 5.3}
Benford's Law 1st digits Distribution:	**{30.1, 17.6, 12.5, 9.7, 7.9, 6.7, 5.8, 5.1, 4.6}**

Hard-working George is quite disappointed. In spite of his dedication to the job and his best effort, first digits for him markedly deviate from Benford with SSD = 23.2, while easy-going Frank unfairly earns a much nicer fit to Benford with SSD = 3.4. The reason for the deterioration in digital fit for George's low growth is due to lack of better calibration of the ratio of last to first elements, not being close to an integral power of ten, being forced to abruptly stop immediately after 170 points as ordered by his boss, and such hasty termination of the series determines the value of the last element - disregarding what happens to exponent difference. Hence for George, the first element is 100000 and the last element is 381058403 with ratio of 381058403/100000, or 3810.58, which is unfortunately not close enough to any integral power of ten, such as 1000 from below or 10000 from above. In other words, that the exponent difference of 3.58 between the first and last elements is not close enough to an integer. For Frank, the first element is 100000 and the last element is $7.236*10^{219}$ with ratio of $7.236*10^{219}/100000$, or $7.236*10^{214}$, which admittedly is also not close to any integral power of ten, such as 10^{214} from below or 10^{215} from above; yet the good fortune of Frank emanates from the extremely high growth rate and the resultant very high order of magnitude of his series which excuses him from any kind of calibration attempt of the first and last elements whatsoever - as in all high growth series.

It should be noted that even though both studies done by George and by Frank trace discrete points on the same continuous 5% curve, and even though both studies are with the same number of data points, namely 170, yet they are indeed of different overall bacterial growth; one spans a shorter time interval and the other spans a much longer time interval; that they are of distinct orders of magnitude from the first to the last elements; and that bacteria grew for Frank much more than it grew for George.

[17] The Constraint of an Integral Exponent Difference for Low Growth

Let us vividly demonstrate with actual numerical examples the importance of the constraint regarding an integral exponent difference for exponential growth series; a constraint derived from the connection of exponential growth series to the k/x distribution as shall be discussed in the next chapter.

The phrase 'exponent difference' refers to the difference between the log values of the minimum (first element) and the maximum (last element). For the exponential series {2, 4, 8, 16, 32, 64, 128, 256, 512, 1024}, exponent difference is LOG(1024) – LOG(2) = 3.0103 – 0.3010 = 2.7093.

A typical real-life manifestation of exponential growth series is a single city growing at a constant rate, where the long stream of official population records by its census office measured at the end of each integral period of time (say each year on December 31) converges to Benford in the limit as the number of periods (years) goes to infinity. In practical terms, for those impatient to wait eternally for an infinite number of years until perfect Benford behavior appears, the rule of thumb for an approximate logarithmic behavior contains two distinct requirements: (**A**) Reasonably large (albeit finite) number of periods which could also be just say 100 or perhaps 500, depending on the desired level of accuracy in terms of obtaining a good fit to Benford. (**B**) The second requirement besides length considerations relates to a very particular relationship between the values of the first element (minimum) and the last element (maximum) of the series, namely that exponent difference LOG(Last) – LOG(First) should be as close as possible to an integer, or equivalently stated, that the ratio Last/First should be as close as possible to an integral power of ten such as 10, 100, 1000, and so forth.

In general, <u>low</u> growth series that adheres closely enough to the above two requirements are close enough to the logarithmic. For very <u>high</u> growth series, where the digital cycle is very short (i.e. passing integral powers of ten points quite frequently), having nearly an integral exponent difference is not really necessary; rather the only necessary condition is to have sufficient number of periods, and for such long series the logarithmic is approximately observed.

The combination of discreteness and finiteness of real-life exponential growth series inevitably ruins perfect logarithmic behavior, and this necessitates the consideration of large enough number of periods in order to get a reasonable fit to Benford in the approximate. As mentioned in Kossovsky (2014) in chapter 62 regarding Corollary I, the integral restriction on exponent difference for the k/x distribution becomes somewhat redundant in practical sense and can be ignored if exponent difference is itself quite large. How large? Well, that depends on the desired level of precision. Certainly, fairly large (non-integral) values of exponent difference over 25 or over 35 say, yield digit distribution that is extremely close to the logarithmic for all practical matters. By extension, such a waiver applies to exponential growth series as well, hence growth series with non-integral but fairly large exponent difference are almost Benford.

A brief explanation of the dispensation or exemption to the usual rule requiring an integral exponent difference for a growth series having say the very large 25.7 exponent difference, is that the vast majority of elements in the series are from the minimum (first) element until that element towards the end completing nearly exactly 25 exponent difference, while only a tiny minority beyond that element until the maximum (last) element are with the very odd 0.7 exponent difference. Since the majority of the elements with that nearly 25 exponent difference constitutes already by themselves a nearly logarithmic series, then the inclusion of a tiny extra minority with 0.7 exponent difference does not manage to ruin the logarithmic-ness of the entire series to any significant degree (by virtue of it being a small minority within the entire series).

For **high growth rates** such as say 60%, an integral power of ten is passed by the series relatively quickly, as in 1.00, 1.60^1, 1.60^2, 1.60^3, 1.60^4, 1.60^5, namely as in 1.00, 1.60, 2.56, 4.10, 6.55, 10.49, so that in just six periods the series expands by one more order of magnitude, and the first digits manage to complete one full cycle. Therefore if one considers say 100 periods for the exponential 60% growth rate, there is no need really to carefully calibrate the first and the last elements in such a way so as to ensure that exponent difference is nearly an integral value. Here there is almost nothing to gain by such calibration; the series is anyhow very close to Benford after say 100 or so elements; and order of magnitude is already very large. Let us learn a lesson or two from two manifestations of such high 60% growth series:

For an exponential 60% growth series with base B = 5 being the quantity at time 0, and considering the **first 96 elements** of the series, we get:

{5.00, 8.00, 12.80, 20.48, 32.77, … , $1.88*10^{19}$, $3.01*10^{19}$, $4.81*10^{19}$, $7.70*10^{19}$, $1.23*10^{20}$}

and exponent difference is not an integral value; $LOG(1.23*10^{20}) – LOG(5.00) = 19.391$; yet 1st significant digits are nearly Benford, coming at {30.2, 17.7, 12.5, 8.3, 8.3, 8.3, 4.2, 6.3, 4.2}, with SSD = 8.8.

Almost nothing is gained now by adjusting the number of periods so as to obtain a near integral exponent difference. To illustrate this point we shall add 3 more elements, thus obtaining nearly an integral value for the exponent difference, and it shall be observed that first digits have barely nudged from their already near perfect logarithmic configuration.

For an exponential 60% growth series with base B = 5 being the quantity at time 0, and considering the **first 99 elements** of the series, we get:

{5.00, 8.00, 12.80, 20.48, 32.77, … , $7.70*10^{19}$, $1.23*10^{20}$, $1.97*10^{20}$, $3.15*10^{20}$, $5.04*10^{20}$}

and exponent difference now is nearly an integral value; $LOG(5.04*10^{20}) – LOG(5.00) = 20.004$; while 1st digits have barely changed, coming at {30.3, 17.2, 13.1, 8.1, 9.1, 8.1, 4.0, 6.1, 4.0}. In fact, SSD = 10.8 which is actually slightly higher, indicating a slightly less perfect fit to Benford!

For this reason the Fibonacci series is always very nearly perfectly Benford so long as one considers sufficient number of elements. Even though the series is defined in terms of additions, it approaches approximately a repeated multiplication process (of the exponential growth series type) very early on, with the golden ratio 1.61803399 as the F factor. Due to its relatively high growth rate of 61.8%, the Fibonacci series passes through an integral power of ten number each 5 terms approximately, thus the requirement of an integral exponent difference can be easily waived and one need not worry at all where we start and where we end, only that enough elements are considered. As a check: $1.618^5 \approx 11 > 10$, so that in 5 terms the cumulative increase is over ten-fold, and an integral power of ten number is passed.

For **low growth rates** such as say 2%, the consideration of merely 100 or 500 elements for example necessitates a very careful selection and calibration of the end points so that exponent difference is nearly of an integral value.

For an exponential 2% growth series with base B = 7 being the quantity at time 0, and carefully and deliberately considering the **first 118 elements** of the series, we get:

{7.00, 7.14, 7.28, 7.43, 7.58, 7.73, … , 65.60, 66.91, 68.25, 69.62, 71.01}

and exponent difference is nearly an integral value; LOG(71.01) – LOG (7.00) = 1.0062; therefore first digits are nearly Benford coming at {29.7, 16.9, 12.7, 9.3, 7.6, 6.8, 6.8, 5.1, 5.1}. SSD value is low 2.1.

On the other hand, mindlessly considering the **first 174 elements** of 2% exponential growth series with base B = 7 being the quantity at time 0, we get:

{7.00, 7.14, 7.28, 7.43, 7.58, 7.73, … , 198.85, 202.83, 206.88, 211.02, 215.24}

and exponent difference is decisively non-integral, coming at LOG(215.24) – LOG(7.00) = 1.4878; therefore even though there are 56 more elements now for this longer series, yet first digits here strongly deviate from Benford coming at {40.2, 13.8, 8.6, 6.3, 5.2, 4.6, 8.0, 6.3, 6.9}. SSD is much higher now, coming at 167.3.

[18] The Connection of Exponential Growth Series to the k/x Distribution

Kossovsky (2014) discusses two essential results regarding exponential growth series:

Proposition V: Deterministic multiplication process of the exponential growth type is characterized by related log density being uniformly distributed (albeit discretely, not continuously).

Proposition VI: Deterministic multiplication process of the exponential growth type is of the k/x distribution albeit in a discrete fashion, having consistent and steady logarithmic behavior all along its entire range. The logarithmic requirement of an integral span on the log-axis (an integral exponent difference) applies to exponential growth series just as it does for any k/x distribution. The approximate [continuous] 'density' of the [discrete] exponential growth series is seen as falling steadily on the right at the constant logarithmic rate all throughout its entire range mimicking the k/x distribution.

Let us vividly demonstrate with an actual numerical example the intimate relationship between exponential growth series and k/x distribution. We choose monthly discrete readings on population that is growing continuously at 5% per year, so that $(F_{MONTHLY})^{12} = F_{YEARLY} = 1.05$, therefore $F_{MONTHLY} = $ 12th root of $1.05 = 1.004074$, namely the very low 0.407% monthly growth. The initial quantity at time 0 is 1, and exactly 567 months (47.25 years) are considered (including the first month before growth begins), after which the quantity has grown to 9.987 (slightly shy of 10). The series is:

1.000	1.004	1.008	1.012	1.016	1.021	1.025	1.029	1.033	1.037	1.041	1.046
1.050	1.054	1.059	1.063	1.067	1.072	1.076	1.080	1.085	1.089	1.094	1.098
1.103	1.107	1.112	1.116	1.121	1.125	1.130	1.134	1.139	1.144	1.148	1.153
1.158	1.162	1.167	1.172	1.177	1.181	1.186	1.191	1.196	1.201	1.206	1.211
1.216	1.220	1.225	1.230	1.235	1.240	1.246	1.251	1.256	1.261	1.266	1.271
1.276	1.281	1.287	1.292	1.297	1.302	1.308	1.313	1.318	1.324	1.329	1.335
1.340	1.346	1.351	1.357	1.362	1.368	1.373	1.379	1.384	1.390	1.396	1.401
1.407	1.413	1.419	1.424	1.430	1.436	1.442	1.448	1.454	1.460	1.465	1.471
1.477	1.483	1.490	1.496	1.502	1.508	1.514	1.520	1.526	1.533	1.539	1.545
1.551	1.558	1.564	1.570	1.577	1.583	1.590	1.596	1.603	1.609	1.616	1.622
1.629	1.636	1.642	1.649	1.656	1.662	1.669	1.676	1.683	1.690	1.696	1.703
1.710	1.717	1.724	1.731	1.738	1.745	1.753	1.760	1.767	1.774	1.781	1.789
1.796	1.803	1.811	1.818	1.825	1.833	1.840	1.848	1.855	1.863	1.870	1.878
1.886	1.893	1.901	1.909	1.917	1.924	1.932	1.940	1.948	1.956	1.964	1.972
1.980	1.988	1.996	2.004	2.012	2.021	2.029	2.037	2.045	2.054	2.062	2.070
2.079	2.087	2.096	2.104	2.113	2.122	2.130	2.139	2.148	2.156	2.165	2.174
2.183	2.192	2.201	2.210	2.219	2.228	2.237	2.246	2.255	2.264	2.273	2.283
2.292	2.301	2.311	2.320	2.330	2.339	2.349	2.358	2.368	2.377	2.387	2.397
2.407	2.416	2.426	2.436	2.446	2.456	2.466	2.476	2.486	2.496	2.506	2.517

2.527 2.537 2.548 2.558 2.568 2.579 2.589 2.600 2.610 2.621 2.632 2.643
2.653 2.664 2.675 2.686 2.697 2.708 2.719 2.730 2.741 2.752 2.763 2.775
2.786 2.797 2.809 2.820 2.832 2.843 2.855 2.866 2.878 2.890 2.902 2.913
2.925 2.937 2.949 2.961 2.973 2.985 2.998 3.010 3.022 3.034 3.047 3.059
3.072 3.084 3.097 3.109 3.122 3.135 3.147 3.160 3.173 3.186 3.199 3.212
3.225 3.238 3.251 3.265 3.278 3.291 3.305 3.318 3.332 3.345 3.359 3.373
3.386 3.400 3.414 3.428 3.442 3.456 3.470 3.484 3.498 3.513 3.527 3.541
3.556 3.570 3.585 3.599 3.614 3.629 3.643 3.658 3.673 3.688 3.703 3.718
3.733 3.749 3.764 3.779 3.795 3.810 3.826 3.841 3.857 3.873 3.888 3.904
3.920 3.936 3.952 3.968 3.984 4.001 4.017 4.033 4.050 4.066 4.083 4.099
4.116 4.133 4.150 4.167 4.184 4.201 4.218 4.235 4.252 4.270 4.287 4.304
4.322 4.340 4.357 4.375 4.393 4.411 4.429 4.447 4.465 4.483 4.501 4.520
4.538 4.557 4.575 4.594 4.612 4.631 4.650 4.669 4.688 4.707 4.726 4.746
4.765 4.784 4.804 4.823 4.843 4.863 4.883 4.903 4.922 4.943 4.963 4.983
5.003 5.024 5.044 5.065 5.085 5.106 5.127 5.148 5.169 5.190 5.211 5.232
5.253 5.275 5.296 5.318 5.339 5.361 5.383 5.405 5.427 5.449 5.471 5.494
5.516 5.538 5.561 5.584 5.606 5.629 5.652 5.675 5.698 5.722 5.745 5.768
5.792 5.815 5.839 5.863 5.887 5.911 5.935 5.959 5.983 6.008 6.032 6.057
6.081 6.106 6.131 6.156 6.181 6.206 6.232 6.257 6.282 6.308 6.334 6.360
6.385 6.411 6.438 6.464 6.490 6.517 6.543 6.570 6.597 6.623 6.650 6.678
6.705 6.732 6.759 6.787 6.815 6.842 6.870 6.898 6.926 6.955 6.983 7.011
7.040 7.069 7.097 7.126 7.155 7.185 7.214 7.243 7.273 7.302 7.332 7.362
7.392 7.422 7.452 7.483 7.513 7.544 7.575 7.605 7.636 7.667 7.699 7.730
7.762 7.793 7.825 7.857 7.889 7.921 7.953 7.986 8.018 8.051 8.084 8.117
8.150 8.183 8.216 8.250 8.283 8.317 8.351 8.385 8.419 8.453 8.488 8.522
8.557 8.592 8.627 8.662 8.697 8.733 8.768 8.804 8.840 8.876 8.912 8.949
8.985 9.022 9.058 9.095 9.132 9.170 9.207 9.244 9.282 9.320 9.358 9.396
9.434 9.473 9.511 9.550 9.589 9.628 9.667 9.707 9.746 9.786 9.826 9.866
9.906 9.946 9.987

The confluence of such meticulous calibration of the first and the last terms where exponent difference is very nearly an integral value [integer 1], together with the abundance of points, ensures an exceptional and very rare fit to Benford, with SSD = 0.03. First digits come out as:

Exponential 0.407% Growth - {30.16, 17.64, 12.35, 9.70, 7.94, 6.70, 5.82, 5.11, 4.59}
Benford's Law First digits − {30.10, 17.61, 12.49, 9.69, 7.92, 6.69, 5.80, 5.12, 4.58}

Let us now construct a histogram of all the quantities falling on (1, 10). Such a construction would omit the time dimension altogether of course. We construct exactly 27 bins for the entire width of (10 – 1) = 9 units, resulting in 9/27 = 0.333 width for each bin. Hence the first bin is on [1.000, 1.333) and there we encounter 71 data points. The second bin is on [1.333, 1.666) and there we encounter 55 data points. For the very last bin on [9.666, 10.000) we encounter only 9 points.

The series of the count of the data points falling within each bin in the entire range is: {71, 55, 45, 38, 33, 29, 26, 23, 21, 20, 18, 17, 16, 15, 14, 13, 13, 12, 12, 10, 11, 10, 10, 9, 9, 8, 9}. Figure 6.31 depicts the histogram.

The series of midpoints on the x-axis for each bin are: {1.167, 1.500, 1.833, 2.167, 2.500, 2.833, 3.167, 3.500, 3.833, 4.167, 4.500, 4.833, 5.167, 5.500, 5.833, 6.167, 6.500, 6.833, 7.167, 7.500, 7.833, 8.167, 8.500, 8.833, 9.167, 9.500, 9.833}.

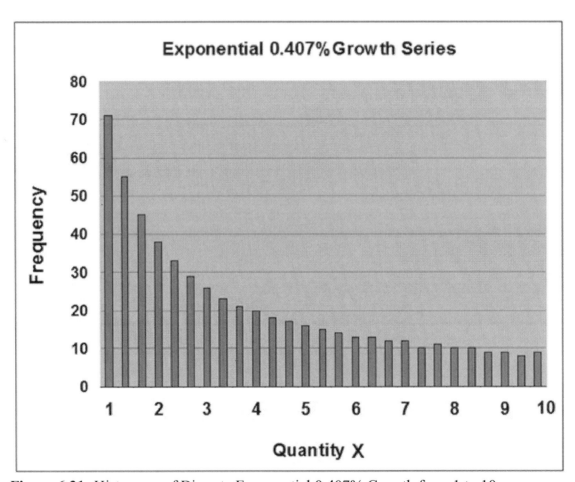

Figure 6.31: Histogram of Discrete Exponential 0.407% Growth from 1 to 10

Let us first roughly check compatibility with k/x distribution. To do this we shall use an essential property of the k/x distribution which states that by doubling quantity x we reduce its frequency (density height) by half. Surely for transition such as: x → 2x the density of k/x always diminishes by half as in: k/x → k/(2x) = (k/x)/2. Very roughly, the two data points of (1.167, 71) and (2.500, 33) on the histogram of the exponential growth series represent approximately the doubling of quantity x, which induces 71 − 33 = 38 reduction in the frequency, namely 38/71 or 53.5% reduction − being comparable approximately to the theoretical 50%. Very roughly, the two data points of (2.167, 38) and (4.500, 18) represent approximately the doubling of quantity x, which induces 38 − 18 = 20 reduction in the frequency, namely 20/38 or 52.6% reduction − being comparable approximately to theoretical 50%. Very roughly, the data points of (4.167, 20) and (8.500, 10) represent the doubling of quantity x, which induces 20 − 10 = 10 reduction in the frequency, namely 10/20 reduction − corresponding nicely to the theoretical 50% expectation.

We shall now look for some k/x distribution that could best fit the histogram of Figure 6.31. Let us find the most fitting parameter k value, one that would minimize the sum of squared errors (SSE) between such an ideal k/x distribution and the histogram of the 0.407% monthly growth series of Figure 6.31 [using the midpoints]. We shall set the derivative of SSE with respect to k equal to zero and solve for k. We therefore obtain:

$$SSE = (k/1.167 - 71)^2 + (k/1.500 - 55)^2 + (k/1.833 - 45)^2 + \ldots + (k/9.833 - 9)^2$$

$$d(SSE)/d(k) = 2(k/1.167 - 71)/1.167 + 2(k/1.500 - 55)/1.500 + 2(k/1.833 - 45)/1.833 + \ldots + 2(k/9.833 - 9)/9.833$$

$$0 = 2(k/1.167 - 71)/1.167 + 2(k/1.500 - 55)/1.500 + 2(k/1.833 - 45)/1.833 + \ldots + 2(k/9.833 - 9)/9.833$$

Dividing both sides of the equation by 2 we get:

$$0 = (k/1.167 - 71)/1.167 + (k/1.500 - 55)/1.500 + (k/1.833 - 45)/1.833 + \ldots + (k/9.833 - 9)/9.833$$

Rearranging terms involving k and those not involving k, we get:

$$0 = (k/1.167)/1.167 + (k/1.500)/1.500 + (k/1.833)/1.833 + \ldots + (k/9.833)/9.833 - (71)/1.167 - (55)/1.500 - (45)/1.833 - \ldots - (9)/9.833$$

$$0 = k*[(1/1.167)/1.167 + (1/1.500)/1.500 + (1/1.833)/1.833 + \ldots + (1/9.833)/9.833] - (71)/1.167 - (55)/1.500 - (45)/1.833 - \ldots - (9)/9.833$$

$$0 = k*[2.67] - 220.28$$

$$k = (220.28)/(2.67) = \mathbf{82.4}$$

Figure 6.32 depicts the excellent fit of 82.4/x distribution with the discrete exponential 0.407% growth series. Points on the line of 82.4/x were constructed as in (x, 82.4/x) using the series of 27 midpoints given earlier for the 27 x values.

82.4 / 1.167 = 70.6   and 71 for the exponential growth series
82.4 / 1.500 = 54.9   and 55 for the exponential growth series
82.4 / 1.833 = 44.9   and 45 for the exponential growth series
82.4 / 2.167 = 38.0   and 38 for the exponential growth series

  … etc. ….

82.4 / 9.833 = 8.4   and 9 for the exponential growth series

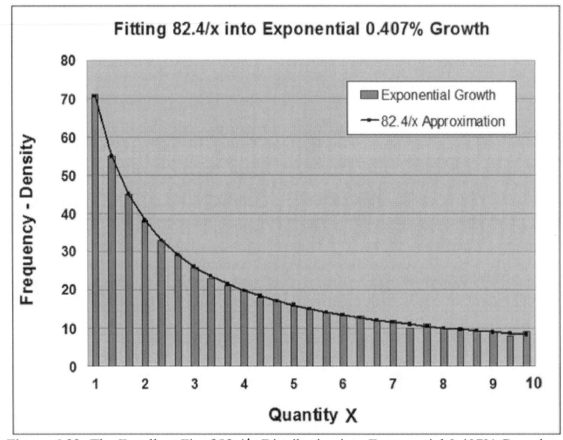

Figure 6.32: The Excellent Fit of 82.4/x Distribution into Exponential 0.407% Growth

One straightforward and decisive mathematical confirmation of the intimate relationship between exponential growth series and the k/x distribution in the context of Benford's Law can be given by the comparison of the expression $X(T) = BF^T$ of the continuous form of exponential growth series and $X(Y) = 10^Y$, Y being the continuous Uniform Distribution, which according to Proposition II in Kossovsky (2014) is of the k/x form. Proposition II and Corollary II state that if Y is uniformly distributed over [R, S], with the length S - R being an integer, then $X = 10^Y$ over $[10^R, 10^S]$ is distributed as in $f(x) = (1/[(S-R)*\ln(10)])*(1/x)$, and it is perfectly logarithmic. Indeed, a moment thought would convinced anybody that the very definition of Benfordness as uniformity of mantissa is also what is seen in the expression $X(Y) = 10^Y$, because whenever X varies between adjacent integral powers of ten, such as from 1 to 10 say, its mantissa Y varies and cycles fully and only once on its entire space of (0, 1).

Let us summarize:

The k/x distribution $\rightarrow$ $X(Y) = 10^Y$
Exponential growth $\rightarrow$ $X(T) = BF^T$

The striking similarity between these two expressions is found in the fact that both are written in an exponential form, with a fixed constant raised to a variable power of the Uniform type. The expression for k/x can be considered as in exponential growth with B = 1 and F = 10.

As discussed in chapter 15 regarding continuous growth, the continuous time T variable for all practical purposes can be discretely selected every minute or every second for the evaluation of the growing quantity as time progresses. Surely time selection must be made fairly, evenly, and uniformly, without focusing on certain time intervals at the expense of other time intervals. By simply choosing to evaluate the series consistently every minute say, fairness and uniformity are ensured by default. The uniform and even dispersions of the projections onto the horizontal time axis of the points in Figures 6.28, 6.29, and 6.30 visually and clearly demonstrate this principle. For example, it would be wrong to take more measurements of the growing quantity in January say each 1/2 a minute, take fewer measurements in February each minute, and take even fewer measurements in March only each 5 minutes.

If the exponent difference of the exponential series $X(T) = F^T$ for a particular T range is approximately of an integral value, then the series is certainly nearly Benford, and multiplying the series by any factor/base B to obtain $X(T) = BF^T$ does not significantly change its logarithmic property as predicated by the Scale Invariance Principle. This implies that the value of B = 1 in the k/x should not affect logarithmic behavior. In addition, surely the high F = 10 value in the k/x case should not preclude logarithmic behavior whatsoever; rather it only implies that growth is with the very high rate of 900% per period. In general, a change in base for the k/x distribution does not change its nature, but such a change in base implies that the end points of the Y range must adjust in order to obtain perfect Benford behavior. Hence $X(Y) = e^Y$ or $X(Y) = 8^Y$ are also Benford, given corresponding proper spans of Y where the ranges lead to perfect logarithmic behavior. There is nothing special in having base 10 for the k/x distribution as in $X(Y) = 10^Y$.

[19] The Last Term of Random Exponential Growth Series is Benford

An innovative explanation of the prevalence of Benford's Law in the physical world has been made relatively recently by **Kenneth Ross** utilizing the collection of the last term of numerous exponential growth series having random growth rate and random initial base.

In Ross (2011) titled "Benford's Law, A Growth Industry", a rigorous mathematical proof is presented showing that a collection of the last term of numerous exponential growth series $\{BF^N\}$ is logarithmic in the limit as N goes to infinity - given that both B and F are randomly selected from suitable Uniform distributions.

The generic expression BF^N signifies exponential growth series, where base B is the initial quantity at time zero, F is the growth factor per period, and N is the number of periods.

Roughly speaking, Ross chooses B and F as in the continuous Uniforms on (1, 10], where all digits 1 to 9 enjoy equitable proportions on the interval, thus avoiding a priori any possible bias against any particular set of digits.

Ross assigns the same value of N (i.e. length) to all the growth series. In other words, Rose is considering the set of growth series, all with equal length of time.

Ross model can be summarized as: $\lim_{N \to \infty} (\text{Uniform}_1)*(\text{Uniform}_2)^N$

One real-life data set that could obvious be relevant to his model is population census data on all the cities and towns in a given country, and which is actually almost always perfectly Benford – except for very small countries with few population centers. Hence, his model can be thought of as a census population snapshot of a large collection of relatively old and well established towns and cities, all being established simultaneously in the same period, and with inter-migration strictly prohibited. In addition, Ross has provided a second model that allows for F to be randomly selected anew for each period, and this model constitutes a more realistic model of cities and towns where the growth rate is generally not constant, but rather varies with the period. Monte Carlo simulations [with a large number of cities] for his first model strongly corroborate his assertion, giving very satisfactory results even after only 20 growth periods. Monte Carlo simulations for his second model show even more rapid convergence, where excellent fit to Benford is obtained even after only 5 or 6 growth periods.

The first model can be succinctly expressed as:

$\text{Uniform}_1[1, 10)*\text{Uniform}_2[1, 10)^N$

The second model can be succinctly expressed as:

$\text{Uniform}_1[1, 10)*[\text{Uniform}_2[1, 10)*\text{Uniform}_3[1, 10)*\ \dots N \text{ times} \dots\ *\text{Uniform}_{N+1}[1, 10)]$

Admittedly, these Uniform distributions are of low POM value of 10. Yet, since the expressions here contain only purely multiplicative terms without any disruptive additions, there exists no tug of war whatsoever between addition and multiplication, and in the limit as N gets large enough and numerous such multiplications are applied, the scheme's POM becomes sufficiently high.

The enormous significance of such logarithmic result for the final elements of random exponential growth series is that it can serve as a logarithmic model for all entities that spring into being gradually via random growth, be it a set of cities and towns with growing populations, rivers forming and enlarging gradually along an incredibly slow geological time scales, biological cells growing, stars and planets formations, and so forth. The potential scope covered by Ross' results is enormous! Hence, instead of having to become a specialist in each and every scientific discipline, laboriously looking for components and physical measurements that may serve as multiplicands within any scientific data set that is Benford, or attempting to fit the physical setup into some data aggregation scheme, Ross lets us skip all this, and obtain extremely general results [for entities that grow] relevant to so many scientific disciplines.

Using the notation B_J for one simulated realization from the random distribution of B, and F_J for one realization from the random distribution of F, his <u>first</u> model can be expressed as the data set $\{B_1F_1^N, B_2F_2^N, B_3F_3^N, B_4F_4^N, \ldots, B_MF_M^N\}$ where N is sufficiently large for the desired convergence and accuracy, and M is the number of simulated cities or entities. Since each B_J and each F_J are realized from the Uniform on the interval [1, 10) and can be expressed as U_I, this can be written more succinctly as: $\{U_1U_2^N, U_3U_4^N, U_5U_6^N, U_7U_8^N, \ldots, U_{2M-1}U_{2M}^N\}$.

For N fixed as 50 for example, his <u>second</u> model can be succinctly expressed as:
$\{U_1U_2U_3\ldots U_{50}U_{51}, \quad U_{52}U_{53}U_{54}\ldots U_{101}U_{102}, \quad U_{103}U_{104}U_{105}\ldots U_{152}U_{153}, \quad \ldots M \text{ times} \ldots \}$

Little reflection is needed to realize that Ross second model can be interpreted as repeated multiplications of random Uniform distributions, and which is distributed almost exactly as Lognormal with high shape parameter and high POM [hence Benford] as predicated by the Multiplicative Central Limit Theorem. His first model can also be interpreted in terms of the well-known general result in the field, namely that any random data set X (logarithmic or non-logarithmic) transformed by raising each value within the data to the Nth power, converges to the logarithmic as N gets large. That is, the set $\{X_1, X_2, X_3, \ldots, X_M\}$ of M realizations from any variable X transformed into $\{X_1^N, X_2^N, X_3^N, \ldots, X_M^N\}$ is logarithmic as N gets large.

In Ross models, both B and F are random variables, while the value of N is identical and constant for all cities. Surprisingly, inverted models where both B and F are fixed as constants being identical for all cities, while only N varies randomly by city, also converge to Benford! In one particular Monte Carlo computer simulation, the model is of a country with 242 cities, all being randomly established at different years - ranging uniformly from 1 to 300 years ago; all having an identical growth rate of 11%; all starting from population of 1 (i.e. a single person). Here we allow for fractional values of persons, since the model represents the generic case of quantitative growth, not necessarily only of integral values. In summarizing the model, the country is said to have cities of varying (random) age; some are very old and established cities, some are not as old, and some are more recent modern cities. This model can be succinctly expressed as $(1)*1.11^{\text{UNIFORM} \{1,2,3, ..., 300\}}$. The variable under consideration is the current snapshot of the populations of all existing 242 cities and towns (only the last elements of the growth series, not including historical population records). Six different simulation runs as well as their average digit distribution came out as follows:

Fixed B & F, Varied N – {27.7, 19.4, 9.9, 10.7, 9.9, 6.6, 7.4, 5.4, 2.9}
Fixed B & F, Varied N – {34.3, 18.6, 9.1, 9.5, 6.6, 5.4, 7.4, 4.5, 4.5}
Fixed B & F, Varied N – {28.5, 24.4, 10.3, 6.6, 7.0, 4.5, 6.2, 5.0, 7.4}
Fixed B & F, Varied N – {28.1, 16.5, 16.1, 7.9, 6.6, 9.1, 6.6, 5.0, 4.1}
Fixed B & F, Varied N – {29.3, 21.5, 12.8, 7.4, 7.4, 7.9, 5.4, 4.1, 4.1}
Fixed B & F, Varied N – {29.3, 17.4, 13.2, 12.4, 4.5, 5.8, 6.6, 7.0, 3.7}

AVG - fixed BF varied N – {29.5, 19.6, 11.9, 9.1, 7.0, 6.5, 6.6, 5.2, 4.5}
Benford's Law 1st digits – {30.1, 17.6, 12.5, 9.7, 7.9, 6.7, 5.8, 5.1, 4.6}

SSD value for this average digit distribution is 6.6, indicating strong conformity to Benford.

In another Monte Carlo computer simulation scheme, the model is of a country with 242 cities; all being established at the same time 1700 years ago (i.e all are of the same age of 1700); all starting from population of 1 (i.e. a single person); all having a fixed but distinct growth rate between 0% and 11% as F is chosen from the continuous Uniform(1.00, 1.11). The variable under consideration is the current snapshot of the populations of all existing 242 cities and towns after 1700 years (the last elements of the growth series, not including historical population records). In summarizing the model, the country is said to have cities of identical age N as well as identical initial population base B, but with distinct random F growth factors. This model can be succinctly expressed as: $(1)*\text{Uniform}(1.00, 1.11)^{1700}$. Seven different simulation runs as well as their average digit distribution came out as follows:

Fixed B & N, Varied F – {36.8, 14.9, 10.3, 11.2, 5.0, 7.0, 7.4, 4.1, 3.3}
Fixed B & N, Varied F – {30.6, 16.9, 16.5, 9.5, 8.7, 2.9, 6.2, 5.0, 3.7}
Fixed B & N, Varied F – {31.8, 20.7, 12.8, 7.9, 6.6, 6.6, 7.4, 3.7, 2.5}
Fixed B & N, Varied F – {31.0, 18.6, 9.5, 9.9, 10.3, 7.9, 4.1, 5.4, 3.3}
Fixed B & N, Varied F – {30.6, 15.3, 13.6, 7.0, 9.9, 6.2, 6.6, 7.4, 3.3}
Fixed B & N, Varied F – {33.5, 13.6, 9.5, 11.2, 9.1, 9.5, 4.1, 5.4, 4.1}
Fixed B & N, Varied F – {27.7, 20.7, 13.2, 9.5, 6.2, 6.2, 7.0, 5.4, 4.1}

AVG - fixed BN varied F – {31.7, 17.2, 12.2, 9.4, 8.0, 6.6, 6.1, 5.2, 3.5}
Benford's Law First digits – {30.1, 17.6, 12.5, 9.7, 7.9, 6.7, 5.8, 5.1, 4.6}

SSD value for this average digit distribution is 4.2, indicating strong conformity to Benford.

Falling well below 1700 years (periods) yields worsening digital results. For example, one run of the above model with only 300 periods yields {34.7, 16.1, 12.4, 8.7, 7.0, 7.4, 4.5, 4.1, 5.0} and its higher SSD value of 28.6 is an indication of diminished conformity to Benford. Yet even this digital configuration is quite similar to the logarithmic overall. A comparison between this scheme and Ross first model points to the fact that fixing base B as a constant (i.e. reducing randomness) requires the system to have many more periods of growth before a convergence to Benford is achieved. Ross first model achieves near-Benford digit configuration quickly even after 20 or so periods, while this model with only factor F varying randomly requires at least 1000 to 2000 periods for convergence.

In another Monte Carlo simulation scheme with all three variables {B, F, N} chosen randomly, results are nearly Benford even with the shorter time span of 100 years/periods. This is so since there is 'more randomness' in the system, or rather because this model is also covered under Ross first model. The model is of a country with 242 cities; all being randomly established at different years ranging from 1 to 100 and selected from the discrete Uniform {1, 2, 3, …, 100}; all having random growth rate between 0% and 14% as F is chosen from the continuous Uniform(1.00, 1.14); all starting from a random population base chosen from the continuous Uniform[1, 10). In summarizing the model, the country is said to have cities of varying random age, random growth rate, and random initial population count. This model can be succinctly expressed as: **Uniform[1, 10)*Uniform(1.00, 1.14)$^{\text{UNIFORM \{1,2,3, …, 100\}}}$**. The variable under consideration is the current snapshot of the populations of all 242 existing cities. Six different simulation runs and their average digit distribution came out as follows:

Varied B, N, F – {33.9, 18.2, 9.5,   7.9,   8.7, 7.0, 6.2, 4.1, 4.5}
Varied B, N, F – {29.8, 15.7, 7.9,   9.1,   9.9, 9.1, 5.8, 7.4, 5.4}
Varied B, N, F – {31.4, 16.5, 12.0,   9.9,   9.1, 7.0, 4.5, 5.0, 4.5}
Varied B, N, F – {27.7, 16.1, 16.1,   8.7,   7.4, 7.0, 6.6, 4.1, 6.2}
Varied B, N, F – {26.0, 17.8, 16.1, 11.2,   5.4, 8.7, 4.1, 6.2, 4.5}
Varied B, N, F – {28.1, 14.0, 10.7, 12.8, 10.3, 8.3, 5.8, 5.8, 4.1}

AVG Varied B, N, F – {29.5, 16.4, 12.1, 9.9, 8.5, 7.9, 5.5, 5.4, 4.9}
Benford's L. 1st digits – {30.1, 17.6, 12.5, 9.7, 7.9, 6.7, 5.8, 5.1, 4.6}

SSD value for this average digit distribution is 4.1, indicating strong conformity to Benford.

Surprisingly, Ross first model where only F and B vary, while N is fixed and identical for all (i.e. cities begin their existence simultaneously at time zero and all are of the same age) yields better results for short N time span such as 20 periods, 100 periods, and such – as compared with this 'more random' model above. The obvious explanation for this apparent paradox is that for a very short time span all the series should start early on [as in Ross model], so that they would have sufficient lengths to even begin resembling growth series. By varying N randomly [as in this last model] we unfortunately allow into the system some very short and immature 'series' of say 3 or 7 years, 'series' which hamper convergence. Empirically, for very long N time span, both models yield the logarithmic equally, and this fact is consistent with this explanation. In addition, another reasonable explanation is that here we restrict the range F to the Uniform(1.00, 1.14) which has a much lower POM value than the POM value of the Uniform(1, 10] for Ross first model.

In contrast to all the successful models above, the model where F and N are fixed as constants while only base B varies randomly, does not converge to Benford in any way. In one Monte Carlo computer simulation scheme, the model is of a country with 242 cities; all being established simultaneously 2500 years ago (i.e. all are of the same ancient age); all having the same 5% growth rate; all starting from a varied population base B randomly selected from the continuous Uniform(0, 100). This 'less random' model can be succinctly expressed as: **Uniform(0, 100)*1.05^{2500}**. The variable under consideration is the current snapshot of the populations of all 242 existing cities after 2500 years. First digit distribution here came out as {10.3, 9.1, 9.5, 14.0, 11.6, 11.2, 14.0, 12.8, 7.4}. Surely if F and N are fixed and are equal for all the cities, then the variation in the final population values is due only to the variable base B, being a random multiplicative factor, hence the Uniform(0, 100)*1.05^{2500} model is equivalent to the model Uniform(0, 100)*Constant, which is of course not Benford at all, but rather of the symmetrical Uniform distribution configuration.

The essential feature leading to Benford convergence in all the above successful models is the multiplicative form in how population [or the generic quantity under consideration] is increasing. If one views [multiplicative] exponential growth as repeated additions, then one sees larger and larger quantities being added as the years pass. A constant 5% population increase implies an addition of 5 at the year when the population is at 100, an addition of 50 at the year when the population is at 1000, and an addition of 500 at the year when the population is at 10000. On a profound level this is almost always the case in the natural world. The larger the city, the more it attracts new inhabitants. A small village attracts very few newcomers. The larger the forming star, the more mass it attracts gravitationally and adds to itself. In sharp contrast, an additive type of growth is one where the quantity being added each year is fixed, not being proportional to the present size of the population. Let us consider the model of a country with 242 cities; all being randomly established at different years ranging uniformly from 1 to 100; all starting from a variable population base B, randomly selected from the continuous Uniform(0, 50); all experiencing fixed and identical additions of quantity D per period. This model can be succinctly expressed as: **Uniform(1, 50) + D*Uniform{1,2,3, …, 100}**. The variable under consideration is the current snapshot of the populations of all 242 existing cities and towns. Three different simulation runs with distinct D values are shown:

Varied B & N, fixed D = 20 Additive – {47.5, 6.2, 8.7, 7.9, 7.4, 5.4, 2.9, 8.7, 5.4}
Varied B & N, fixed D = 33 Additive – {38.8, 33.5, 12.4, 1.7, 2.5, 2.1, 3.7, 2.1, 3.3}
Varied B & N, fixed D = 8 Additive – {17.4, 10.3, 11.2, 13.2, 14.5, 14.0, 11.6, 6.2, 1.7}

Certainly this model is not Benford at all, being a linear combination of sorts of two Uniform distributions.

Acknowledgement:

The late mathematician **Ralph Raimi** who has eloquently written some of the first mathematically rigorous articles on Benford's Law in the 1960's and 1970's has also written about the existence of such anomalous series, using the term 'reentrant series' - in an article titled "The Peculiar Distribution of First Digit", Scientific America, 1969. The author is still quite content to re-invent this old wheel by himself and to add several new features to it. The author wishes to convey his strong sense of affinity and rapport with Raimi for thinking along the same lines in the quest for a thorough understanding of the behavior of exponential growth series in the context of Benford's Law. A lasting impression for the author was a 2-day visit to Raimi's residence in Rochester, New York, in April 2013, discussing Benford's Law in general, and in particularly the discovery of Digital Development Pattern which Raimi enthusiastically endorsed. Funnily, anomalous (reentrant) series were never discussed during the entire visit!? Raimi passed away on January 2, 2017, at the age of 92.

The author wishes to thank the mathematician **Kenneth Ross** for his correspondence regarding the two logarithmic growth models of his article.

SECTION 7

INNOVATIVE FORENSIC FRAUD DETECTION APPLICATIONS IN BENFORD'S LAW

[1] Forensic Digital Analysis in Data Fraud Detection Applications

Checking data for authenticity has become feasible with applications in the field of Benford's Law which predicts that the first significant digit on the left-most side of numbers in real-life data is proportioned between all possible 1 to 9 digits approximately as in LOG(1 + 1/digit), so that low digits occur much more frequently than high digits. In addition, a patented algorithm employing Digital Development Pattern [chapter 15 section 1] has been invented by the author to detect fraud in data sets for which Benford's Law does not apply, or when fraudsters are already aware of the law and attempt to calibrate digits in fake data according to the law so as to make it appear genuine. Digital Development Pattern is found when partial digital configurations are examined on mini sub-intervals between integral powers of ten points, and where a pronounced rise in digital skewness becomes apparent as focus shifts from low values to high values.

Benford's Law is widely used nowadays for detecting fraud in reported data, especially in accounting and financial data, a practice which has started very modestly around 1995, and then accelerated in use ever since. Forensic digital analysis is now the standard practice of any respectable auditor; being used in addition to all other standard auditing procedures and techniques. Many governments around the world nowadays routinely apply Benford's Law in their tax offices and revenue departments in order to detect tax evasion.

More recently, Benford's Law is being applied in attempts to detect manipulations and fraudulent activities regarding the counting of electoral votes. This later use of Benford's Law may appear quite silly and useless whenever representative democracy in and of itself is used to deceive and manipulate the public, avoiding the more direct and participatory form of democracy such as constant referendums, the bane of all politicians and dictators.

Benford's Law is also applied as a check on fraud in publication of scientific research regarding experiments involving numerical data. If the dishonest scientist fudges or exaggerates the numerical results obtained in laboratory experiments or scientific data so as to support the main thesis or purported claim of the paper, digit configuration typically appears as non-Benford and this in principle could lead to the detection of fraud.

Let us illustrate the most basic and simplest application of Benford's Law, namely comparing 1st digits configuration of accounting data to the set of the Benford proportions. Figure 7.1 depicts 20 hypothetical amounts representing revenue data in units of dollars, for each of the five different companies, Honda, Exxon, Sony, Nike, and Canon. Four out of these five companies are totally innocent of any financial wrongdoing, and only one company appears to be dishonest, strongly suspected of providing fake data so as to reduce tax payment, or in order to appear financially sound and profitable to investors and stock market traders.

There is no way other than the magical or the miraculous to detect any unusual or suspicious financial activities here by merely focusing on the numbers themselves. Yet, focusing on the proportions of the [first] digits within the set of reported numbers for any given company could possibly help in detecting unusual digital patterns and anomalies.

Can the reader find that one particular company which is being suspected of tax and financial fraud by merely scrutinizing those numbers in Figure 7.1 and before any further readings or peeks at the next tables and figures which reveal the answer?! For those readers who give up all too quickly, Figure 7.2 could be useful, as it serves as a digital tool emphasizing 1st digits by showing them in bold and black color.

Honda	Exxon	Sony	Nike	Canon
347.21	68.44	765.3	20.45	607.28
17.75	7.77	18.50	129.57	26.52
120.68	1.65	107.71	10.50	349.81
33.98	388.5	8.93	307.25	47.03
60.53	20.55	251.23	12.50	131.13
120.5	47.33	33.95	3.98	7541.25
8.55	97.4	12.50	1980.50	204.29
43.99	120.55	530.52	56.28	166.73
387.23	7.54	290.55	560.66	59.02
492.72	603.34	60.77	78.31	30.57
5.62	100.23	122.45	409.81	402.24
51.29	87.34	30.22	924.05	176.21
102.85	54.77	1332.81	85.76	32.26
18.33	30.44	12.25	104.46	970.25
128.76	80.11	1277.50	13.34	113.63
27.45	93.05	46.52	670.50	106.02
76.44	80.50	309.90	18.84	55.25
301.23	702.25	54.30	208.78	270.67
106.56	30.41	2.43	127.66	15.25
298.88	27.33	40.70	20.54	23.98

Figure 7.1: Hypothetical Accounting Data for Five Companies

Honda	Exxon	Sony	Nike	Canon
347.21	68.44	765.3	20.45	607.28
17.75	7.77	18.50	129.57	26.52
120.68	1.65	107.71	10.50	349.81
33.98	388.5	8.93	307.25	47.03
60.53	20.55	251.23	12.50	131.13
120.5	47.33	33.95	3.98	7541.25
8.55	97.4	12.50	1980.50	204.29
43.99	120.55	530.52	56.28	166.73
387.23	7.54	290.55	560.66	59.02
492.72	603.34	60.77	78.31	30.57
5.62	100.23	122.45	409.81	402.24
51.29	87.34	30.22	924.05	176.21
102.85	54.77	1332.81	85.76	32.26
18.33	30.44	12.25	104.46	970.25
128.76	80.11	1277.50	13.34	113.63
27.45	93.05	46.52	670.50	106.02
76.44	80.50	309.90	18.84	55.25
301.23	702.25	54.30	208.78	270.67
106.56	30.41	2.43	127.66	15.25
298.88	27.33	40.70	20.54	23.98

Figure 7.2: First Digits of the Accounting Data are Emphasized in Bold Black Color

The answer to the challenge posed to the reader is given by Figure 7.3 which shows the first digits count for each of the five companies. Figure 7.4 depicts the digital proportions for each of the five companies. This simple forensic digital analysis tells us in no uncertain terms that Exxon is highly suspected of fraud.

Digit	Honda	Exxon	Sony	Nike	Canon
1	7	3	7	8	6
2	2	2	3	3	4
3	4	3	3	2	3
4	2	1	2	1	2
5	2	1	2	2	2
6	1	2	1	1	1
7	1	3	1	1	1
8	1	3	1	1	0
9	0	2	0	1	1

Figure 7.3: First Digits Count of the Hypothetical Accounting Data

Digit	Honda	Exxon	Sony	Nike	Canon
1	35%	15%	35%	40%	30%
2	10%	10%	15%	15%	20%
3	20%	15%	15%	10%	15%
4	10%	5%	10%	5%	10%
5	10%	5%	10%	10%	10%
6	5%	10%	5%	5%	5%
7	5%	15%	5%	5%	5%
8	5%	15%	5%	5%	0%
9	0%	10%	0%	5%	5%

Figure 7.4: First Digits Proportions of the Hypothetical Accounting Data

One should keep in mind that nearly all accounting and financial data strongly and almost universally obey Benford's Law, and that there is almost no exception to this rule. Salary amounts which are typically of very low order of magnitude as well as data sets with very few entries [i.e. small-size data sets with less than about 100 values] are practically the only counter examples to Benford's Law in accounting and finance data. Here, first digits of Exxon are approximately uniform and even. The first digits proportions of Honda, Sony, Nike, and Canon, are all with an overall pattern of digital skewness, favoring low digits, and discriminating against high digits, and which is consistent with the pattern of almost all accounting data. These four companies certainly appear to be honest, and no expectation whatsoever is made for such small data sets the size of only 20 data points to achieve digital configuration very close to Benford.

[2] Digital Development Pattern – 1st Digits Basis

Benford's Law relates to the overall digit distribution of the entire data set in question. A closer digital scrutiny within smaller sub-intervals reveals a very consistent pattern of an approximate digital equality on the left part of the x-axis for low values; the logarithmic digital configuration roughly around the center for the bulk of the data; then severe digital inequality in favor of low digits on the extreme far right for high values, and which is skewer and ever more uneven than the logarithmic configuration itself. Overall, when digital configurations of all the regions on the left, right, and center are aggregated, the logarithmic is encountered as predicated by the law.

This very particular differentiation in digital configuration along the entire range is called 'Digital Development Pattern', and it is found in almost all random data sets (Benford and non-Benford), with only extremely rare exceptions, making it by far more prevalent and universal than even Benford's Law itself! Yet, this pattern can only be found under a partition of the entire range along adjacent Integral Powers of Ten (IPOT) pairs of points, such as (0.01, 0.1), (0.1, 1), (1, 10), (10, 100), (100, 1000) and so forth. For deterministic data types such as exponential growth series this pattern does not exist; and instead the logarithmic configuration is found consistently and equally everywhere throughout the entire range. Nonetheless, these deterministic cases are very rare exceptions indeed.

For example, a close examination of mini digital configurations between IPOT points for the US census data on the population of all its cities and towns in 2009 clearly reveals a definite development pattern [detailed in section 1 chapters 15 and 20, and shown in Figure 1.13].

The definition of skewness over and above the Benford configuration is given by the observed [combined] percents of digit 1 and digit 2 minus the Benford default sum of 47.7% for both digits. In other words, the proportion of digits 1 and 2 in the observed data set for the sub-interval under consideration, over and above the rightful legal allocation of 47.7% for digits 1 and 2 as expected by the law. This measure is called '**Excess Sum digits 1 & 2**' and abbreviated as **ES12**.

ES12 = [observed % of digit 1 + observed % of digit 2] – [theoretical % allocation for 1 and 2]
ES12 = [observed % of digit 1 + observed % of digit 2] – [$100*\log(1 + 1/1) + 100* \log(1 + 1/2)$]
ES12 = [observed % of digit 1 + observed % of digit 2] – [47.7%]

In Kossovsky (2014) chapter 43, the meaning and motivation for this definition are discussed. In a nutshell, extraordinary digital skewness manifests itself by lending digits 1 and 2 greater proportions than their typical combined 47.7% portion of the logarithmic configuration, and that occurs at the expense of all the other higher digits which would then earn lower proportions than their typical proportions of the logarithmic configuration.

Figure 7.5 depicts ES12 values for almost all the IPOT intervals of the US 2009 population data, except for the 1,000,000 to 10,000,000 sub-interval on the far right, due to scarcity of points there and the consideration of this sub-interval as an outlier. The Figure 7.6 visually depicts the Digital Development Pattern of the US 2009 population data.

Left Point	1	10	100	1,000	10,000	100,000
Right Point	10	100	1,000	10,000	100,000	1,000,000
	===	===	=====	======	=======	========
Digit 1	14.8	5.3	19.1	37.3	46.0	62.9
Digit 2	7.4	8.1	17.4	19.7	20.2	17.6
Digit 3	3.7	7.0	13.6	11.6	10.9	6.0
Digit 4	7.4	9.2	11.5	8.6	6.4	4.1
Digit 5	7.4	11.5	9.9	6.3	5.8	3.0
Digit 6	14.8	13.9	8.8	5.3	3.8	3.0
Digit 7	7.4	13.9	7.6	4.3	2.8	1.5
Digit 8	14.8	17.0	6.1	4.0	2.3	1.1
Digit 9	22.2	14.1	6.0	2.9	1.7	0.7
	-------	-------	-------	-------	-------	-------
ES12	-25%	-34%	-11%	9%	19%	33%
% of Data	0.14%	5.5%	42.0%	37.3%	13.6%	1.4%

Figure 7.5: ES12 for IPOT Sub-Intervals – US Population 2009 Census

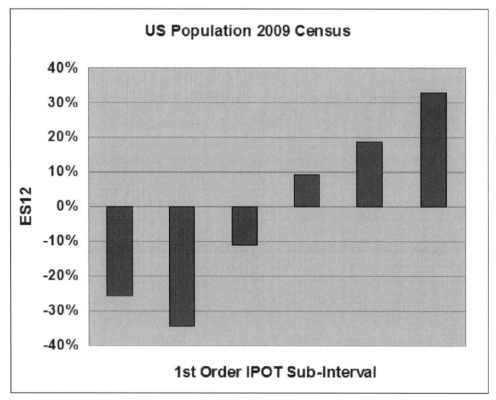

Figure 7.6: Digital Development Pattern – US Population 2009 Census

[3] Detecting Fraud in Data via Digital Development Pattern

Nowadays, increasing number of accountants, auditors, economists, and businessmen, are becoming aware of Benford's Law and the possibility of forensically testing data for authenticity via digital analysis. This awareness trend will surely accelerate in the future. As a consequence, those sophisticated and well-educated potential cheaters who are already aware of Benford's Law, might then alter the way they cheat and concoct their numbers in such a way as to arrive approximately at the Benford proportions for the fraudulent data they provide. How could then the professional statistician or auditor detect such sophisticated fraud when cheaters are carefully calibrating digits in fake data according to the law so as to make it appear authentic?

To overcome this challenge, it is necessary to forensically test digital configurations on local sub-intervals, to verify that Digital Development Pattern does indeed exist in provided data. Surely, even very bright, sophisticated, and well-read cheaters would nevertheless be unaware of this inner digital pattern of development associated with Benford's Law, and therefore wouldn't be able to properly concoct real data with all the intricate and hidden inner patterns associated with it. Naively, the cheater would concoct data without any consideration to development, focusing exclusively on the resultant digital configuration for the entirety of the data.

Such half-baked imitation of Benford's Law and partially-sophisticated falsification of data could potentially result in three basic development scenarios.

(1) Local digit configurations are consistently Benford-like throughout the entire data in all localities, mistakenly mimicking deterministic-like data types. Figure 7.7 depicts one such possible result in fraudulent data reporting regarding accounting and financial data which are almost always inherently of the random flavor. Here, data invention was carefully performed so that the aggregated digit distribution [the last column on the right] comes out very close to the expected Benford proportions. What appears extremely odd in the table of Figure 7.7 is the consistent digital configurations locally, which is very similar to the case of deterministic exponential growth series. The differentiated portions of overall data falling within each sub-interval is actually quite proper and in the same manner as for almost all random data types, where most of the data lies in the central region, with only modest data portions falling around the left and right edges.

Note: the aggregated digit proportions for the entire data set in the last column are weighted averages, taking percent of data into account. For example, the 30.6% proportion for digit 1 is calculated as: $0.292*0.046 + 0.312*0.352 + 0.306*0.475 + 0.297*0.127 = 0.306$.

(2) Meaningless, zigzag, chaotic, and pattern-less style, lacking any development or trend. Figure 7.8 depicts one such possible result in fraudulent data reporting regarding accounting and financial data where the expected development is nowhere to be found. Here as well, data invention was carefully performed and calibrated so that the aggregated digit distribution [the last column on the right] comes out very close to the expected Benford proportions.

From: To:	0.1 1	1 10	10 100	100 1,000	Entire Range
1	29.2%	31.2%	30.6%	29.7%	30.6%
2	16.8%	16.5%	18.5%	19.4%	17.8%
3	12.9%	13.5%	11.7%	13.3%	12.6%
4	9.1%	9.6%	9.3%	9.2%	9.4%
5	8.3%	7.7%	8.3%	7.4%	8.0%
6	7.3%	6.8%	7.1%	6.2%	6.9%
7	5.8%	5.3%	5.5%	5.5%	5.4%
8	6.4%	4.9%	5.1%	5.2%	5.1%
9	4.2%	4.5%	3.9%	4.1%	4.2%
% of Data	4.6%	35.2%	47.5%	12.7%	100.0%

Figure 7.7: Fake Data Concocted to Appear as Benford – Unusually Steady

From: To:	10 100	100 1,000	1,000 10,000	10,000 100,000	Entire Range
1	21.3%	44.1%	16.7%	33.6%	30.8%
2	18.4%	20.1%	15.2%	21.8%	18.6%
3	11.1%	11.3%	14.1%	13.7%	12.6%
4	10.7%	6.2%	12.0%	8.8%	9.1%
5	9.6%	4.5%	11.1%	8.3%	7.9%
6	8.6%	4.5%	8.4%	6.1%	6.5%
7	7.9%	3.9%	7.3%	3.1%	5.3%
8	6.3%	2.9%	8.2%	2.5%	5.0%
9	6.1%	2.7%	7.0%	2.1%	4.4%
% of Data	10.5%	39.4%	33.6%	16.5%	100.0%

Figure 7.8: Fake Data Concocted to Appear as Benford – Zigzag and Chaotic

(3) An inverse development style, where local digit configurations start out as highly skewed on the left for low values, and terminate with an approximate digital equality on the right for high values. Figure 7.9 depicts one such possible (but very rare) result in fraudulent data reporting regarding accounting and financial data where the expected development is strangely found in reversed order. Here as well, data invention was carefully performed and calibrated so that the aggregated digit distribution [the last column on the right] comes out very close to the expected Benford proportions.

From: To:	1 10	10 100	100 1,000	1,000 10,000	Entire Range
1	38.9%	31.5%	26.2%	12.3%	29.1%
2	22.4%	18.5%	14.7%	11.8%	17.4%
3	12.3%	14.1%	13.5%	11.3%	13.2%
4	9.3%	8.5%	12.1%	10.5%	9.9%
5	6.2%	7.3%	9.1%	11.4%	8.1%
6	4.1%	6.8%	8.2%	11.2%	7.2%
7	3.3%	5.9%	6.0%	11.1%	6.0%
8	2.3%	4.2%	5.2%	10.1%	4.8%
9	1.2%	3.2%	5.0%	10.3%	4.2%
% of Data	19.9%	38.9%	29.1%	12.1%	100.0%

Figure 7.9: Fake Data Concocted to Appear as Benford – Inverse Development

These three examples of Figures 7.7, 7.8, and 7.9, are cases associated with crystal clear conclusions, all pointing to suspicious data structure; and with high level of confidence that fraudulent reporting is involved. It should be noted that in some very particular [honest and real] cases, the data set appears with muddled and somewhat indecisive development structure while still showing an overall positive increase in digital skewness; such as when the overall differential in skewness between the left-most sub-interval and the right-most sub-interval is milder than the sharp polarity in developments seen in Figures 7.5 and 7.6 of US population data. At times, development polarity is sharp enough, but it is interrupted by a reversal in one rebellious sub-interval where skewness is temporarily decreasing, then immediately returning back to the overall positive development with increasing skewness. Such temporary setbacks in digital development are [usually] normal and should not cause any suspicion of fraud.

What is to be made of these muddled and fuzzy cases? How does one certify a particular accounting data set as probably fraudulent and another data set as probably honest; and yet without incurring the accusation of being subjective? In other words: can digital development be quantified, analysis automated, and conclusions stated objectively?

This author had conceived of the possibility of applying Digital Development Pattern in forensic digital analysis, and subsequently faced this exact dilemma; namely the need to construct some objective criterion, with clear threshold values and cutoff points differentiating between the honest and the fraudulent, between the normal and the anomalous. Forced by circumstances to invent, the author then created an algorithm and arrived at an automated forensic development tests. While the algorithm is based in part on considerations in theoretical statistics, it is also firmly rooted in reality via the empirical examinations of digital developments in a large variety of real-life data sets, all of which enable one to properly perform the comparison of digital development of any given data set to the average digital development of numerous real-life (honest) data sets. Subsequently the U.S. Patent Office had granted the author a patent in fraud detection methods. Curious readers, zealous and strict auditors, patent thieves who copy other people's work, and others, are referred to detailed readings about the algorithm in the U.S. Patent Office link provided in the Reference, as well as in Kossovsky (2014) chapters 33, 43, 82, 83, and 84.

The other challenge in this context for statisticians and auditors - is the detection of fraud in accounting, financial, and economics-related data types which do not obey Benford's Law in the first place, such as payroll/salary data, other data sets with low order of magnitude, and the few other exceptions to the rule. How could the statistician forensically check for fraud when the data is actually expected not to obey Benford's Law?! Here again, the answer lies simply in forensically testing digital configurations on local sub-intervals, to verify that Digital Development Pattern does indeed exist in provided data. This is so since accounting and financial data sets that do not conform to Benford's Law should still show development pattern.

Digital Development Pattern is more prevalent than Benford's Law itself and practically all accounting and financial data (Benford and non-Benford) should come with this development pattern - with the pronounced exception of exponential growth series, such as a bank account frozen for many decades or centuries, earning either fixed or variable/random interest rate, and where withdrawals and deposits are not permitted. The absence of development pattern in forensic digital tests should trigger suspicion of fraudulent reporting on the part of the statistician or the auditor, and this would merit further forensic investigations and additional checks.

The manner of how digital configuration develops over the entire range of data is akin to a hidden digital signature [code] inscribed deep within the standard Benford signature [code] of $LOG(1 + 1/d)$, and as such it provides a magnificent tool in data forensic fraud detection work.

[4]   First Order Digits Versus Second Order Digits

The first leading digit is the first non-zero digit of a given number appearing on the left-most side.

419   → digit 4
0.005161   → digit 5
8   → digit 8
-8   → digit 8
3,221,955   → digit 3
-0.233879   → digit 2

Benford's Law also describes a distribution for the second order digits. For example, the 2nd leading digit (second digit from the left) of the number 72 is digit 2, of 0.00347 it's digit 4, and of 2,178,393 it's digit 1. It is noted that for the 2nd order the digit 0 must be included, since all ten digits {0, 1, 2, 3, 4, 5, 6, 7, 8, 9} are used.

729   → digit 2
0.0039524   → digit 9
2,874,912   → digit 8
-0.817392   → digit 1
508   → digit 0

The 2nd order proportions among the digits are by far more equal than those of the 1st order. In other words, the 1st order is by far skewer with severe digital inequality in comparison to the much milder digital differences of the 2nd order. Second order digits distribution (unconditional probabilities) according to Benford's Law as well as the 1st order digits distribution are given in the tables of Figure 7.10.

Digit	Probability
1	30.1%
2	17.6%
3	12.5%
4	9.7%
5	7.9%
6	6.7%
7	5.8%
8	5.1%
9	4.6%

Digit	Probability
0	12.0%
1	11.4%
2	10.9%
3	10.4%
4	10.0%
5	9.7%
6	9.3%
7	9.0%
8	8.8%
9	8.5%

Figure 7.10: First Digits Versus Second Digits in Benford's Law

Figure 7.11 depicts the chart of the distributions of the 1st and the 2nd digital orders superimposed, in order to emphasize the stark contrast between them.

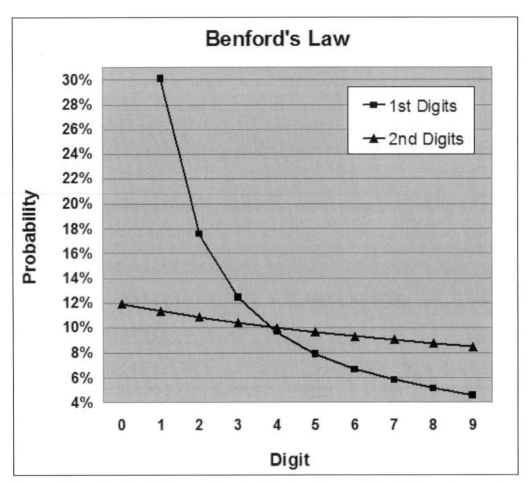

Figure 7.11: Benford's Law – Chart of First and Second Digits Superimposed

[5] Digital Development Pattern – 2nd Digits within 1st Digits Basis

The above discussion about Digital Development Pattern focused exclusively on the 1st order digits, although all this can be considered and performed on the 2nd order digits as well, with only few and minor modifications.

Whenever 2nd order digital development analysis is performed, it is suggested to stick to a partition along adjacent IPOT as well, just as it is done for the 1st order test – assuming that there are plenty of IPOT sub-intervals in the data set. In other words, the suggestion is to partition the entire range along 1st order basis of IPOT points while testing 2nd order development, as opposed to the more natural scheme of partitioning the entire range along the much shorter 2nd order basis of the 2nd order cycles while testing the 2nd order development. This suggestion is given in spite of the apparent dichotomy in mixing 1st order themes with the 2nd order themes. It should be noted that the full cycle of the second order digits is much shorter, being merely 1/9 the length of the full cycle of the 'local' first order. For example, for the first order whole cycle on the sub-interval [1, 10): from **7.0 to 8.0** all second order digits possibilities get a chance to manifest themselves, with digit zero 2nd leading on [7.0, 7.1), digit one 2nd leading on [7.1, 7.2), digit two 2nd leading on [7.2, 7.3), and so forth. The same can be said about the range from **3.0 to 4.0**, where all second order digits possibilities get a chance to manifest themselves there, with digit zero 2nd leading on [3.0, 3.1), digit one 2nd leading on [3.1, 3.2), digit two 2nd leading on [3.2, 3.3), and so forth.

Within (1, 10), 2nd order cycles are 1-unit long.
Within (10, 100), 2nd order cycles are 10-unit long.
Within (100, 1000) 2nd order cycles are 100-unit long.

The reason for suggesting to test the 2nd order along the 1st order basis of [1, 10), [10, 100), and such, is due to the interdependencies of the orders; being that 2nd order probabilities depend on 1st order probabilities; that 2nd digits' proportions are more skewed in favor of 2nd-order low digits whenever 1st-order digits are also low. By selecting IPOT sub-intervals in which all 1st order digits are equally represented such as [1, 10) or [10, 100), by extension we allow all 2nd order digits equal opportunity to fully express themselves and their logarithmic property as well. Kossovsky (2014) chapter 100 discusses the dependencies between the orders and provides some concrete examples.

Clearly, {0, 1, 2, 3, 4} should be considered low 2nd digits, and {5, 6, 7, 8, 9} high 2nd digits. Hence a reasonable definition of skewness over and above the 2nd order [unconditional] Benford configuration on any given sub-interval is the observed percent of numbers being led by digits {0, 1, 2, 3, 4} minus the Benford default sum of 54.7% for {0, 1, 2, 3, 4} 2nd order digits.

ES04 = [observed % of 2nd order digits {0, 1, 2, 3, 4}] – [theoretical % allocation for digits 0 to 4]
ES04 = [observed % of 2nd order digits {0, 1, 2, 3, 4}] – [12.0% + 11.4% + 10.9% + 10.4% + 10.0%]
ES04 = [observed % of 2nd order digits {0, 1, 2, 3, 4}] – [54.7%]

This 2nd order skewness measure, to be called '**Excess Sum digits 0 to 4**' and abbreviated as **ES04**, varies from + 45.3% to -54.7%.

Figure 7.12 depicts the 2nd order ES04 values within 1st-digit-basis IPOT sub-intervals for the 2009 US population data. The left-most sub-interval on (1, 10) with only 0.14% of overall data was omitted here so as to avoid distortion from outliers in the more delicate [less skewed] 2nd order case. Certainly a clear 2nd order development pattern is seen here of increasing skewness as focus moves to the right. Figure 7.13 depicts this 2nd order development pattern for the data on US population centers.

Left Point	10	100	1,000	10,000	100,000
Right Point	100	1,000	10,000	100,000	1,000,000
	===	=====	======	========	=========
Digit 0	9.7	9.9	13.0	13.8	24.0
Digit 1	8.7	10.6	12.3	12.1	11.6
Digit 2	10.7	10.7	11.3	13.0	14.2
Digit 3	10.1	10.5	10.7	10.6	8.6
Digit 4	11.8	10.1	10.1	10.4	9.0
Digit 5	10.1	9.6	9.4	8.7	7.9
Digit 6	10.1	10.0	9.2	9.6	5.2
Digit 7	9.9	9.3	8.1	8.6	7.1
Digit 8	9.7	9.8	8.4	6.9	6.0
Digit 9	9.1	9.3	7.4	6.5	6.4
	--------	--------	--------	--------	--------
ES04	-4%	-3%	3%	5%	13%
% of Data	5.5%	42.0%	37.3%	13.6%	1.4%

Figure 7.12: ES04 of 2nd Order within 1st Order Basis – US Population Data

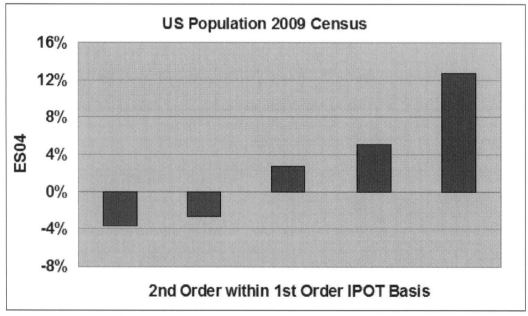

Figure 7.13: Development Pattern of 2nd Order within 1st Order - US Population Data

[6] Digital Development Pattern – Purely 2nd Order Basis

When the entire data set spans merely one order of magnitude, or less, such as when data falls on (10, 100), or on (284, 2375) for example, neither the 1st digital order comparisons of skewness (i.e. ES12) nor the 2nd order comparisons of skewness along the 1st order basis (i.e. ES04) can be performed, since there is only one or less such sub-interval to consider. Nonetheless, digital development pattern can still be seen clearly in such cases via the purely 2nd order digital cycles, totally disregarding the 1st order and the dependencies of the 2nd order on the 1st order. This is done by dividing the entire short range into sub-intervals spanning the 2nd order cycles, followed by the examination of the 2nd order configuration of skewness of ES04 on each of these sub-intervals. For example, data falling on the short range (10, 80) is not even of 1 whole order of magnitude, and its upper bound of 80 does not even reach the next IPOT point of 100. Here we focus purely on the 2nd order and subdivide the range into 7 sub-intervals relating to the 2nd order, namely into {[10, 20), [20, 30), [30, 40), [40, 50), [50, 60), [60, 70), [70, 80)}; then ES04 is evaluated on each of these sub-intervals. A development pattern of increasing ES04 in the aggregate is expected to be observed – assuming that the data is authentic and not fraudulently concocted.

How could this development analysis work well in spite of the distortion arising from the dependencies between the orders? The answer is that here the effects of digital development on 2nd order configurations is by far stronger and more influential than the effects of the more mild differentiation in digital 2nd order configurations due to the dependency on the 1st order.

The effects of digital development on the 2nd digits is caused by the dramatic differentiation in the log histogram of the data itself, which sharply rises on the left and precipitously falls on the right, quickly and dramatically changing course from the minimum value to the maximum value (both of which are not that much far apart from each other, since order of magnitude is small).

More specifically, the purely 2nd order basis development analysis is highly effective because the dependencies between the orders are easily overwhelmed by the differentiation in log position on the curve along the values of log-axis, which are much sharper and more dramatic for data sets with very low order of magnitude.

Care must be taken to ensure that these cycles expand by a factor of 10 at each encounter of IPOT point. Thus some sub-intervals within the entire comparison scheme are wider, while some are narrower, and this disparity in width does not distort in any way the comparison, rather it facilitates it! For example, for data falling on (60, 300), the sub-intervals for the purely 2nd order development pattern are: {[60, 70), [70, 80), [80, 90), [90, 100), [100, 200), [200, 300)}. Here the 4 left-most sub-intervals are of 10-unit each, while the 2 right-most sub-intervals are of 100-unit each, and this disparity in width induces and facilitates the proper 2nd order skewness comparison between these six sub-intervals.

This additional method of examining development via the purely 2nd order basis is a highly valuable and indispensable tool in digital forensic analysis in the context of fraud detection whenever data is of low order of magnitude! Three case studies of the purely 2nd order basis development analysis shall be given in the next three chapters.

[7] Case Study I: State of Oklahoma Payroll Data

The State of Oklahoma Payroll data of the Department of Human Services for the 1st quarter of 2012 shall be considered; see Kossovsky (2014) chapter 138 for details. This smaller data set regarding salaries (payroll) is markedly different from the very large data set of the State of Oklahoma (general) expenses discussed in chapter 20 section 2. Only those 2189 rows from the column 'Amount' pertaining to the Department of Human Services are considered. On the face of it, the data from its minimum value of **1.16**, to its maximum value of **13562.50**, spans multiple IPOT sub-intervals and comes with a very large order of magnitude. Yet in reality, when outliers and edges are taken into consideration, as they should be, a much narrower range and lower effective order of magnitude of only 0.8 is found, and the data set is decisively not Benford. First order digits proportions come out as {12.9, 41.2, 25.4, 8.3, 5.2, 2.8, 1.3, 1.8, 1.0}.

Such low effective order of magnitude precludes using IPOT sub-intervals for either the 1st digits basis or for the 2nd digits within the 1st digits basis. Only an analysis based on the purely 2nd order cycles can be used here in order to demonstrate the existence of digital development for this random data set.

The entire data set – including those disruptive outliers - is initially partitioned between all the purely 2nd order cycles, namely as in:

{

[1, 2), [2, 3), [3, 4), [4, 5), [5, 6), [6, 7), [7, 8), [8, 9), [9, 10),

[10, 20), [20, 30), [30, 40), [40, 50), [50, 60), [60, 70), [70, 80), [80, 90), [90, 100),

[100, 200), [200, 300), [300, 400), [400, 500), [500, 600), [600, 700), [700, 800), [800, 900), [900, 1000),

[1000, 2000), [2000, 3000), [3000, 4000), [4000, 5000), [5000, 6000), [6000, 7000), [7000, 8000), [8000, 9000), [9000, 10000),

[10000, 20000)

}

The range from the minimum of 1.16 to 1000 is thinly spread, containing only about 16% of overall data. The range from 6000 to the maximum of 13562.50 is even more thinly spread, containing only about 2% of overall data. It is decided here for this data set that any sub-interval having less than 2% of overall data shall be considered as potentially disruptive to the analysis and discarded as an outlier, thus leaving only the relevant core sub-intervals of {[1000, 2000), [2000, 3000), [3000, 4000), [4000, 5000), [5000, 6000)} to be considered, having data proportions of {9.7%, 39.3%, 23.4%, 6.1%, 3.2%}, and representing in total 81.7% of overall data. Calculating the 2nd order digital configuration on each sub-interval via ES04 values yields {-19.8%, -9.8%, +2.1%, +8.7%, +29.6%} as the vector of the purely 2nd order development pattern, as shown in Figures 7.14 and 7.15. This vector shows a decisive trend of increasing skewness and strongly confirms the expectation of such differentiated 2nd order behavior along the core range of the data. Further empirical data analyses on a variety of real-life data is necessary in order to arrive at strict cutoff points - as was laboriously done for the 4 forensic tests in Kossovsky (2014) chapter 43. The discussion here merely illustrates the general tendency of

ES04 to increase between the purely 2nd order cycles, signifying development pattern, and that such generic pattern should be found in all honest random data sets.

| Left   Border | 1,000 | 2,000 | 3,000 | 4,000 | 5,000 |
Right Border	2,000	3,000	4,000	5,000	6,000
Digit 0	13.2%	9.2%	8.2%	14.2%	2.9%
Digit 1	2.4%	6.4%	12.1%	9.7%	40.0%
Digit 2	9.0%	7.7%	14.6%	14.9%	20.0%
Digit 3	5.7%	13.0%	8.4%	14.2%	12.9%
Digit 4	4.7%	8.7%	13.5%	10.4%	8.6%
Digit 5	8.5%	12.7%	7.4%	4.5%	1.4%
Digit 6	8.5%	9.9%	6.6%	9.7%	8.6%
Digit 7	1.9%	8.9%	19.5%	6.0%	0.0%
Digit 8	8.5%	12.1%	2.1%	11.2%	1.4%
Digit 9	37.7%	11.5%	7.4%	5.2%	4.3%
% of Overall Data	*9.7%*	*39.3%*	*23.4%*	*6.1%*	*3.2%*
ES04	**-19.8%**	**-9.8%**	**2.1%**	**8.7%**	**29.6%**

Figure 7.14: Purely 2nd Order ES04 Values - Oklahoma Payroll Data

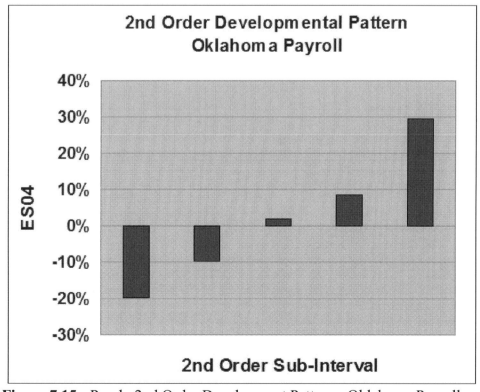

Figure 7.15: Purely 2nd Order Development Pattern - Oklahoma Payroll

[8] Case Study II: U.S. County Area Data

Forensic digital analysis shall be performed on US County Area data. The data pertains to areas of all the 3,143 counties in the US. Data can be downloaded from the US Census website http://www.census.gov/support/USACdataDownloads.html#LND where "Land Area" is selected with the choice of LND01.xls for data downloads. The data on column X called LND110210D is selected for analysis. It is necessary to eliminate from the data the following items: (a) 50 aggregated values of 50 states, (b) area for the District of Columbia, (c) total area of the entire country, (d) 3 entries with zero as area; resulting in areas for 3,143 proper counties. Data is reported in units of Square Miles. Some examples are shown below:

Yukon-Koyukuk County in Alaska has the largest area - 145,504.8 square miles
North Slope County in Alaska has the 2nd largest area -   88,695.4 square miles
Otoe County in Nebraska has the median area size -   615.6 square miles
Falls Church County in Virginia has the 2nd smallest area -   2.5 square miles
Lexington County in Virginia has the smallest area -   2.0 square miles

Certainly in this case, fraud detection is not an issue. The US Census Office is trusted in providing honest data, having no incentive whatsoever to cheat. Yet, this case is quite instructive and useful in demonstrating some relevant procedures and ideas in Forensic Digital Analysis.

The 1st order digit distribution is:
US County Area Census - {16.2, 10.0, 10.7, 15.8, 15.2, 10.4, 8.6, 7.1, 5.9}
Benford's Law 1st Digits - {30.1, 17.6, 12.5, 9.7, 7.9, 6.7, 5.8, 5.1, 4.6}

The 2nd order digit distribution is:
US County Area Census   - {13.5, 11.1, 10.2, 10.0, 8.5, 9.3, 9.4, 9.8, 8.6, 9.7}
Benford's Law 2nd Digits - {12.0, 11.4, 10.9, 10.4, 10.0, 9.7, 9.3, 9.0, 8.8, 8.5}

All in all, US County Area data set certainly fails the Benford test. The fact that higher orders digit distributions do indeed behave highly logarithmically does not excuse the data set in its entirety, since such selective compliance with the law is extremely typical in most non-Benford data types - as discussed in Kossovsky (2014) chapter 71 titled "The Near Indestructibly of Higher Order Distributions". The conceptual explanation for the digital behavior of US County Area data is that this data set is too narrowly focused on a limited range of values without sufficient order of magnitude. County boundaries within each state are man-made and artificial; legal declarations done in a way so as to give approximately similar geographic areas to each county. State officials would rarely assign any county huge land areas (at the expense of others), nor would they usually assign any county very tiny area. Therefore the spread of the data is too small, focused narrowly on certain magnitudes. The extreme cases shown above [Alaska and Virginia] with deceptive high order of magnitude of supposedly about 4.9 (with the minimum of 2.0 and the maximum of 145,504.8) are merely four examples at the extreme edges of the data set, truly constituting outliers on the very margin of data, while the bulk of the data is spread over a much narrower range. Applying the general logarithmic rule of Core Physical Order of Magnitude (CPOM) discussed in chapter 11 section 1, namely that (90th percentile)/(10th percentile) must be larger than 100, or at least not below 50 or 30, we obtain:

90th percentile = 1843.1
10th percentile = 286.0
CPOM = $P_{90\%}/P_{10\%}$ = (1843.1)/(286.0) = 6.5

Since CPOM should be over 50 or at least over 30 at a minimum for an expectation of logarithmic behavior in a given data set, such low CPOM value of 6.5 clearly explains why US County Area data set is not Benford at all.

Such low [effective] order of magnitude precludes using either the 1st digits basis or the 2nd digits within the 1st digits basis for the detection of digital development pattern. Instead, the purely 2nd order approach is performed here by dividing the core of the range into sub-intervals spanning whole 2nd order cycles, followed by the examination of the 2nd order measure of skewness of ES04 on each of these sub-intervals.

For the U.S. County Area data set, the range below 100, with only 1.8% of overall data is ignored as unmistakable outlier. Also the thinly spread range above 3000 containing only 5.4% of overall data is ignored as an outlier. More specifically, the whole range of the 2nd-digit-cycle sub-intervals above 3000 is data-anemic; where each sub-interval there contains less than 2% of overall data, therefore they are considered as outliers, not to be incorporated into the analysis.

The purely 2nd order partition for the core of the range is then:

{
  [100, 200), [200, 300), [300, 400), [400, 500), [500, 600), [600, 700), [700, 800),
  [800, 900), [900, 1000), [1000, 2000), [2000, 3000)
}

The skewness vector of ES04 values for these 11 sub-intervals depicted in Figure 7.16 came out as {-26.4, -10.6, -8.5, +3.4, -14.4, +3.9, +3.1, -14.2, +8.9, +16.5, +5.3}, with "%" sign omitted. An overall decisive trend of increasing ES04 is seen in Figure 7.17, confirming the expected digital development trend, albeit not in a very consistent and monotonic manner.

Left Border	100	200	300	400	500	600	700	800	900	1,000	2,000
Right Border	200	300	400	500	600	700	800	900	1,000	2,000	3,000
Digit 0	8.7%	6.9%	8.3%	14.7%	9.5%	13.2%	12.4%	9.0%	18.8%	25.5%	17.4%
Digit 1	2.2%	6.9%	8.7%	11.6%	7.3%	10.9%	17.8%	6.7%	15.3%	17.3%	9.6%
Digit 2	5.4%	9.6%	8.7%	10.5%	6.6%	9.5%	12.0%	9.0%	13.1%	12.8%	12.2%
Digit 3	5.4%	11.2%	9.7%	13.4%	8.8%	13.2%	8.1%	8.6%	9.7%	8.2%	13.0%
Digit 4	6.5%	9.6%	10.8%	8.0%	8.1%	11.8%	7.4%	7.1%	6.8%	7.4%	7.8%
Digit 5	9.8%	14.4%	9.4%	9.1%	9.9%	8.9%	9.7%	7.1%	8.0%	5.9%	12.2%
Digit 6	14.1%	11.7%	8.7%	8.0%	16.1%	6.3%	8.5%	12.9%	5.7%	5.6%	11.3%
Digit 7	12.0%	8.5%	9.7%	8.2%	17.6%	7.2%	8.1%	13.3%	9.1%	7.1%	5.2%
Digit 8	17.4%	13.3%	10.8%	6.2%	9.0%	7.9%	6.2%	9.0%	6.3%	6.6%	6.1%
Digit 9	18.5%	8.0%	15.2%	10.2%	7.0%	11.2%	9.7%	17.1%	7.4%	3.6%	5.2%
% of Data	2.9%	6.0%	8.8%	14.3%	14.4%	9.7%	8.2%	6.7%	5.6%	12.5%	3.7%
ES04	-26.4%	-10.6%	-8.5%	3.4%	-14.4%	3.9%	3.1%	-14.2%	8.9%	16.5%	5.3%

Figure 7.16: Table of the Purely 2nd Order ES04 Values - US County Area

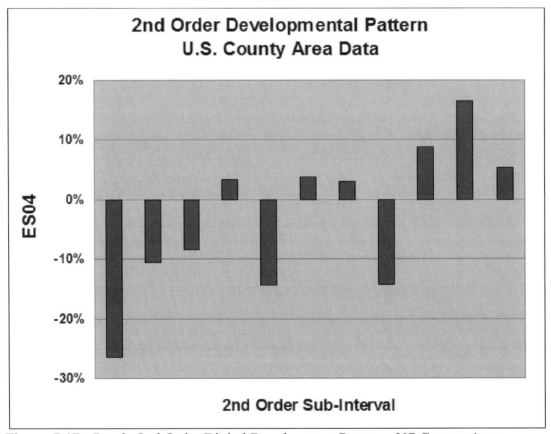

Figure 7.17: Purely 2nd Order Digital Development Pattern - US County Area

[9] Case Study III: The Lognormal Distribution

The Lognormal distribution is frequently found in accounting and financial data, as well as in scientific and physical data sets. For this reason it is important that the purely 2nd order basis development is confirmed also in the Lognormal case. For Lognormal distribution with shape parameter value of 1 or higher, digits are nearly perfectly Benford, and the range of its log histogram is wide enough, namely having high order of magnitude of over 3 or 4 for the distribution itself. For Lognormal distribution with shape parameter value much lower than 1, digits are not logarithmic for two reasons: (1) the range of its log histogram is too narrow and order of magnitude of the distribution itself is very low; (2) its histogram is not skewed at all or not skewed enough.

Yet, even for Lognormal distribution with low shape parameter where its entire range spans barely one order of magnitude or even less, thus precluding any possible development comparison on the 1st order basis or on the 2nd order within the 1st order basis, digital development pattern can still be detected via the purely 2nd order configurations along that narrow range. For example, Lognormal with shape parameter 0.45 and location parameter 6.45 is decisively not Benford, yet - its digital development pattern can be clearly detected!

Note: Monte Carlo computer simulations here are done via 35,000 realized values.

Simulated results regarding its 1st order digit distribution yields:

Lognormal(6.45, 0.45) 1st Digits - {15.4, 4.8, 10.5, 14.7, 15.4, 13.7, 11.0, 8.3, 6.3}
Benford's Law 1st Digits Order   - {30.1, 17.6, 12.5, 9.7, 7.9, 6.7, 5.8, 5.1, 4.6}

Simulated results regarding its 2nd order digit distribution yields:

Lognormal(6.45, 0.45) 2nd Digits - {13.0, 11.5, 10.6, 10.1, 9.7, 9.2, 9.2, 8.8, 9.1, 8.8}
Benford's Law 2nd Digits Order   - {12.0, 11.4, 10.9, 10.4, 10.0, 9.7, 9.3, 9.0, 8.8, 8.5}

About 99.9% of data here falls within **(141, 2698)**, and this indicates that order of magnitude is quite low. CPOM measure shows that the main body of the data comes with an even lower order of magnitude, namely that 80% of the core data falls within the narrower range of **(354, 1124)**.

90th percentile = 1124
10th percentile = 354
CPOM = $P_{90\%}$/$P_{10\%}$ = (1124)/(354) = 3.2

Since CPOM should be over 50 or at least over 30 at a minimum for an expectation of logarithmic behavior in a given data set, such low CPOM value of 3.2 clearly explains why Lognormal(6.45, 0.45) is not Benford at all.

Therefore, here there is practically not even a single proper IPOT 'sub-interval' to work with; and digital development pattern cannot be detected by way of 1st order digits basis or by way of 2nd order performed along 1st order basis. Indeed, only the purely 2nd order development pattern can be detected here. The range shall be partitioned here quite liberally in this case of the Lognormal to include the wider range of (141, 2698) where 99.9% of data falls.

The reason for the wider and more liberal consideration of range here springs from reduced fear of outliers and disruptive edges, being that a very smooth, predictable, and abstract distribution is under consideration here. Also, the fact that plenty of realized/simulated values are being used, namely 35,000 points in total, ensures more smoothness and continuity. Thus the purely 2nd order sub-intervals to be considered here are as follows:

{
  [100, 200), [200, 300), [300, 400), [400, 500), [500, 600), [600, 700), [700, 800),
  [800, 900), [900, 1000), [1000, 2000), [2000, 3000)
}

The vector of ES04 values for these 11 sub-intervals came out as (with '%' sign omitted): {-38.8, -22.9, -11.5, -7.1, -3.8, -3.6, -0.7, -0.7, -1.6, +30.0, +29.4}, and it is depicted in the table of Figure 7.18. This vector shows a decisive trend of increasing ES04 skewness as seen in Figure 7.19. This result strongly confirms the expectation of digital development pattern, and which is being detected here via the differentiated purely 2nd order behavior of the data.

Left Border	100	200	300	400	500	600	700	800	900	1,000	2,000
Right Border	200	300	400	500	600	700	800	900	1,000	2,000	3,000
Digit 0	0.0%	4.1%	7.2%	9.3%	9.9%	11.2%	10.6%	11.7%	11.0%	29.4%	29.0%
Digit 1	0.5%	4.9%	8.2%	9.4%	10.5%	10.8%	11.4%	11.0%	11.8%	21.8%	26.1%
Digit 2	2.7%	6.4%	8.8%	9.0%	10.1%	10.4%	10.6%	9.5%	10.9%	15.5%	11.4%
Digit 3	6.6%	7.6%	9.1%	9.7%	10.3%	9.6%	11.9%	10.9%	9.9%	10.4%	9.7%
Digit 4	6.0%	8.8%	9.9%	10.2%	10.1%	9.2%	9.7%	10.9%	9.6%	7.7%	8.0%
Digit 5	10.4%	10.8%	10.4%	10.3%	10.3%	9.9%	9.3%	9.6%	11.5%	5.5%	8.0%
Digit 6	7.7%	12.3%	10.8%	10.4%	9.6%	10.3%	9.6%	10.1%	9.3%	3.9%	3.4%
Digit 7	16.5%	14.6%	12.0%	11.0%	9.7%	10.3%	9.1%	9.3%	9.8%	2.5%	2.3%
Digit 8	20.9%	14.9%	11.0%	10.0%	9.4%	8.9%	8.9%	8.2%	8.1%	2.1%	1.1%
Digit 9	28.6%	15.6%	12.6%	10.7%	10.1%	9.5%	9.1%	8.8%	8.2%	1.3%	1.1%
% of Data	0.5%	4.4%	10.9%	14.8%	15.4%	13.3%	11.0%	8.4%	6.2%	14.6%	0.5%
ES04	-38.8%	-22.9%	-11.5%	-7.1%	-3.8%	-3.6%	-0.7%	-0.7%	-1.6%	30.0%	29.4%

Figure 7.18: Table of the Purely 2nd Order ES04 Values - Lognormal(6.45, 0.45)

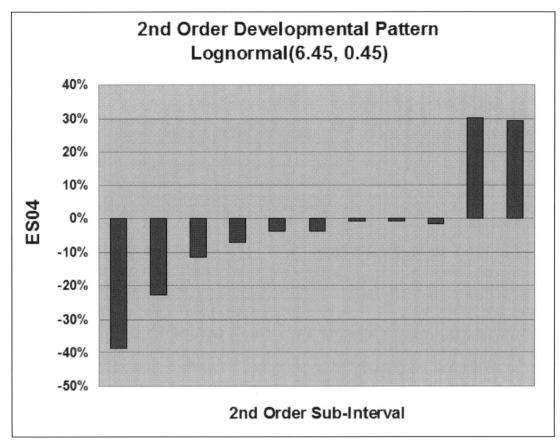

Figure 7.19: Purely 2nd Order Development Pattern - Lognormal(6.45, 0.45)

Figure 7.20 depicts the (decimal) log histogram of Lognormal(6.45, 0.45).

Leading Digits Inflection Point (LDIP) is defined and discussed in Kossovsky (2014) chapter 81, as well as in section 3 here. Very briefly, LDIP is such that to the left of it digital configuration is milder than the Benford configuration, and to the right of it digital configuration is more skewed and severe than the Benford configuration. Clearly, Digital Development Pattern revolves around LDIP.

As discussed in Kossovsky (2014) chapter 81, theoretical LDIP on the (decimal) log histogram of any Lognormal distribution is at LOG(e)*(Location), calculated here as (0.434)*(6.45) = 2.8, and the existence of this pivotal log value is demonstrated empirically via the simulations as seen in Figure 7.20, where to the left of approximately 2.8 log value, log histogram is rising, and to the right of it, log histogram is falling.

On the data histogram itself of any Lognormal distribution, theoretical LDIP is at $e^{Location}$, calculated here as $e^{6.45} = 633$. Indeed, as can be seen in the table of Figure 7.18, the empirical results via the simulations for the purely 2nd order cycles show that to the left of approximately 633, ES04 is decisively negative, and to the right of it, beginning from 700, ES04 is either a very low negative number near zero, or decisively positive. Hence, consistent with its nature, LDIP of 633 had proved itself to be a pivotal point also for the purely 2nd order development pattern.

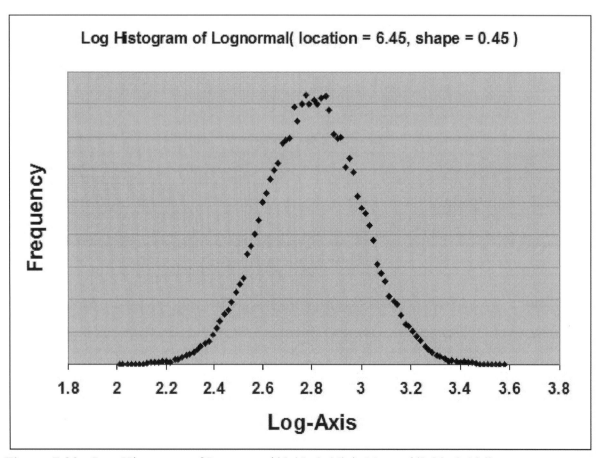

Figure 7.20: Log Histogram of Lognormal(6.45, 0.45) is Normal(2.80, 0.194)

[10] Order Dependencies Induce Mild Distortions in the Purely 2nd Order Pattern

Let us demonstrate the mild distortions in ES04 measurements arising from the dependencies between the orders. The table in Figure 7.21 provides the conditional probabilities of the 2nd order digits for any given 1st order occurrence.

This table should be contrasted with Benford's Law for the unconditional 2nd order digits distribution of {**12.0, 11.4, 10.9, 10.4, 10.0, 9.7, 9.3, 9.0, 8.8, 8.5**}.

Further discussion and more details about the dependencies between the digital orders are provided in Kossovsky (2014) chapter 100.

Given that 1st digit is	Probabilities for 2nd digit:									
	0	1	2	3	4	5	6	7	8	9
1	13.8	12.6	11.5	10.7	10.0	9.3	8.7	8.2	7.8	7.4
2	12.0	11.5	11.0	10.5	10.1	9.7	9.3	9.0	8.7	8.4
3	11.4	11.0	10.7	10.4	10.1	9.8	9.5	9.3	9.0	8.8
4	11.1	10.8	10.5	10.3	10.1	9.8	9.6	9.4	9.2	9.1
5	10.9	10.7	10.4	10.3	10.1	9.9	9.7	9.5	9.4	9.2
6	10.7	10.5	10.4	10.2	10.1	9.9	9.8	9.6	9.5	9.3
7	10.6	10.5	10.3	10.2	10.1	9.9	9.8	9.7	9.5	9.4
8	10.5	10.4	10.3	10.2	10.0	9.9	9.8	9.7	9.6	9.5
9	10.5	10.4	10.3	10.2	10.0	9.9	9.8	9.7	9.6	9.5

Figure 7.21: Conditional Probabilities of Second Order Digits Given First Order Digit

As can be deduced from the table in Figure 7.21, 2nd order digits are skewer (in favor of low digits) whenever the 1st digit is low, and are more equal whenever the 1st digit is high. In other words, the 1st and the 2nd orders positively correlate in this limited sense. Therefore, ES04 should be a bit higher than its usual level whenever the 1st digit is low, and ES04 should be a bit lower than its usual level whenever the 1st digit is high. The crucial implication here for forensic analysis is that on the interval (10, 20) for example where the 1st digit is always 1, skewness for the 2nd digits and its measure of ES04 tend to be higher than on the interval (80, 90) for example where the 1st digit is always 8 - excluding other factors influencing ES04 such as Digital Development Pattern which could counteract, overcome, and even overwhelm this dependency tendency, enough to give (80, 90) even higher ES04 value than for (10, 20).

[11] Distortions in Purely 2nd Order Pattern for k/x Due to Dependencies

The k/x distribution defined over (1, 1000) shall be forensically analyzed via the purely 2nd order basis. As discussed in previous sections, the log histogram (density) of the k/x distribution is flat and uniform; hence Digital Development Pattern does not exist here. Rather, the Benford configuration is consistent and constant throughout the entire range. Hence, ES04 is not swayed at all by development, and on the face of it, should in principle be constant and totally stable throughout the entire range. Yet in reality ES04 is not stable! It does fluctuate indeed due to the dependencies of the 2nd order on the 1st order. Computer realizations of k/x over (1, 1000) were generated, yielding 42,858 realized values. The Program starts by generating methodically a long series of uniformly-spaced log values from 0 to 3 in tiny and even steps of 0.00007 increments. For example, the first 8 log values that it generates are {0.00000, 0.00007, 0.00014, 0.00021, 0.00028, 0.00035, 0.00042, 0.00049}, and so forth. The last 8 log values that it generates are {2.99950, 2.99957, 2.99964, 2.99971, 2.99978, 2.99985, 2.99992, 2.99999}. The program then generates k/x = $10^{\text{UNIFORM-LOG-SERIES}}$ values. This entire data set of 42,858 values is partitioned between all the relevant purely 2nd order cycles, namely as in:

{
  [1, 2), [2, 3), [3, 4), [4, 5), [5, 6), [6, 7), [7, 8), [8, 9), [9, 10),

  [10, 20), [20, 30), [30, 40), [40, 50), [50, 60), [60, 70), [70, 80), [80, 90), [90, 100),

  [100, 200), [200, 300), [300, 400), [400, 500), [500, 600), [600, 700), [700, 800), [800, 900), [900, 1000)
}

The vector of ES04 values for these 27 sub-intervals came out as (with '%' sign omitted):

{
  +3.8, +0.3, -1.1, -1.9, -2.4, -2.7, -3.1, -3.3, -3.4,
  +3.8, +0.3, -1.1, -1.9, -2.4, -2.8, -3.0, -3.1, -3.3,
  +3.8, +0.3, -1.1, -1.9, -2.4, -2.8, -3.0, -3.3, -3.5
}

Figure 7.22 depicts this purely 2nd order sequence of ES04 values for k/x over (1, 1000).

None of these 27 sub-intervals should be considered as an outlier or data-anemic in any sense whatsoever, and in fact none came out as data-anemic. This is so since this is not some random real-life physical data, but rather computer-generated data set. Moreover, this computer scheme directly generates values in a methodical and totally non-random fashion, not applying the Monte Carlo simulation method.

Since Digital Development Pattern does not exist here, there is nothing to counteract these intrinsic fluctuations due to the dependencies between the orders. This is the result of having a log density that is uniformly distributed and perfectly flat, thus totally lacking development.

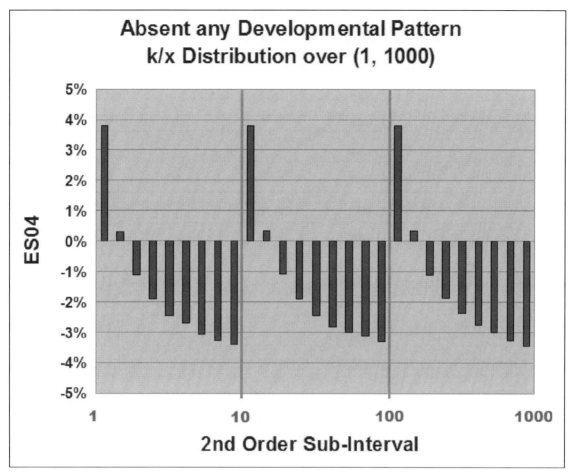

Figure 7.22: Purely 2nd Order Digital Fluctuations Due to Dependencies – k/x on (1, 1000)

The empirical fluctuations seen in Figure 7.22 correspond almost perfectly to the theoretical variation in skewness for the 2nd order digits according to the dependency on the 1st order digits as seen in Figure 7.21. These computerized results of the k/x case can be easily checked against the calculations of the theoretical expectations of ES04 values on each sub-interval.

For example, according to the table of Figure 7.21, the conditional probabilities of the 2nd digits given that the 1st digit is **1** are shown in the top row:

{13.8%, 12.6%, 11.5%, 10.7%, 10.0%, 9.3%, 8.7%, 8.2%, 7.8%, 7.4%}.

Calculations are: ES04 = [13.8 + 12.6 + 11.5 + 10.7 + 10.0] – [54.7] = [58.6] – [54.7] = **+3.9**

For example, according to the table of Figure 7.21, the conditional probabilities of the 2nd digits given that the 1st digit is **9** are shown in the bottom row:

{10.5%, 10.4%, 10.3%, 10.2%, 10.0%, 9.9%, 9.8%, 9.7%, 9.6%, 9.5%}.

Calculations are: ES04 = [10.5 + 10.4 + 10.3 + 10.2 + 10.0] – [54.7] = [51.4] – [54.7] = **– 3.3**

[12] Distortions in Purely 2nd Order Pattern for US Population Due to Dependencies

The data set on the US Population 2009 census survey shall be revisited, and development along the purely 2nd order pattern shall be examined. As was done for the this data set regarding its 2nd order analysis along 1st order basis of Figure 7.12, here too for the purely 2nd order analysis we shall omit all the points below 10 and all the points over 1,000,000, considering them as outliers and such. Hence the entire data set is to be partitioned between all the purely 2nd order cycles from 10 to 1,000,000, namely as in:

{

[10, 20), [20, 30), [30, 40), [40, 50), [50, 60), [60, 70), [70, 80), [80, 90), [90, 100),

[100, 200), [200, 300), [300, 400), [400, 500), [500, 600), [600, 700), [700, 800), [800, 900), [900, 1000),

[1000, 2000), [2000, 3000), [3000, 4000), [4000, 5000), [5000, 6000), [6000, 7000), [7000, 8000), [8000, 9000), [9000, 10000),

[10000, 20000), [20000, 30000), [30000, 40000), [40000, 50000), [50000, 60000), [60000, 70000), [70000, 80000), [80000, 90000), [90000, 100000),

[100000, 200000), [200000, 300000), [300000, 400000), [400000, 500000), [500000, 600000), [600000, 700000), [700000, 800000), [800000, 900000), [900000, 1000000)

}

Figure 7.23 depicts the chart of the purely 2nd order sequence of ES04 values. Visually, there is a decisive global or overall trend of increasing ES04 values, albeit in a somewhat irregular pattern, with roughly 5 or 6 zigzag cycles. The last 7 sub-intervals from 300,000 to 1,000,000 are quite data-anemic containing only 0.08%, 0.06%, 0.04%, 0.04%, 0.02%, 0.02%, 0.01% proportions of overall data respectively, and this fact may explain their outlier-like erratic behavior on the right-most part of the chart. The chart seen in Figure 7.23 is derived from the confluence of two forces acting upon 2nd order digital configuration, namely, the strong, dominant, and global Digital Development Pattern, as well as the weak and local Order Dependency Pattern, resulting in the actual shape of the chart. Figure 7.23 is certainly compatible with what Figure 7.13 and Figure 7.22 jointly imply. Obviously, Figure 7.13 which performs the 2nd order digits analysis on the 1st order basis of IPOT points is much preferred to Figure 7.23 which performs the analysis on the purely 2nd order basis. This is so because in Figure 7.13 the unmistakable upward trend of development is clearly visible, but in Figure 7.23 that trend is not very clearly observed as it is confounded a bit by the Order Dependency Pattern. This is exactly the motivation behind the forensic rule [discussed earlier in chapter 5] that prefers ES04 analysis of the 2nd order to be performed on the 1st order basis and not on the purely 2nd order basis - whenever there are at least 3 IPOT sub-intervals available for such analysis. Surely, in cases where there is only 2 or less IPOT sub-intervals available for analysis, such as in the 3 case studies above, the purely 2nd order basis is the only proper and useful forensic test to execute.

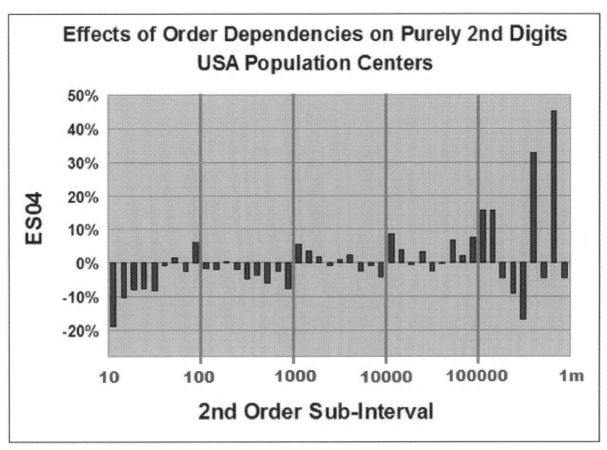

Figure 7.23: Purely 2nd Order Digital Development Pattern – USA Population

Around the central part of the chart in Figure 7.23, the transition around IPOT points 1000 and 10000 strongly remind us of the zigzag pattern of Order Dependency Pattern seen in Figure 7.22. It is mostly around the left and right edges that Digital Development Pattern totally overwhelms and dominates digital configurations.

In Figure 7.19 regarding the purely 2nd order analysis of Lognormal(6.45, 0.45), there is a sharp and pronounced rise in ES04 for the last 2 sub-intervals of [1000, 2000) and [2000, 3000). Surely one can reasonably point out that this nicely coincides or being reinforced by the Order Dependency Pattern as we leave [900, 1000) which is the least skewed 2nd-digits-wise, and enter [1000, 2000) which is the most skewed 2nd-digits-wise. Yet, since Order Dependency Pattern is mild in comparison to the overwhelming and strong influence of Digital Development Pattern, this explanation is not very convincing. As seen in Figure 7.22, that quantum jump in ES04 values due to Order Dependency Pattern as IPOT points of 10 and 100 are passed is just 7.2%, while the sharp rise in Figure 7.19 from [900, 1000) into [1000, 2000) is a whopping 31.6%, and this sharp rise is mostly caused by Digital Development Pattern, with only minor assistance from Order Dependency Pattern.

Figure 7.24 depicts the (decimal) log histogram the 2009 US Population data. Leading Digits Inflection Point (LDIP) seems to be around 2.9 on the log-axis, and around $10^{2.9}$ or simply 795 approximately on the x-axis itself. The sharp contrast seen in Figure 7.24 between the sharply rising region on the left of LDIP and the precipitously falling region on the right of LDIP implies that development is quite intense when the focus is around the values of 795 or so, and that there is no central region here where log histogram is at least just locally horizontal-like and development is momentarily relaxing its overbearing and strong influence on digital configuration.

The existence of LDIP around 795 implies that the part of the data to the left of it should be milder and less skewed than Benford, and that the part of the data to the right of it should be of extreme inequality and even skewer than Benford. All this can be confirmed by staring at the table of Figure 7.5 for ES12 and at the table of Figure 7.12 for ES04 (both of which pertain to the data set of US population census). Flexibility is required in thinking along very round and generic numbers; hence instead of wrongly focusing on the fanatically exact value of 795, one should focus on the round and generic number of 1000. Clearly in Figure 7.5 as well as in Figure 7.12, to the left of 1000 Excess Sums are negative (i.e. milder digital configuration), while to the right of 1000 Excess Sums are positive (i.e. extreme digital inequality) – all of which is consistent and in harmony with the definition and the meaning of LDIP.

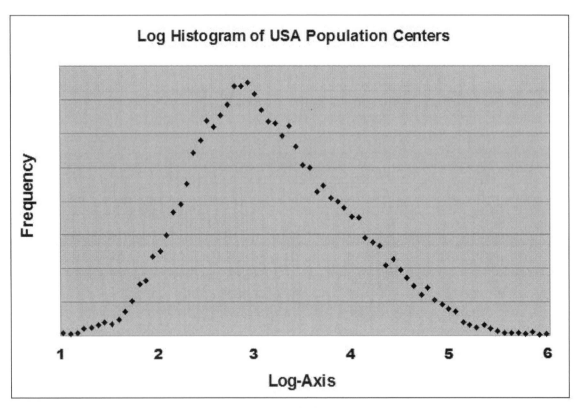

Figure 7.24: Log Histogram of 2009 US Population Centers Data

[13] Development Pattern Overwhelms Order Dependency in Data with Low Order of Magnitude

What could be seen as puzzling perhaps is the fact that for the data sets of Case Studies I, II, III, ES04 values of development on the purely 2nd order basis came out with smooth and very decisive pattern, without any apparent interference from Order Dependency Pattern - as visually seen in Figures 7.15, 7.17, 7.19. They all lack zigzags and fluctuations except for the minor ones seen in Figure 7.17; and even those slight fluctuations are mostly not centered around IPOT points and thus are not associated with Order Dependency Pattern. One wonders why Order Dependency Pattern couldn't manifest itself at all in these three cases, not even mildly.

Case Studies I, II, III are all with exceedingly low order of magnitude of just about 1, lacking a variety of IPOT sub-intervals, and as a consequence development pattern cannot be forensically checked by comparing digital configurations of the 1st digits or of the 2nd digits on the 1st order basis. Is it simply by fortunate coincidences and rare luck that we were able to verify development via the purely 2nd order for these three particular data sets - and without any distortions due to order dependencies? No, it's not luck, and these results do not spring from any rare coincidences, rather they are derived indeed from some very general, intrinsic, and profound statistical forces and tendencies which apply to all data sets with low order of magnitude.

Three general facts regarding digit configurations and log histograms of data need to be stated and acknowledged before an explanation for the above puzzling results could be given.

The first fact is the mathematical basis of Digital Development Pattern discussed in chapter 19 section 1 which states that log density of practically all random data is almost always an upside-down U-like curve in the spirit of Related Log Conjecture of chapter 18 section 1, namely that it starts from the bottom on the log-axis itself and that it ends all the way down on the log-axis as well. Figure 1.16 enables the visualization of the generic sketch of practically all log histograms of random data.

The second fact is about the relationship between the slope or steepness of log histogram and digit configuration - as discussed in chapter 16 section 1 on mantissa. The concept of the 'slope' is similar to the one in Calculus, but it's not identical, as the focus here is on relative proportions of areas under the curve which depend not only on the 'slope' but also on height of the histogram above the log-axis. Figure 1.15 enables the visualization of this relationship, where a falling log histogram implies severe digital skewness, while a rising log histogram implies more equal and almost even digital configuration. What was not explicitly stated in that chapter [but expected to be intuitively grasped by the reader] is that the steeper the fall in the log histogram the more severe is digital skewness, and that the steeper the rise in the log histogram the more equal digit configuration is. So much so, that an extremely sharp rise in log histogram could imply inverse skewness where high first digits {7, 8, 9} are more frequent than low first digits {1, 2, 3}.

The third fact is that the range on the log histogram signifies the order of magnitude of the data, so that if log of data spans (2.5, 6.5) on the log-axis, then order of magnitude is $6.5 - 2.5 = 4$. Data set with low order of magnitude has a narrow span on the log-axis, while data set with high order of magnitude has a wide span.

Figure 7.25, depicts two superimposed log histograms for comparison, one with low order of magnitude and one with high order of magnitude. The details of these two log histograms shall be discussed later, but the explanation given now revolves around the generic shapes of the log histograms in Figure 7.25. In a nutshell, the explanation lies in the fact that when order of magnitude is very small, log histogram then necessarily curves sharply over about only 1 or so log-units; it rises very sharply on the left [steep slope], and then it falls precipitously on the right [steep slope]; quickly and dramatically changing course from the minimum value to the maximum value - both of which are not that much far apart from each other, since order of magnitude is small. As a consequence, Digital Development Pattern truly dominates digital configuration and it totally overwhelms Order Dependency Pattern to the point of completely eclipsing it. When order of magnitude is high enough, say over 3 or over 4 log units, log histogram then necessarily curves around much more gently [gradual slope], and therefore Digital Development Pattern cannot totally overwhelm Order Dependency Pattern – resulting in some minor reversals and zigzag patterns centered around IPOT points.

In order to clearly demonstrate the principles involved, ES04-development analyses on the purely 2nd order basis for two distinct Lognormal distributions shall be compared.

Lognormal with location parameter 6, and shape parameter 2.0 – high order of magnitude.
Lognormal with location parameter 6, and shape parameter 0.6 – low order of magnitude.

These two distributions have an identical density form, and they have an identical median value since both are with the same location parameter of 6, yet they have distinct orders of magnitude due to their distinct shape parameter values of 2.0 and 0.6. Exactly 30,000 realizations shall be obtained from each Lognormal, ensuring that both have the same number of data points as well. Hence these two distributions are comparable in almost every aspect, and the main distinction between them is having distinct orders of magnitude. Let us now calculate the generic expression for the decimal log density of any Lognormal as follows:

$$\text{Lognormal(location, shape)} = e^{\text{Normal(location, shape)}}$$

$$\log_{10}(\text{Lognormal(location, shape)}) = \log_{10}(e^{\text{Normal(location, shape)}})$$

Utilizing the logarithmic identity $\text{LOG}_A(X) = \text{LOG}_B(X) / \text{LOG}_B(A)$ for the right hand side of the equation we get:

$$\log_{10}(\text{Lognormal(location, shape)}) = \log_e(e^{\text{Normal(location, shape)}})/\log_e 10$$

$$\log_{10}(\text{Lognormal(location, shape)}) = \text{Normal(location, shape)}/\log_e 10$$

$$\log_{10}(\text{Lognormal(location, shape)}) = (\log_{10}e)*\text{Normal(location, shape)}$$

Hence the decimal logarithm of the Lognormal is simply the generating Normal distribution adjusted by a scale multiplicative factor of $\log_{10}(e) = 0.43429$.

The decimal logarithm of Lognormal(6, 2.0) is Normal(6*[0.43429], 2.0*[0.43429]).
The decimal logarithm of Lognormal(6, 2.0) is **Normal(2.6, 0.87)**.

The decimal logarithm of Lognormal(6, 0.6) is Normal(6*[0.43429], 0.6*[0.43429]).
The decimal logarithm of Lognormal(6, 0.6) is **Normal(2.6, 0.26)**.

Figure 7.25 depicts the decimal log histograms of Lognormal(6, 2.0) and Lognormal(6, 0.6) superimposed on the relevant log-axis range. Visually attempting to ascertain the orders of magnitude of the two Lognormals by estimating the total log range of each log histogram gives approximately the values of 4.5 and 1.5 respectively.

The three-sigma empirical rule states that for many reasonably symmetric unimodal distributions, almost all of the points lie within three standard deviations of the mean. More specifically, the 99.73% empirical rule for the Normal distribution states that 99.73% of the values lie within three standard deviations of the mean.

Hence the approximate calculations of the order of magnitude of Lognormal(6, 2.0) viewed via its log density of Normal(2.6, 0.87) yield the estimated full range of (2.6 – 3*0.87, 2.6 + 3*0.87), namely (–0.01, 5.21), thus order of magnitude is around (5.21) – (–0.01) = 5.22.

Hence the approximate calculations of the order of magnitude of Lognormal(6, 0.6) viewed via its log density of Normal(2.6, 0.26) yield the estimated full range of (2.6 – 3*0.26, 2.6 + 3*0.26), namely (1.82, 3.38), thus order of magnitude is around (3.38) – (1.82) = 1.56.

In reality, when the left and right extreme edges (outlier-like points) are excluded from the calculations of the effective order of magnitude - as they should be - the core data is of somewhat lower Physical Order of Magnitude (POM) value.

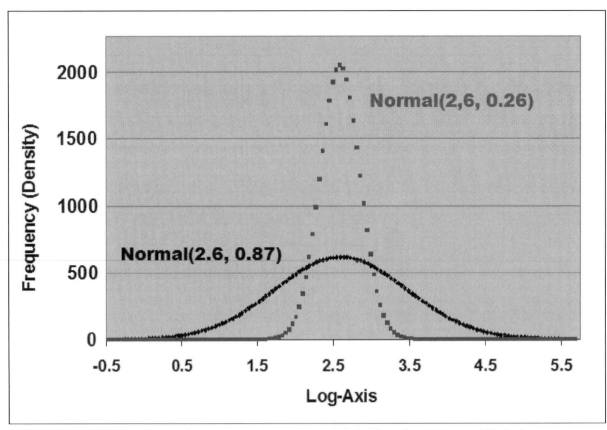

Figure 7.25: Decimal Log Histograms of Lognormal(6, 2.0) and Lognormal(6, 0.6)

The confluence of the facts that each of the two histograms in Figure 7.25 contains 30,000 points, namely that both have the same data size, and that both must start and must end on the log-axis itself, geometrically implies that the histogram with lower order of magnitude of narrower range is tall in statue, and thus it comes with extreme slopes and steepness; while the histogram with higher order of magnitude of wider range is short in statue, and thus it comes with gentle slopes and mild steepness.

More generally, the conversion of histograms into full fledged statistical densities involves adjusting the vertical y-axis so that total area of each density sums to 1, and this feature drives the point that densities of lower order of magnitude with narrower span on the log-axis, must be quite tall in statue along the vertical axis [so that total area sums to 1], and as a consequence they necessarily come with extreme slopes and steepness; while densities of higher order of magnitude with wider span on the log-axis, must be quite short in statue along the vertical axis [so that total area sums to 1], and as a consequence they necessarily come with gentle slopes and mild steepness.

Let us now perform development analysis of ES04 on the purely 2nd order basis for Lognormal(6, 2.0) of high order of magnitude.

We shall omit 40 points below 1 - constituting only 0.13% of the entire 30,000 data set, considering them as outliers. We shall also omit 9 points above 400000 - constituting only 0.03% of the entire 30,000 data set, considering them as outliers. Hence the entire data set is to be partitioned between all the purely 2nd order cycles from 1 to 400000, a range containing 29,951 data points - constituting 99.84% of the entire data set, namely as in:

{
  [1, 2), [2, 3), [3, 4), [4, 5), [5, 6), [6, 7), [7, 8), [8, 9), [9, 10),

  [10, 20), [20, 30), [30, 40), [40, 50), [50, 60), [60, 70), [70, 80), [80, 90), [90, 100),

  [100, 200), [200, 300), [300, 400), [400, 500), [500, 600), [600, 700), [700, 800), [800, 900), [900, 1000),

  [1000, 2000), [2000, 3000), [3000, 4000), [4000, 5000), [5000, 6000), [6000, 7000), [7000, 8000), [8000, 9000), [9000, 10000),

  [10000, 20000), [20000, 30000), [30000, 40000), [40000, 50000), [50000, 60000), [60000, 70000), [70000, 80000), [80000, 90000), [90000, 100000),

  [100000, 200000), [200000, 300000), [300000, 400000)
}

Figure 7.26 depicts the chart of the sequence of ES04 values on the purely 2nd order basis. Visually, there is a decisive global or overall trend of increasing ES04 values, signifying development, yet, due to the fact that the distribution is of relatively high order of magnitude, the pattern is a bit irregular, being partially influenced and confounded locally by the Order Dependency Pattern, causing roughly 4 zigzag cycles.

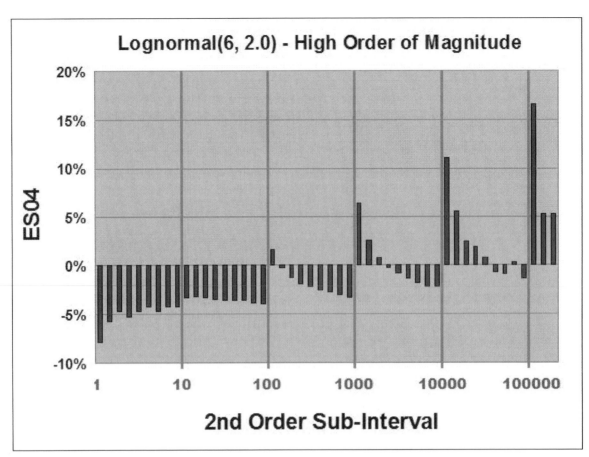

Figure 7.26: Fluctuations Due to Dependencies in the Purely 2nd Order Development Pattern

Let us now perform development analysis of ES04 on the purely 2nd order basis for Lognormal(6, 0.6) of low order of magnitude.

We shall omit 7 points below 50 - constituting only 0.02% of the entire 30,000 data set, considering them as outliers. We shall also omit 13 points above 3000 - constituting only 0.04% of the entire 30,000 data set, considering them as outliers. Hence the entire data set is to be partitioned between all the purely 2nd order cycles from 50 to 3000, a range containing 29,980 data points - constituting 99.93% of the entire data set, namely as in:

{

  [50, 60), [60, 70), [70, 80), [80, 90), [90, 100),

  [100, 200), [200, 300), [300, 400), [400, 500), [500, 600), [600, 700), [700, 800), [800, 900), [900, 1000),

  [1000, 2000), [2000, 3000)

}

Figure 7.27 depicts the chart of the sequence of ES04 values on the purely 2nd order basis. Visually, there is a decisive global as well as local development trend of increasing ES04 values, almost consistently so, and almost without reversals or zigzags, except the minor one around 100. This is so due to the fact that the distribution is of relatively low order of magnitude and where Digital Development Pattern almost totally overwhelms Order Dependency Pattern.

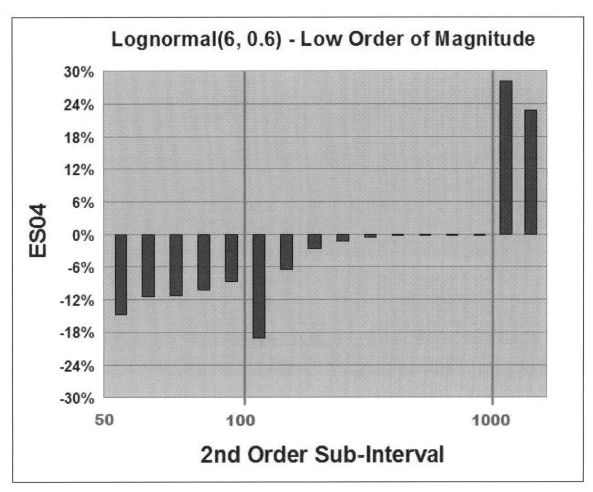

Figure 7.27: Smooth and Consistent Pattren of the Purely 2nd Order Digital Development

In summary, the lesson learnt from these two distinct yet similar Lognormal distrinbutions is that relatively high order of magnitude allows Order Dependency Pattern to manifest itself with few gentle reversals or zigzags, slightly confounding the visualization of development; while relatively low order of magnitude causes Digital Development Pattern to overwhelm Order Dependency Pattern to the extent that development can be seen clearly and decisively without even minor reversals or zigzags. The forensic analyst is quite fortunate that this feature of the purely 2nd order digital development analysis exists, as it nicely compensates for the total inability to perform any development analysis on the 1st digit basis whenever data is of low order of magnitude.

SECTION 8

ORDER OF MAGNITUDE AND BENFORD'S LAW

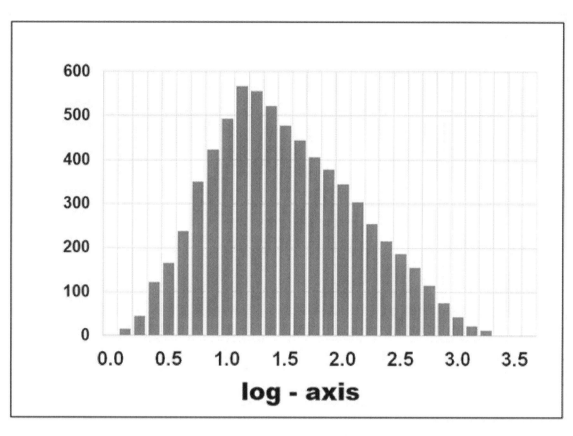

[1] Correlation between Order and Magnitude and Benford Behavior

Four distinct demonstrations of the strong correlation between Benford behavior and order of magnitude of data shall be presented here. This is accomplished via the explorations of four distinct data types generated from Benford random distributions and Benford random processes, each with a wide variety of orders of magnitude, so that digital results can be tested against the Benford requirement that order of magnitude of data should be at least 3 in the approximate. The four demonstrations are: **(1)** The Lognormal Distribution, **(2)** Random Partition Process, **(3)** Random Multiplication Process, and **(4)** Chain of Two Uniform Distributions.

(1) The first demonstration of the strong correlation between Benford's Law and order of magnitude of data is provided via several computer simulations of the Lognormal distribution involving a wide variety of orders of magnitude. Only the shape parameter of the Lognormal distribution determines its order of magnitude, while its location parameter is irrelevant and thus it is being fixed at the value of **5.0** in all of these computer simulations.

One of the generic expressions of the Lognormal is as follows:

$$\text{Lognormal(location, shape)} = e^{\text{Normal(location, shape)}}$$

This implies that the Lognormal distribution can be obtained by taking Euler's number e to the power of the random Normal distribution. The 'location' parameter of the Normal is actually its mean, and the 'shape' parameter of the Normal is actually its standard deviation (s.d.), hence this can be written as:

$$\text{Lognormal(location, shape)} = e^{\text{Normal(mean, s.d.)}}$$

with the understanding that 'mean' = 'location', and that 's.d.' = 'shape'. Since s.d. is the only factor (i.e. parameter) controlling the variability of the Normal, it follows that the variability of the Lognormal (i.e. order of magnitude) is controlled exclusively by the s.d. of the generating Normal, namely by the shape parameter of the Lognormal.

For these Lognormal computer simulations, a wide variety of shape parameter values are used, from 0.2 to 1.4 in small even steps of 0.05. Order of magnitude is determined after the elimination of the top 1% of values and the bottom 1% of values, so as not to let extreme computer simulation values (outliers) influence or distort order of magnitude; and this path is chosen due to the realization that such distortions could easily be very significant. Digital configuration on the other hand is determined for the entire data set coming out of the computer simulation without any elimination of what might be considered extreme values; and this path is chosen due to the realization that any distortion from potential outliers could never be significant at all [since they could constitute at most only 2% of all digits].

Simulation for each shape parameter value is performed with 30,000 random realizations.

Figure 8.1 depicts simulation results for these Lognormal distributions regarding resultant digital configuration together with their SSD measure as well as resultant order of magnitude. Figure 8.2 depicts the scatter chart of digit 1 proportions versus order of magnitude.

What should also be acknowledged here [although it's not visible in the figures] is that while the shape parameter increases from 0.2 to 1.4 causing OOM to increase, skewness is also increasing, from the nearly symmetrical Normal-like curve when shape is around 0.2, to a highly positively skewed curve when shape is around 1.4 and where histogram comes with a tail falling to the right overall, except for a brief, temporary, and minor rise on the very left side of the x-axis near the origin. Clearly, this transformation from symmetry to asymmetry also contributes decisively to the convergence of first digits to the Benford configuration. The confluence of increasing order of magnitude as well as increasing quantitative skewness as the shape parameter increases leads to full Benford convergence.

Shape	OOM	Digit 1	Digit 2	Digit 3	Digit 4	Digit 5	Digit 6	Digit 7	Digit 8	Digit 9	SSD
0.20	0.4	90.6%	6.9%	0.0%	0.0%	0.0%	0.0%	0.1%	0.5%	1.8%	4196.7
0.25	0.5	82.8%	11.2%	0.3%	0.0%	0.0%	0.1%	0.6%	1.5%	3.5%	3204.2
0.30	0.6	74.3%	15.4%	0.9%	0.0%	0.1%	0.4%	1.4%	2.8%	4.7%	2310.0
0.35	0.7	67.3%	17.5%	1.8%	0.3%	0.4%	1.2%	2.4%	3.8%	5.3%	1689.7
0.40	0.8	61.0%	18.9%	3.3%	0.7%	0.9%	2.0%	3.2%	4.4%	5.6%	1196.6
0.45	0.9	55.8%	19.2%	4.6%	1.8%	1.7%	2.6%	3.8%	4.9%	5.7%	847.3
0.50	1.0	51.5%	19.2%	5.8%	2.7%	2.6%	3.3%	4.3%	5.0%	5.6%	595.1
0.55	1.1	47.7%	19.0%	7.1%	3.6%	3.4%	4.0%	4.8%	5.0%	5.4%	405.4
0.60	1.2	42.9%	19.7%	8.3%	4.9%	4.1%	4.4%	5.1%	5.2%	5.4%	229.2
0.65	1.3	41.0%	18.4%	9.6%	5.9%	4.9%	4.8%	5.1%	5.3%	5.0%	155.0
0.70	1.4	38.1%	18.6%	10.1%	6.8%	5.6%	5.2%	5.3%	5.4%	5.0%	87.0
0.75	1.5	36.5%	18.1%	10.5%	7.4%	6.2%	5.8%	5.3%	5.2%	5.0%	54.6
0.80	1.6	34.8%	18.3%	10.9%	8.2%	6.5%	5.9%	5.4%	5.2%	4.9%	29.9
0.85	1.7	33.6%	18.0%	11.3%	8.3%	7.1%	6.0%	5.6%	5.2%	4.9%	16.7
0.90	1.8	32.7%	17.8%	11.9%	8.8%	7.4%	5.9%	5.7%	4.9%	4.8%	9.0
0.95	1.9	31.6%	18.0%	11.9%	8.9%	7.4%	6.5%	5.9%	5.2%	4.6%	3.6
1.00	2.0	31.2%	17.9%	12.4%	9.2%	7.7%	6.6%	5.6%	4.9%	4.6%	1.7
1.05	2.1	31.1%	17.8%	12.2%	9.4%	7.7%	6.7%	5.4%	5.2%	4.7%	1.4
1.10	2.2	30.2%	17.8%	12.7%	9.2%	8.0%	6.4%	5.8%	5.1%	4.8%	0.5
1.15	2.3	30.2%	17.3%	12.5%	9.7%	7.9%	6.8%	6.0%	5.1%	4.5%	0.1
1.20	2.4	30.2%	17.7%	12.1%	10.0%	8.0%	6.8%	5.7%	5.0%	4.6%	0.3
1.25	2.5	30.4%	17.7%	12.4%	9.5%	8.0%	6.9%	5.8%	5.0%	4.5%	0.3
1.30	2.6	30.1%	17.6%	12.7%	9.8%	7.8%	6.9%	5.8%	5.0%	4.3%	0.2
1.35	2.7	30.2%	17.9%	12.7%	9.7%	7.8%	6.3%	5.6%	5.1%	4.7%	0.4
1.40	2.8	30.0%	17.4%	12.5%	9.7%	8.1%	6.9%	5.9%	5.1%	4.4%	0.1
2.00	4.1	30.2%	17.6%	12.4%	9.6%	8.0%	6.6%	5.9%	5.0%	4.6%	0.1
3.00	6.1	30.0%	17.7%	12.4%	9.9%	7.9%	6.8%	5.6%	5.3%	4.6%	0.1
Benford's Law:		**30.1%**	**17.6%**	**12.5%**	**9.7%**	**7.9%**	**6.7%**	**5.8%**	**5.1%**	**4.6%**	**0**

Figure 8.1: Lognormal Distributions – Approaching Benford Configuration as OOM Nears 2.5

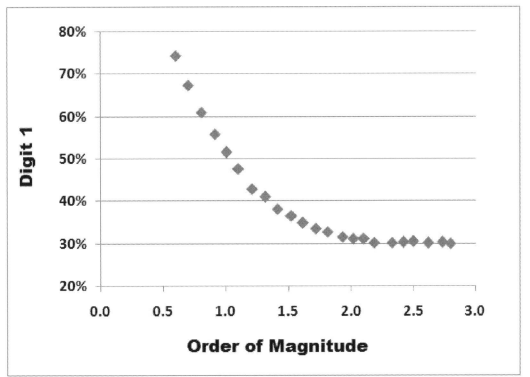

Figure 8.2: Lognormal Distributions – Digit 1 Approaches LOG(1 + 1/1) namely 0.30

(2) The second demonstration of the strong correlation between Benford's Law and order of magnitude of data is provided via several computer simulations of a particular random partition process. Practically all random partition processes point to positive quantitative skewness as well as to strong digital tendency towards the Benford configuration for the resultant set of parts/pieces – given that plenty of parts/pieces are created in the process, and that partition is done freely in a truly random fashion without too many constraints (such as insisting on integral values for the parts/pieces, or limiting sizes to be below a certain maximum or above a certain minimum). The strong connection between Benford's Law and random partition processes is examined in details in Section 3.

One such partition process is the random breaking up of an imaginary big rock in multiple stages into much smaller imaginary pieces, and this is termed 'Random Dependent Partition' or 'Random Rock Breaking'. This process actually follows a strict partition procedure with exact and carefully executed stages. In the first stage the rock is broken into 2 pieces using a random pair of percentage values, such as 23% and 77% for example. Then the second stage starts with the orderly breaking of each of the 2 pieces in a random fashion using two new random pairs of percentage values, resulting in 4 pieces altogether. In the third stage, each of the 4 pieces is broken in a random fashion into two pieces, using four new random pairs of percentage values for the four pieces, resulting in 8 pieces altogether, and so forth.

In one computer simulation of Random Rock Breaking, the process starts with an original quantity of 100 kilograms, and which is broken up in 14 stages using simulated values from the Uniform(0, 1) to decide on random percents for any piece at any stage. The process ends with 16,384 much smaller pieces. At each stage, order of magnitude increases while digits are converging ever closer to the Benford configuration. Figure 8.3 depicts the snapshots of the entire process, showing for each stage, the stage number, the number of resultant pieces, OOM, digit distribution, and SSD measure. Figure 8.4 depicts the scatter plot of SSD versus OOM. A logarithmic scale is used for the SSD vertical axis. Here, even though OOM rapidly passes the required threshold of 3 after the 5th stage, yet digits are not quite Benford because the quantitative structure of the data is of extra skewness, over and above that of the Benford level, and it's only after the 11th or 12th stage that quantitative structure moderates significantly and assumes the Benford style as skewness becomes less intense and nearly as of the Benford level.

#	Pieces	OOM	Digit 1	Digit 2	Digit 3	Digit 4	Digit 5	Digit 6	Digit 7	Digit 8	Digit 9	SSD
0	1	0.0	100.0%	0.0%	0.0%	0.0%	0.0%	0.0%	0.0%	0.0%	0.0%	5633.9
1	2	0.2	0.0%	0.0%	50.0%	0.0%	0.0%	50.0%	0.0%	0.0%	0.0%	4735.7
2	4	0.5	25.0%	50.0%	0.0%	25.0%	0.0%	0.0%	0.0%	0.0%	0.0%	1653.9
3	8	1.6	50.0%	25.0%	0.0%	12.5%	0.0%	0.0%	0.0%	12.5%	0.0%	831.1
4	16	2.6	31.3%	18.8%	18.8%	6.3%	6.3%	6.3%	12.5%	0.0%	0.0%	148.6
5	32	4.4	28.1%	15.6%	15.6%	12.5%	3.1%	12.5%	3.1%	0.0%	9.4%	138.6
6	64	4.9	35.9%	9.4%	14.1%	15.6%	6.3%	3.1%	6.3%	6.3%	3.1%	158.6
7	128	5.7	35.9%	15.6%	14.1%	10.2%	10.2%	5.5%	3.9%	3.1%	1.6%	63.8
8	256	7.2	29.3%	20.7%	10.2%	9.4%	8.2%	5.9%	7.4%	5.9%	3.1%	21.9
9	512	7.4	34.4%	15.4%	10.7%	9.2%	10.0%	5.9%	4.7%	6.1%	3.7%	34.1
10	1024	7.5	30.0%	17.1%	11.3%	11.3%	8.6%	6.1%	5.3%	5.6%	4.8%	5.7
11	2048	7.9	29.5%	18.8%	13.1%	10.3%	7.5%	5.8%	6.0%	4.8%	4.2%	4.0
12	4096	8.1	30.7%	17.2%	12.4%	9.4%	7.3%	6.8%	6.1%	5.2%	4.9%	1.2
13	8192	8.4	29.2%	17.5%	12.4%	9.5%	7.9%	7.4%	6.5%	5.3%	4.3%	1.8
14	16384	8.6	29.9%	17.7%	12.6%	9.7%	7.9%	7.1%	5.7%	5.0%	4.4%	0.3
Benford's Law:			30.1%	17.6%	12.5%	9.7%	7.9%	6.7%	5.8%	5.1%	4.6%	0.0

Figure 8.3: Random Partition Process – Approaching Benford as OOM Increases

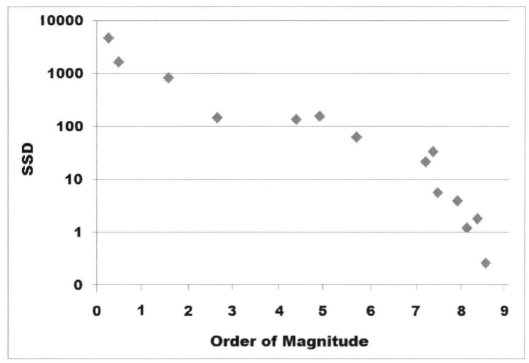

Figure 8.4: Random Partition Process – SSD Approaches 0 as OOM Nears 9

(3) The third demonstration of the strong correlation between Benford's Law and order of magnitude of data is provided via several computer simulations of a particular random multiplication process. As mentioned in Section 2, practically all multiplication processes, either random or deterministic, lead to positive quantitative skewness as well as to strong digital tendency towards the Benford configuration. Indeed, for multiplication processes, order of magnitude reigns supreme, as it is the only factor determining skewness and Benford behavior! Resultant OOM here is the confluence of two factors. The first factor is the individual OOM of the multiplicands themselves; and the second factor is the number of multiplicands involved, as OOM increases with each new multiplication. The only exception here is a multiplication process with limited number of multiplicands, and where each multiplicand is with very low OOM, resulting in a bit higher OOM for the final product, and yet perhaps not sufficiently close to the 3 OOM threshold. All multiplication processes - no matter how they are arranged and which multiplicands are involved - demand resultant OOM of almost the **exact level of 3** to arrive at the nearly perfect Benford configuration yielding SSD below 1.

The Monte Carlo computer simulations here would involve the Normal(5, 1.5), namely the Normal distribution with mean 5 and standard deviation 1.5. First, the Normal(5, 1.5) itself is simulated 30,000 times, without any multiplicative involvement, and then OOM, first digit configurations, and SSD are measured and recorded for these 30,000 realizations.

411

Then two independent realizations from the Normal(5, 1.5) are multiplied by each other, and this is repeated 30,000 times, followed by the recording of OOM, digit configurations, and SSD of these 30,000 products. This is followed by multiplying three independent realizations from the Normal(5, 1.5), repeated 30,000 times, and results are recorded. As more Normals are multiplied the process tends to the perfect Benford digital configuration and towards more quantitative skewness, while OOM level increases decisively and monotonically. Figure 8.5 depicts the snapshots of the process showing OOM, digit distribution, and SSD measure for each additional Normal joined as multiplicand. Figure 8.6 depicts the scatter plot of SSD versus OOM. A logarithmic scale is used for the SSD vertical axis.

Process	OOM	Digit 1	Digit 2	Digit 3	Digit 4	Digit 5	Digit 6	Digit 7	Digit 8	Digit 9	SSD
Normal(5, 1.5)	0.75	1.9%	6.8%	15.7%	24.8%	25.2%	16.1%	7.3%	1.9%	0.3%	1570.8
2 Normals	1.07	29.5%	35.0%	20.8%	7.5%	2.3%	1.1%	1.0%	1.3%	1.5%	486.2
3 Normals	1.32	46.4%	13.5%	4.8%	4.5%	5.3%	5.7%	6.5%	6.8%	6.6%	382.8
4 Normals	1.53	22.1%	11.6%	12.8%	12.2%	11.4%	9.5%	8.1%	6.8%	5.6%	135.2
5 Normals	1.71	26.2%	22.0%	16.2%	11.3%	7.7%	5.6%	4.4%	3.7%	3.1%	58.6
6 Normals	1.90	35.4%	18.7%	10.6%	7.3%	6.3%	5.9%	5.7%	5.2%	4.9%	42.1
7 Normals	2.08	29.5%	14.9%	11.7%	9.9%	8.4%	8.0%	6.5%	5.8%	5.3%	12.0
8 Normals	2.24	27.7%	18.3%	13.7%	10.5%	8.5%	6.8%	5.9%	4.6%	4.2%	8.7
9 Normals	2.40	31.5%	18.6%	12.3%	9.3%	7.3%	6.1%	5.4%	4.8%	4.7%	4.1
10 Normals	2.57	30.6%	17.1%	11.7%	9.3%	8.1%	6.9%	6.1%	5.4%	4.8%	1.6
11 Normals	2.74	29.5%	17.6%	12.8%	10.0%	8.0%	6.8%	5.7%	5.0%	4.6%	0.6
12 Normals	2.84	30.1%	18.2%	12.6%	9.7%	7.6%	6.5%	5.4%	5.3%	4.6%	0.7
13 Normals	3.02	29.9%	17.7%	12.5%	9.4%	8.0%	6.7%	5.9%	5.2%	4.7%	0.2
Benford's Law:		30.1%	17.6%	12.5%	9.7%	7.9%	6.7%	5.8%	5.1%	4.6%	0.0

Figure 8.5: Random Multiplication Process – Approaching Benford Exactly as OOM Reaches 3

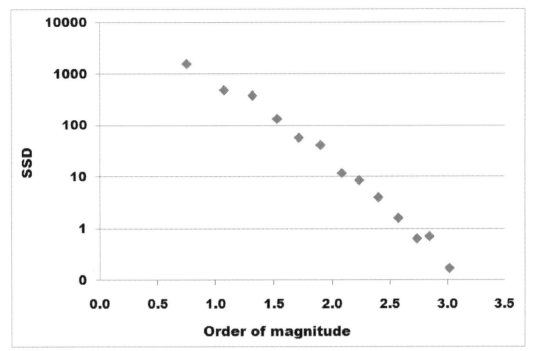

Figure 8.6: Random Multiplication Process – SSD Approaches 0 as OOM Reaches 3

NOTE: The demonstrations involving the Lognormal Distribution and Random Multiplication Processes are closely related, and in the limit they are actually identical. This is so since the Lognormal distribution can be obtained by taking Euler's number e to the power of the random Normal distribution, and this together with the statement of the Central Limit Theorem (CLT) imply that the Lognormal distribution can be represented as a process of repeated multiplications of a random variable, namely as:

$$\text{Lognormal} = e^{[\text{ Normal }]} = e^{[X_1 + X_2 + X_3 + \dots + X_N]}$$
$$\text{Lognormal} = e^{X_1} e^{X_2} e^{X_3} \dots e^{X_N}$$

This result, namely that repeated multiplications of a random variable is Lognormal in the limit as N gets large is called the Multiplicative Central Limit Theorem.

(4) The fourth demonstration of the strong correlation between Benford's Law and order of magnitude of data is provided via several computer simulations of a particular format of chains of distributions. As mentioned in Section 1 of this book, as well as in Kossovsky (2014), chapters 54, 102, 103, practically all chains of distributions are nearly Benford. Orthodox statistical theory deals with distributions where parameters are static and fixed, while chains of distributions are statistical constructs where parameters are not fixed but are rather randomly derived from other statistical distributions. Chains of distributions almost always point to positive quantitative skewness as well as to strong digital tendency towards the Benford configuration – assuming that the number of sequences [parametrical dependencies] are at least 3 or 4, and preferably over 5, involving either the location parameter or the scale parameter, or both. Chaining the shape parameter typically does not lead to Benford behavior in any way.

15 Monte Carlo simulations schemes are performed, each with 20,000 realizations from the Uniform(3, Uniform(7, M)); M = {11, 14, 17, 20, 23, 26, 29, 32, 35, 38, 41, 44, 47, 50, 53}. For example, for M = 14, we start with 20,000 realizations from the Uniform(7, 14), which are obtained, recorded, and called the set {R_j; $1 \leq j \leq 20,000$}. Then each individual R_j value is used in another single simulation (i.e. realization) of the Uniform(3, R_j), which is the final or ultimate value of the chain. Figure 8.7 depicts the 15 distinct simulations of these 15 chains of distributions for each M value, showing in details OOM, digit distribution, and SSD measure. Clearly, the larger the value of M, the wider is the spread of the values of the chain on the x-axis, and the higher the order of magnitude of the chain. Figure 8.8 depicts the scatter plot of SSD versus OOM. A logarithmic scale is used for the SSD vertical axis. Interestingly here, lower OOM value of only 1.1 approximately was enough to obtain digital results fairly close to Benford (almost).

It should be noted that further increases in the value of M beyond 53 do not yield better digital results and greater fit to Benford, rather SSD values start to oscillate, retreating a bit from Benford, then returning closer to it and where SSD again approaches the 15 level (which is the lowest possible SSD level here for this particular chain format).

Chain	OOM	Digit 1	Digit 2	Digit 3	Digit 4	Digit 5	Digit 6	Digit 7	Digit 8	Digit 9	SSD
U(3, U(7, 11))	0.53	1.7%	0.0%	17.4%	17.0%	17.7%	17.2%	14.4%	9.6%	5.1%	1495.9
U(3, U(7, 14))	0.62	11.8%	0.0%	14.7%	14.3%	14.2%	14.5%	13.1%	10.0%	7.5%	857.8
U(3, U(7, 17))	0.70	21.6%	0.0%	12.6%	12.4%	12.3%	12.5%	11.7%	8.9%	7.9%	502.9
U(3, U(7, 20))	0.76	28.7%	0.0%	11.2%	11.2%	11.4%	11.1%	10.2%	8.8%	7.4%	387.6
U(3, U(7, 23))	0.82	33.6%	1.6%	9.7%	10.1%	10.4%	10.2%	9.1%	8.3%	7.1%	323.9
U(3, U(7, 26))	0.87	36.1%	4.5%	9.0%	9.0%	8.9%	9.4%	8.8%	7.8%	6.5%	249.4
U(3, U(7, 29))	0.92	37.0%	8.5%	8.6%	8.2%	8.5%	8.0%	7.9%	7.0%	6.3%	160.8
U(3, U(7, 32))	0.96	36.5%	11.2%	8.8%	7.8%	7.9%	7.7%	7.3%	6.8%	6.1%	106.7
U(3, U(7, 35))	0.99	37.0%	13.7%	8.7%	7.5%	7.3%	6.8%	6.9%	6.4%	5.9%	87.3
U(3, U(7, 38))	1.03	35.2%	15.5%	10.1%	6.8%	7.1%	7.2%	6.5%	6.1%	5.5%	48.1
U(3, U(7, 41))	1.06	34.8%	16.5%	11.8%	6.5%	6.8%	6.4%	6.3%	5.7%	5.1%	36.4
U(3, U(7, 44))	1.09	34.0%	17.4%	13.3%	6.8%	6.1%	6.4%	5.9%	5.4%	4.7%	27.3
U(3, U(7, 47))	1.12	32.7%	17.6%	14.1%	7.4%	6.3%	5.9%	5.9%	5.3%	4.9%	17.7
U(3, U(7, 50))	1.14	32.1%	18.0%	14.8%	8.4%	5.9%	5.7%	5.8%	4.6%	4.7%	16.3
U(3, U(7, 53))	1.16	31.4%	17.8%	15.2%	9.2%	5.8%	5.8%	5.2%	5.1%	4.4%	15.3
Benford's Law:		**30.1%**	**17.6%**	**12.5%**	**9.7%**	**7.9%**	**6.7%**	**5.8%**	**5.1%**	**4.6%**	**0.0**

Figure 8.7: Chains of Two Uniforms – Approaching Approx Benford Behavior as OOM > 1.1

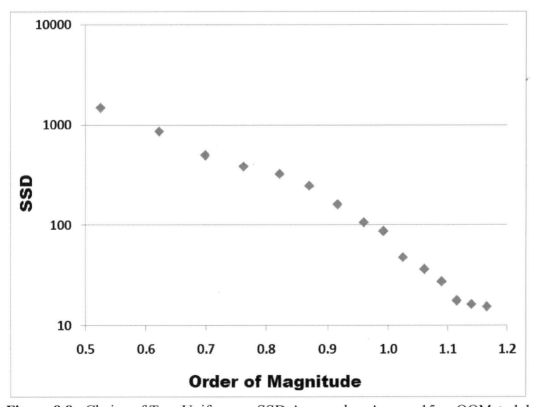

Figure 8.8: Chains of Two Uniforms – SSD Approaches Approx 15 as OOM > 1.1

415

[2] The Universality of the Order of Magnitude Rule Across All Data

Why should order of magnitude be high for Benford behavior? On the face of it, there is no obvious connection between these two distinct measures. And why should the threshold value or cutoff point be approximately 3 for all data types? The answer to these questions can be found if one is willing to let go of his or her vista of raw numbers and of first digits distributions and focus instead on the logarithmically-transformed values of the data, and thus to consider densities and histograms of the log of data.

The explanation regarding the universality of the large order of magnitude rule, in a nutshell, is that order of magnitude of data is actually the range on the log-axis of the data's log histogram, and it so happened that a wide spread of approximately 3 units on the log-axis induces some very intrinsic and profound geometrical considerations in favor of Benford first digits configuration. This principle is what unites and what synchronizes diverse data sets in requiring at least 3 orders of magnitude approximately for Benford behavior.

The fact that order of magnitude is indeed the length of the range on the log-axis for the log histogram of data springs directly from its definition:

$$\text{Order of Magnitude} = \text{LOG}_{10}(\text{Max}/\text{Min}) = \text{LOG}_{10}(\text{Max}) - \text{LOG}_{10}(\text{Min})$$

[3] Related Log Conjecture and the Quest for High Order of Magnitude

In this chapter it will be shown how the typically <u>curved</u> histogram of the logarithm of data can yield <u>straight</u> and uniform fractional log density, although for this to occur it is necessary to have wide span on the log-axis of at least 3 units, namely that it is necessary to have large order of magnitude. Since uniformity of fractional log is required for Benford behavior, it follows that large order of magnitude of data of approximately 3 is a requirement for such digital behavior.

The histogram of the logarithm of k/x distribution is flat, horizontal, and uniform. As it happened, real-life data is almost never as in k/x distribution. Indeed, the k/x distribution is rarely found in real-life data. Practically all real-life data sets come with curved log histogram. Chapters 16, 17, 18, and 19 of Section 1 discuss these issues.

Some profound philosophical and intuitive considerations explain this fact of life. Could we really believe that log of real-life random data would abruptly pop out of nowhere very high on the vertical frequency/count axis exactly at one point (say 1), only to continue steadily, exactly, and stubbornly with the same flat density level all throughout, and then finally just as abruptly to terminate that high at another point further on (say 2)?!

Those who know temperamental Mother Nature well and are familiar with the chaotic and leisurely way she works, know that she would probably never bother to create such regimented and inflexible data in the world. Mother Nature would more likely bend and twist the log density around, letting it rise and fall gradually, consistent with her free-spirit attitude and her strong dislike of regimentation. Gradualism in how quantities in the real world are found is not only intuitively believed, but also supported by plenty of subliminal empirical personal experiences. Can one imagine weights of a large random group of people that starts abruptly at 45 kilos, having no records below that cutoff value, and continues with the same (numerical or log) count/frequency until it reaches 120 kilos where it abruptly terminates?!

There exists almost no data in the physical world or in relation to economics, financial, demographic and census measures that comes with anything resembling uniform and horizontal log histogram! Just about the only exception is Exponential Growth Series which indeed come with that rare flat and uniform log density as in the k/x case.

But all this appears paradoxical, because as discussed in Section 1 chapters 16 and 18, Benford's Law implies uniformity of mantissa, namely the uniformity of the fractional part of the logarithm, and if logarithm itself is not typically uniform, so then how do we ever manage to obtain Benford in the real world?

In order to resolve our dilemma it is necessary to consider the possibility of obtaining Benford in round log curves. Indeed, the alternative to a linear and flat log density is simply an upside-down U-like curve where the rising part on the left approximately or exactly offsets the falling part on the right, leaving the central part of approximately flat log density as the best representative of the entire log curve. A conjecture is raised claiming that sufficiently wide range on the log-axis [i.e. high order of magnitude] would lead to nearly perfect or approximate cancelation and offsetting effects, given that log density starts from the very bottom on the log-axis itself and that it also terminates all the way down on the log-axis as well. This is coined as "Related Log Conjecture". It should be noted that the focus here is on the <u>fractional</u> part of the logarithm, which is the aggregation or superimposition of all the log histogram occurrences on the sub-ranges between the integers of the log-axis.

Figure 8.9 facilitates the visualization behind Related Log Conjecture. In this graph, log is defined on (1, 7); it is quite symmetrical; it rises and falls gradually; it starts and terminates on the log-axis itself, touching it; and it spans a wide range of 6 units on the log-axis. Overall resultant digits configuration is calculated by averaging out the contributions from the 6 mini log configurations on the 6 different sub-intervals standing between integral log-points.

Let us focus on the two gray areas of (2, 3) and (5, 6). It is quite plausible to argue that the confluence of these two log densities - superimposed one upon the other – yields uniformity of resultant (fractional) log density. The same comparisons and conclusions can be made between the regions (1, 2) and (6, 7) as well as between the regions (3, 4) and (4, 5). Since such moderations and balancing acts exist separately within each of the three pairs of regions, overall (fractional) log density is seen as being uniform in the aggregate, and Benford's Law is confirmed!

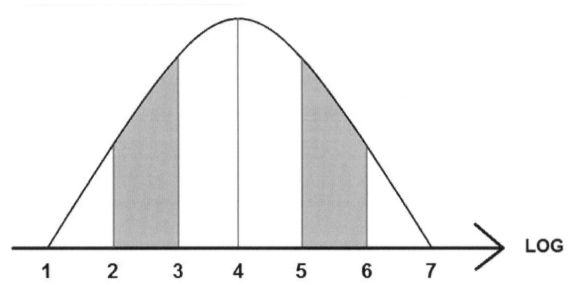

Figure 8.9: The Rising portion on the Left Cancels and Offsets the Falling Portion on the Right

The remarkable and highly surprising flexibility, versatility, and near-universality of this geometrical tendency can hardly be overstated or exaggerated. Assuming wide enough range on the log-axis and a strict adherence to the rule of starting and terminating on the log-axis itself, then log histogram does not have to be nicely aligned with integral values as in Figure 8.9, nor does it have to be nicely symmetrical as in Figure 8.9, and it can even have abrupt turns, and it can assume unusual shapes! Often though, it 'costs' a bit more in terms of extra log-axis range (i.e. order of magnitude) to be asymmetrical, to be misaligned along the integers, and to have odd shapes and sharp turns, but that cost is fairly 'cheap', usually no more than say ≈ 0.5 or even less of extra log-axis territory (i.e. extra OOM), and indeed numerous real-life data sets are perfectly able and are even enthusiastic to pay such low price in order to enter the prestigious club of Benford data. Some illustrative examples of the flexibility of the conjecture in many odd cases can be found in Kossovsky (2014) chapter 68 titled "The Remarkable Malleability of Related Log Conjecture".

The **crux** of Related Log Conjecture is that it demands sufficiently wide range on the log-axis (i.e. high order of magnitude), so that all these cancelation and offsetting effects are carried out meaningfully in relation to all the 9 different fractional log-axis compartments of the digits [see Figure 1.15 of Section 1]. When the range on the log-axis is very narrow, all these cancelation and offsetting effects are carried out only in relation to one, two, or maybe three different fractional log-axis compartments of digits, and then Benford cannot be found.

Figure 8.10 depicts the useless and ineffective cancelation and offsetting effects of data with the narrow span of (0.0, 0.5) log-axis. Here, the cancelations are mostly between compartment [0.000, 0.301) of digit 1 and compartment [0.301, 0.477) of digit 2, which does not lead to anywhere near the Benford configuration of course. Figure 8.10 yields roughly 60% digit 1, around 38% digit 2, and roughly 2% digit 3, but 0% for the digits 4 to 9, and so it is decisively non-Benford. The same terrible fate would befall other 0.5-OOM data sets focused on very short log-axis ranges, such as say (0.5, 1.0) or (0.75, 1.25), and so forth.

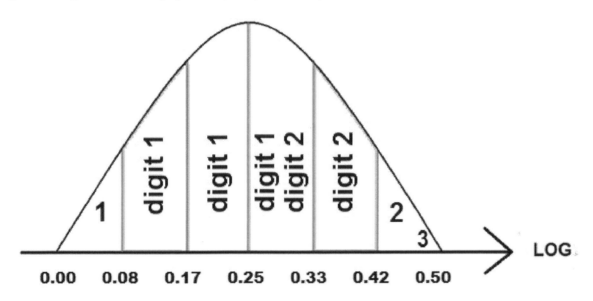

Figure 8.10: No Meaningful Cancelations and Offsetting Effects for Short Ranges (Low OOM)

[4] Curvy-Closure Log Pattern – Nearly Universal Across All Data

Figure 8.11 depicts a hypothetical log histogram curve of the typical random data type. On the left side, log is rising, and digital equality prevails in the approximate, or else low digits obtain only slight advantage over high digits. Around the center, log reaches its peak, it is approximately flat, and the Benford configuration is found locally in the approximate, albeit very briefly and for only very short sub-range around the peak. Finally, on the right side, log is falling, low digits strongly dominate, and extreme digital inequality prevails, much more so than in the usual Benford configuration. It should be noted that the variety of digital configurations as described above assumes wide span on the log-axis.

419

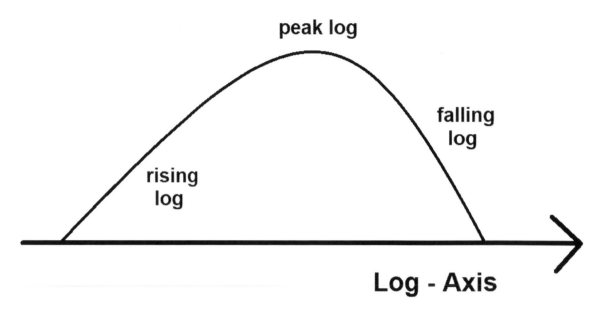

Figure 8.11: Typical Log Histogram of Typical Random Data – Rising, Peaking, then Falling

The generic shape of the chart in Figure 8.11 is extremely typical in almost all log curves of random real-life data sets - in Benford data as well as in non-Benford data. Such geometrical description of the curve is coined as the "**Curvy-Closure**" property of log histogram. The term "**Curvy**" reflects the fact that it curves around, usually gradually and smoothly, but not always so, as it could be on occasions rough and abruptly changes its direction, yet, even in such cases, it is almost always being done in an approximately continuous way, so that the curve of the log is without significant breaks or gaps, and that it does not have truly disconnected parts. The term "**Closure**" reflects the fact that it focuses and warps itself thoroughly and tightly around the log-axis; that it literally 'closes' the log-axis; starting from the log-axis itself and terminating there as well; leaving no opening.

If nearly all real-life data sets are endowed with the Curvy-Closure property for the histogram of their logarithm, then all they need in order to have the Benford property is simply sufficient range on the log-axis, namely they must have high order of magnitude of over 3 approximately.

In reality they also need to have quantitative skewness, but not of the too gentle type sloping less than Benford, nor of the too steep type falling on the right precipitously, but rather just the right type of skewness at the Benford level. Such proper rate of quantitative fall in the data histogram is automatically guaranteed, assuming large order of magnitude and given that the log histogram is smooth and continuous, and that it starts and ends on the log-axis itself.

Hence, exceptions and counter examples to the order of magnitude rule exist, such as the Normal Distribution or the Uniform Distribution, which are of the symmetrical quantitative configuration, totally lacking skewness, and their log histogram is quite odd, therefore no matter how high their order of magnitude happened to be, they are never Benford.

[5] Curvy-Closure Log Pattern – Confirmed in Generic Benford Processes

Empirical examinations performed by the author with numerous real-life data sets (prior to the publication of Kossovsky (2014) "Benford's Law: Theory, the General Law of Relative Quantities, and Forensic Fraud Detection Applications") thoroughly and consistently confirmed the Curvy-Closure property of log histograms. In this chapter, this property shall be empirically examined again with the 4 data sets relating to the 4 abstract processes and abstract distributions discussed in chapter 1 of this section, all of which are of the Benford type. Surely, only the ultimate and fully converging model with the largest OOM (i.e. final stage or final process) shall be considered, namely: Lognormal with the highest shape parameter of 1.4; the pieces after the last 14th stage in random partitions; the multiplication of all 13 Normals; and the widest chain with M = 53. Figure 8.12 depicts the log histograms of these four processes and distributions, and it strongly confirms the Curvy-Closure log assertion above. Readers should be able to visually 'read' clearly and directly the value of the order of magnitude of each data set via the width of the span on the log-axis of the respective log chart.

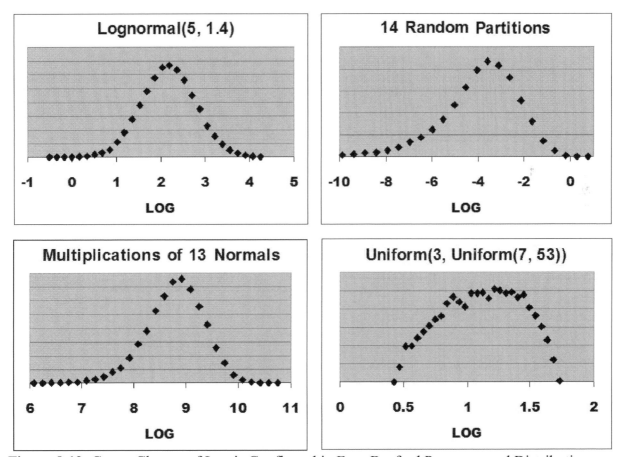

Figure 8.12: Curvy-Closure of Log is Confirmed in Four Benford Processes and Distributions

[6] Curvy-Closure Log Pattern – Confirmed in Real-Life Benford Data

Let us also empirically check the Curvy-Closure log assertion with four real-life Benford data sets. Figure 8.13 depicts the log histograms of these four data sets, namely: US Census on the population in all its 19,509 cities and towns in 2009 discussed in Section 7, see Kossovsky (2014) Chapter 13 for more details; time in seconds between all the 19,451 global earthquake occurrences in 2012, see Kossovsky (2014) Chapter 11 for more details; the price list or catalog of Canford Audio PLC containing 15,194 items for sale in 2013, see Kossovsky (2014) Chapter 44 for more details; and US market capitalization of 5,486 registered companies as of January 1, 2013, see Kossovsky (2014) Chapter 28 for more details. Orders of magnitude of the above four data sets were calculated as: 3.6 for Population, 2.8 for Earthquakes, 4.1 for Canford, 5.9 for Market Capitalization, and these values are easily corroborated for each data set via the reading of the width of the span on the log-axis of the respective log chart.

Clearly, these four real-life data sets strongly confirm the Curvy-Closure log assertion above.

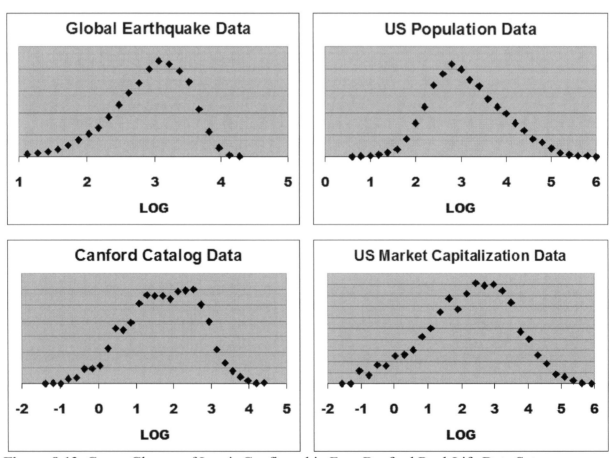

Figure 8.13: Curvy-Closure of Log is Confirmed in Four Benford Real-Life Data Sets

[7] Curvy-Closure Log Pattern – Confirmed in Non-Benford Real Data

Empirical examinations of many Non-Benford data by the author also thoroughly and consistently confirmed the Curvy-Closure property of log histograms. The inescapable conclusion is then that the Curvy-Closure log pattern is even more ubiquitous and more universal than the Benford digital phenomenon itself!

Most typically, real-life Non-Benford data sets only suffer from lack of sufficient order of magnitude (i.e. lack of sufficient width on the log-axis) and which precludes these intrinsic geometrical cancelations and offsetting effects from manifesting themselves meaningfully. A minority of real-life Non-Benford data sets suffer from symmetry or from having gentle and moderate skewness so that their data histogram is not skewed enough and thus digits are not Benford, even though they might come with a large order of magnitude. In some rare cases Non-Benford data sets with high order of magnitude suffer from severe skewness, over and above the Benford level of skewness, and thus their digits are not Benford.

Figure 8.14 depicts the log histograms of US County Area data and of State of Oklahoma payroll data relating to the Department of Human Services for the 1st quarter of 2012. See Kossovsky (2014) Chapter 45 for details about US County data. The Oklahoma data can be found on their website: https://data.ok.gov/Finance-and-Administration/State-of-Oklahoma-Payroll-Q1-2012/dqi7-zvab. Only those 2377 rows from the column 'Amount' pertaining to the Department of Human Services are considered. These two data sets are decisively not Benford.

Clearly, even these two Non-Benford data sets strongly confirm and validate the Curvy-Closure log assertion above. Part of the reason they do not comply with Benford's Law is that their order of magnitude is too low, and the other part is the incomplete or partial skewness of their data histogram. The histogram of US County data looks a bit symmetrical, while the histogram of Oklahoma data does not look symmetric or asymmetric, but rather an odd mixture of both, and quite confusing.

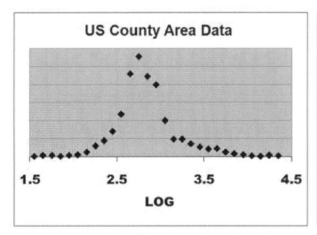 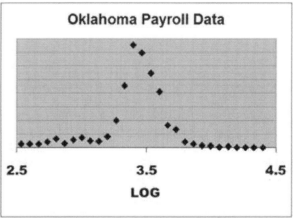

Figure 8.14: Curvy-Closure is Confirmed in Non-Benford Real-Life Data – Range is Too Short

SECTION 9

ROBUST ORDER OF MAGNITUDE OF DATA

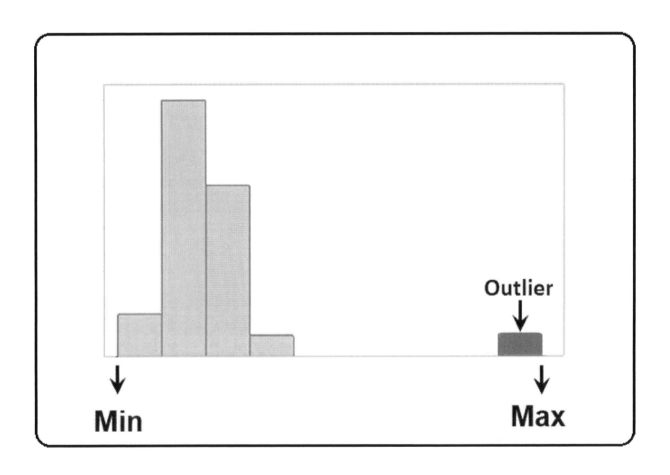

[1] Core Order of Magnitude of Data (COM)

Order of magnitude is defined as the log of the ratio of the maximum value to the minimum value. The data set is assumed to contain only positive numbers greater than zero.

Order of Magnitude (OOM) = $\text{LOG}_{10}(\text{Maximum}/\text{Minimum})$

However, occurrences of outliers have the potential to distort and exaggerate the value of order of magnitude, thus leading to numerous problems and paradoxes. Trust in the entire phenomenon of Benford's Law as well as faith in the validity of its data forensics applications often erodes during the encounters of data sets which are not inherently Benford due to low order of magnitude but which possess distorting outliers greatly exaggerating the order of magnitude and causing them to be considered as Benford – thus misleading researchers to declare the encounters of contradictions and paradoxes in this entire field of study.

In order to establish firm and robust guidelines regarding variability of data, it is necessary to eliminate any possible outliers on the left for small values and on the right for big values. This is accomplished by narrowing the focus exclusively onto the core 80% part of the data. This brutal purge eliminates without mercy any malicious and misleading outliers as well as any innocent and proper data points which happened to stray just a little bit away from the core part of the data. The measure shall be called Core Order of Magnitude and it is defined as follows:

Core Order of Magnitude (COM) = $\text{LOG}_{10}(P_{90\%}/P_{10\%})$

The definition simply reformulates OOM by substituting the 10th percentile (in symbols $P_{10\%}$) for the minimum, and by substituting the 90th percentile (in symbols $P_{90\%}$) for the maximum.

Core Physical Order of Magnitude (CPOM) was defined in Section 1 as $P_{90\%}/P_{10\%}$, hence Core Order of Magnitude = $\text{LOG}_{10}(\text{Core Physical Order of Magnitude})$.

By strictly applying the 10th and 90th percentiles for all data sets, the procedure avoids vague and fuzzy definitions and statistical arguments about outliers. Surely, the price paid for such universality in the procedure is the occasional eliminations of innocent data points near the edges which are actually very much part of the data, but are being swept away by the crude cleansing method of COM.

The two essential prerequisites or conditions for data configuration with regards to compliance with Benford's Law are: (1) having sufficient variability with order of magnitude of at least three; and (2) overall positive skewness, where the histogram is falling to the right in the aggregate.

As it happened, the vast majority of real-life physical and abstract data sets are indeed such! Most data sets come with sufficient variability and are generally positively skewed in the aggregate, and as a consequence Benford's Law is confirmed in most data sets.

Let us state the Benford requirement of high variability in terms of OOM and COM:

$$LOG_{10}(\text{Maximum}/\text{Minimum}) > 3 \qquad\qquad OOM > 3$$

Actually, even lower OOM values such as 2.5 or even 2.0 are expected to yield approximately Benford, but falling below 2.0 does not bode well for getting anywhere near the Benford distribution.

The above prerequisite for compliance totally ignores the thorny issue of outliers and edges, and in that sense it is too simplistic and even completely erroneous for some data sets. Hence, using the COM qualification is essential in judging whether or not a given data set is expected or not expected to comply with Benford's Law. The proper qualification for expectance of compliance with the law in the approximate - obtained via extensive empirical studies - is then as follows:

$$LOG_{10}(P_{90\%}/P_{10\%}) > 2 \qquad\qquad COM > 2$$

Actually, even lower COM values such as 1.7 or even 1.5 are expected to yield approximately Benford, but falling below 1.5 does not bode well for getting anywhere near the Benford distribution.

How do we determine the above rule? This is not being argued mathematically or statistically, but rather empirically. When a large number of positively skewed data sets are easily available for computer analysis, they are classified according to their Benford compliance, and their order of magnitude is evaluated, then they are ready for correlation analysis. These data sets should come with a diversity of Benford compliance; where some are deemed nearly perfectly Benford with only tiny deviation from the $LOG(1 + 1/d)$ proportions; some are just slightly off; some are with a moderate deviation; and some are decisively non-Benford. What is consistently and almost universally observed here is that order of magnitude of less than 1 is correlated with a decisive non-Benford digit behavior, and as order of magnitude increases toward the 2 level, digital configuration steadily becomes closer to the Benford proportions, and finally attaining a near perfect Benford configuration as we approach the 3 to 4 level.

Here are some real-life Benford data examples:

US 2009 Census on population in all 19,509 cities - {29.4, 18.1, 12.0, 9.5, 8.0, 7.0, 6.0, 5.3, 4.6}
Time between 19,451 Global Earthquakes in 2012 - {29.9, 18.8, 13.5, 9.3, 7.5, 6.2, 5.8, 4.8, 4.2}
Canford Audio PLC 2013 Price List 15,194 items - {28.8, 17.7, 14.2, 9.2, 8.1, 7.0, 5.3, 5.1, 4.6}
US 2013 Market Capitalization, 5,486 companies -{30.1, 18.0, 12.6, 9.7, 8.1, 6.5, 5.9, 4.6, 4.4}
Benford's Law   **-{30.1, 17.6, 12.5, 9.7, 7.9, 6.7, 5.8, 5.1, 4.6}**

Orders of magnitude of the above four data sets respectively are **{3.6, 2.8, 4.1, 5.9}**.

These values were obtained after the elimination of the top 1% of values and the bottom 1% of values in each data set so as not to let extreme values (outliers) influence or distort order of magnitude. In other words, neither OOM nor POM were used; instead a measure somewhere in between them was chosen, a measure which is much closer to OOM than to POM.

[2] Outliers Elimination Method Via Tukey's Inner Fences

Another robust technique which could be considered superior to the usage of the Core Order of Magnitude is simply the elimination of all perceived outliers prior to the measurement of the order of magnitude. This technique might be considered superior in terms of preserving almost all of the core data points, gently purging only those that truly appear as outliers, instead of blindly and ruthlessly purging all points below $P_{10\%}$ regardless, and all points above $P_{90\%}$ regardless. This technique potentially establishes a new minimum and a new maximum which might emerge after the elimination of all perceived outliers, and this is followed by the re-evaluation of the order of magnitude, namely as $LOG_{10}($ New Maximum/New Minimum $)$; while applying the same compliance criterion, namely that this newly evaluated OOM should be at least **3** for expectation of very good compliance with Benford's Law.

The interquartile range (IQR) is often used to decide on outliers in data. Outliers X_i are eliminated in the standard way for extreme values such that:

$$X_i > Q_3 + \mathbf{1.5}^*(Q_3 - Q_1) \qquad \text{for } X_i > Q_3$$
$$X_i < Q_1 - \mathbf{1.5}^*(Q_3 - Q_1) \qquad \text{for } X_i < Q_1$$

$Q_1 = P_{25\%}$ the 25th percentile, and it is called the first quartile.
$Q_2 = P_{50\%}$ the 50th percentile, and it is called the second quartile (the median).
$Q_3 = P_{75\%}$ the 75th percentile, and it is called the third quartile.

$IQR = (Q_3 - Q_1)$

This is known in the literature as '**Tukey's inner fences**' employing the distance of **1.5*IQR**, as opposed to '**Tukey's outer fences**' employing the distance of **3.0*IQR**.

Tukey's fences technique is superior to the COM technique in that it does not involve the elimination of so many 'normal' or 'essential' parts of the core data. On the other hand, the advantage of the COM technique is in its ability to definitely eliminate all outliers regardless of any data structure, never missing on any outlier whatsoever, while the weakness of Tukey's fences technique is that at times it fails to detect what might be thought of as apparent outliers still positioned inside the fences.

[3] Outliers Elimination Method Via a Modified Version of Tukey's Fences

The above suggestion regarding Tukey's inner fences proves a bit too liberal in the sense that it occasionally allows some disruptive data points residing not too far from the core data to be included, and this liberal inclusion in turns contributes to a larger observed measure of order of magnitude than warranted in the context of Benford's Law.

Thus the utilization of a modified version of Tukey's inner fences is suggested via (0.75)*IQR slack above Q3 and below Q1, instead of the standard (1.5)*IQR slack, constituting inner-most fences more tightly bound around the core part of the data for the formation of this truly robust measure of the order of magnitude of data in this digital context.

For this (0.75)*IQR rule, outliers X_i are eliminated for extreme values such that:

$$X_i > Q_3 + \mathbf{0.75}^*(Q_3 - Q_1) \qquad \text{for } X_i > Q_3$$
$$X_i < Q_1 - \mathbf{0.75}^*(Q_3 - Q_1) \qquad \text{for } X_i < Q_1$$

These limits shall be called "**Tukey's inner-most fences**"

Paul Velleman, a statistician at Cornell University in New York, was a student of John Tukey, the renowned statistician and researcher who invented the boxplot and the 1.5*IQR Rule. When he asked Tukey "Why 1.5?", Tukey answered "Because 1 is too small and 2 is too large". Perhaps here in the context of Benford's Law compliance and the unwarranted influence of outliers on the order of magnitude measure, when we are asked "Why 0.75?" we should answer "Because 0.50 is too small and 1.00 is too large".

[4] Case Study I – Artificially Concocted 50 Data Points

In this example, 50 data points, sorted low to high, and shown below, shall be examined:

2	23	24	25	26	27	28	29	32	33
33	33	34	36	37	38	38	39	40	41
42	47	48	50	51	52	53	55	56	57
59	60	63	67	68	75	76	77	78	79
80	84	86	91	94	103	107	114	**213**	**567**

There are three 'apparent outliers' here, namely {2, 213, 567} shown above in bold font for emphasis. Initially, these three data points are named 'outliers' in a totally subjective and intuitive way, without employing any rigorous statistical machinery or any strict rule. Figure 9.1 depicts the histogram of the above data set, showing the three apparent outliers as thick black marks, as well as the 10th and 90th percentile points utilized in the definition of COM.

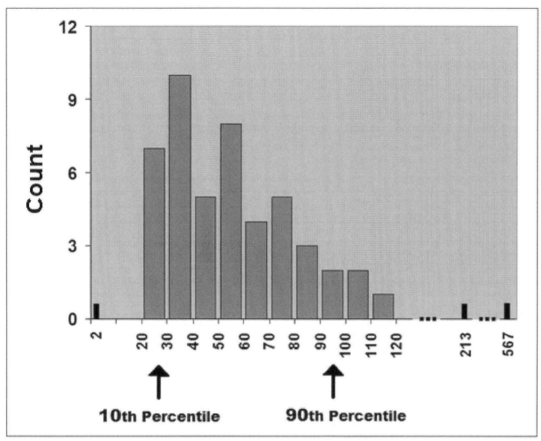

Figure 9.1: COM Variability Measure Unnecessarily Eliminates Essential Data Points

Examining 1st order digit distribution shows a drastic deviation from Benford:

Data Set of 50 Data Points - { 6.0, 18.0, 20.0, 10.0, 18.0, 8.0, 10.0, 6.0, 4.0}
Benford's Law 1st Digits - {30.1, 17.6, 12.5, 9.7, 7.9, 6.7, 5.8, 5.1, 4.6}

Calculating OOM mindlessly without any worries whatsoever about possible distortions from potential outliers, leads to OOM = LOG((567)/(2)) = 2.5. Supposedly, such relatively high order of magnitude which is almost near the compliance threshold of 3, combined with positive skewness, should show sufficient or an approximate conformity to Benford, yet, this data set is decisively non-Benford!?

Let us employ the more robust measure of COM. The 10th percentile is 26.9 thus pointing to the value 27 as the new minimum. The 90th percentile is 94.9 thus pointing to 94 as the new maximum, hence COM = LOG($P_{90\%}$ / $P_{10\%}$) ≈ LOG((94)/(27)) = 0.5, and such low COM value indicates the expectation of decisive non-compliance with Benford. Our dilemma is then satisfactorily resolved.

Admittedly, this elimination of all data points below the 10th percentile and all data points above the 90th percentile clearly cuts into essential parts of the data.

This is akin to a malignant cancer surgery where the surgeon is keen on making sure that no cancer cells whatsoever are left in area, and that remission is to be avoided at all cost, therefore the surgeon is cutting some more all around the tumor, even in the healthy tissue and eliminates cells immediately surrounding it. For the data set above, the crude outlier-surgery of COM removes also the innocent points {23, 24, 25, 26, 103, 107, 114} which are actually authentic part of the data. Yet, this is the price some statisticians would be willing to pay in order to standardize the procedure, instead of endlessly arguing about which data points truly constitute outliers and which are not.

The alternative 'Outliers Elimination Method' would simply eliminate all outliers before any calculation of the simple OOM measure. It would then apply the newly observed minimum and maximum values of the purified and cleansed data set in OOM calculations.

$Q_1 = 34.5$ $\qquad$ $Q_3 = 76.8$

$IQR = Q_3 - Q_1 = 76.8 - 34.5 = 42.3$

In summary, outliers X_i are eliminated in the standard way for extreme values such that:

$X_i > 76.8 + 1.5*(42.3)$

$X_i < 34.5 - 1.5*(42.3)$

Hence here Tukey's inner fences are:

$X_i > 140.1$ $\quad$ and $\quad$ $X_i < -28.9$

This method correctly eliminates the two high value points of {213, 567} which are apparently outliers, yet it does not eliminate the low value point of {2} which perhaps (on the face of it) appears to be an outlier nonetheless. If we trust our subjective judgment that data point 2 is an outlier, then Tukey's fences seem as not robust enough in eliminating all outliers.

With the elimination of {213, 567} as outliers, the new maximum is 114, while point 2 still serves as the minimum, therefore the newly calculated measure is OOM = LOG(Max / Min) = LOG((114)/(2)) = 1.8. Yet, this somewhat lower value of OOM (which is not very much lower than say the 2.5 or the 2.0 levels) does not really explain why this data set deviates so drastically from Benford. Our dilemma is not fully resolved. Our glass is half empty and half full.

It should be noted here that Tukey's LEFT fence penetrates well below the 0 origin and deep within negative territory, even though all the values in the data set are positive.

Figure 9.2 depicts Tukey's inner fences together with the histogram of the above data set.

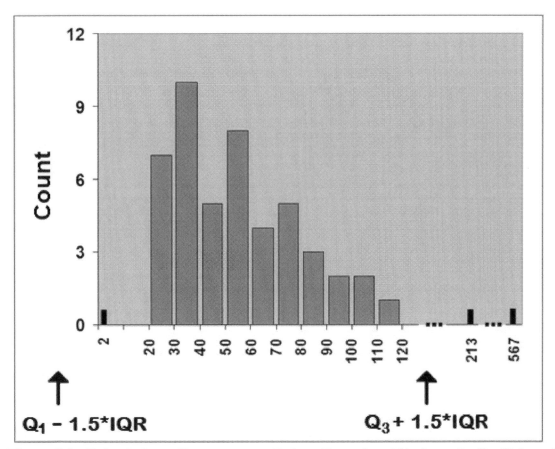

Figure 9.2: Tukey's Inner Fences are not Robust Enough to Eliminate Outlier Value of 2

The alternative (0.75)*IQR slack rule yields tighter fences around the core data as follows:

$X_i > 76.8 + (0.75)*(42.3)$

$X_i < 34.5 - (0.75)*(42.3)$

Hence modified Tukey's inner-most fences are:

$X_i > 108.4$ and $X_i < 2.8$

This method correctly eliminates all apparent outliers, and data point 2 is now excluded, although the method has also excessively eliminated the point {114} of the core data.

With the elimination of {2, 114, 213, 567} as outliers, the new maximum is 107, the new minimum is 23, therefore the newly calculated measure is OOM = LOG(Max / Min) = LOG((107)/(23)) = 0.7, and such low OOM value indicates the expectation of a decisive non-compliance with Benford. Our dilemma is then satisfactorily resolved. Figure 9.3 depicts modified Tukey's inner-most fences together with the histogram of data.

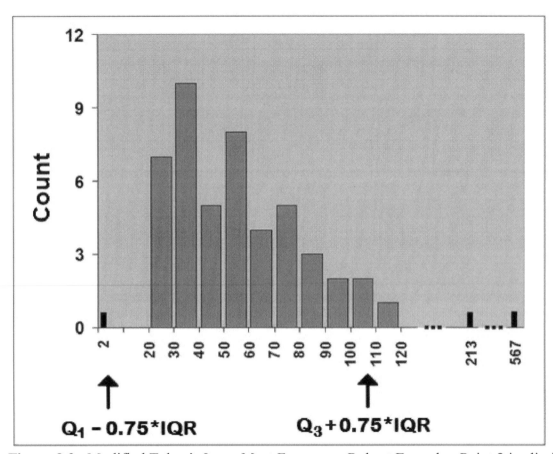

Figure 9.3: Modified Tukey's Inner-Most Fences are Robust Enough – Point 2 is eliminated

Conclusion - 50 Points Data:

(1) COM measure applying $P_{10\%}$ and $P_{90\%}$ with its value of 0.5, proved successful in explaining the decisive non-compliance of this data set with Benford's Law, although it unnecessarily eliminated many essential data points.

(2) Tukey's inner fences leading to an exaggerated OOM value of 1.8, failed to explain why this data set does not comply at all with Benford's Law. Tukey's inner fences seem a bit too liberal here by allowing one apparent outlier to be included, namely data point {2}.

(3) Modified Tukey's inner-most fences of the (0.75)*IQR slack rule leading to the realistic OOM value of 0.7, fully and decisively explain the strong non-compliance of this data set with Benford's Law, and properly eliminated all apparent outliers including {2}, although admittedly the method also excessively eliminated {114} of the core data.

In trying to justify the <u>intuitive</u> or <u>apparent</u> inclination to declare points {2, 213, 567} as outliers, Figure 9.4 is used. It depicts the additive progression (distances) and the multiplicative progression (factors) of the ordered data. Clearly, the unusually long distances of {21, 99, 354} as well as the unusually big factors of {11.50, 1.87, 2.66} relating to these 3 special points are extraordinary long and big relative to the much shorter distances and smaller factors of the core data!

Data	Distances	Factors	Data	Distances	Factors
(2)	===	===	52	1	1.020
23	21	11.500	53	1	1.019
24	1	1.043	55	2	1.038
25	1	1.042	56	1	1.018
26	1	1.040	57	1	1.018
27	1	1.038	59	2	1.035
28	1	1.037	60	1	1.017
29	1	1.036	63	3	1.050
32	3	1.103	67	4	1.063
33	1	1.031	68	1	1.015
33	0	1.000	75	7	1.103
33	0	1.000	76	1	1.013
34	1	1.030	77	1	1.013
36	2	1.059	78	1	1.013
37	1	1.028	79	1	1.013
38	1	1.027	80	1	1.013
38	0	1.000	84	4	1.050
39	1	1.026	86	2	1.024
40	1	1.026	91	5	1.058
41	1	1.025	94	3	1.033
42	1	1.024	103	9	1.096
47	5	1.119	107	4	1.039
48	1	1.021	114	7	1.065
50	2	1.042	(213)	99	1.868
51	1	1.020	(567)	354	2.662

Figure 9.4: Distances and Factors – Potentially Differentiating Data Points as Outliers

NOTE: The author has chosen the simple and round value of 0.75 as in 3/4 as a reasonable 'rule of thumb'; although admittedly the exact choice of 0.75 is not a rule set in stones. In any case, decisive and consistent empirical examinations in this context of Benford's Law tell us in no uncertain terms that here Tukey's standard (1.5)*IQR slack rule simply does not work! It's too liberal and too inclusive, and therefore 1.5 must be substantially reduced to be effective in eliminating those outliers that greatly exaggerate order of magnitude in our digital sense.

NOTE: Even though the symmetrical Standard Normal Distribution is considered a non-Benford case, it may teach us a lesson or two about the 0.75 versus 1.50 threshold choice.
Tukey's inner fences of the (1.50)*IQR rule established for the Standard Normal are –2.70 and +2.70, and they enclose 99.30% of the 'core' data, while eliminating only 0.70% of data on the very margin of the range, and all this can be considered as being too inclusive and too liberal.
Tukey's inner-most fences of the (0.75)*IQR rule established for the Standard Normal are –1.688 and +1.688, and they enclose 90.9% of the truly core data, and strictly eliminate 9.1% of data on the left and right edges of the range, ensuring the elimination of any possible outliers.

[5] Case Study II – US County Area Data

US County Area data pertains to areas of all 3143 counties in the USA as of the year 2013. See Kossovsky (2014) Chapter 45 for details about this data set.

Data is reported in units of Square Miles. Some extreme examples are shown below:

Yukon-Koyukuk County in Alaska has the <u>largest</u> area -   145,504.79 square miles
North Slope County in Alaska has the 2nd largest area -   88,695.41 square miles
Otoe County in Nebraska has the <u>median</u> area size -   615.63 square miles
Falls Church County in Virginia has the 2nd smallest area -   2.50 square miles
Lexington County in Virginia has the <u>smallest</u> area -   2.00 square miles

The 1st order digit distribution is:

US County Area Census   - {16.2, 10.0, 10.7, 15.8, 15.2, 10.4, 8.6, 7.1, 5.9}
Benford's Law 1st Digits - {30.1, 17.6, 12.5, 9.7, 7.9, 6.7, 5.8, 5.1, 4.6}

Clearly, this data set is not Benford at all, yet order of magnitude is high enough, indicating an expectation for a full compliance with the digital law (assuming positive skewness).

Order of magnitude here is:

$$OOM = LOG_{10}(Maximum/Minimum) = LOG_{10}(145504.8/2.0) = 4.9$$

The answer to this dilemma is that these two extreme cases shown above [Alaska and Virginia], constituting the maximum and the minimum, erroneously point to deceptive high order of magnitude. These are merely two examples at the extreme ends of the data set, truly constituting outliers on the margin of data, while the core of the data is spread over a much narrower range. Applying the more robust measure of Core Order of Magnitude, we get a more accurate picture of the data structure and are able to acknowledge the much narrower variability of this data set.

$$Core\ Order\ of\ Magnitude = LOG_{10}(P_{90\%}/P_{10\%}) = LOG_{10}(1845.7/286.0) = 0.8$$

This clearly explains why US County Area data set is not Benford at all. Our dilemma is then satisfactorily resolved.

The alternative 'Outliers Elimination Method' would simply eliminate all outliers before any calculation of the originally-defined OOM measure. It would then apply the newly observed minimum and maximum values of the purified and cleansed data set in OOM calculations.

$Q_1 = 430.7$ $Q_3 = 924.0$

$IQR = Q_3 - Q_1 = 924.0 - 430.7 = 493.2$

In summary, outliers X_i are eliminated in the standard way for extreme values such that:

$X_i > 924.0 + 1.5*(493.2)$

$X_i < 430.7 - 1.5*(493.2)$

Hence here Tukey's inner fences are:

$X_i > 1663.8$ and $X_i < -309.1$

The minimum is the positive county area 2.0 for Lexington County in Virginia, which is kept, not being considered as an outlier. The new maximum just below the fence of 1663.8 is the county area 1659.1 for Grant county in North Dakota. Therefore, robust variability according to Tukey's inner fences is between 2.0 and 1659.1, and we get:

$OOM = LOG_{10}(1659.1/2.0) = LOG_{10}(829.6) = 2.9$

Hence Tukey's inner fences method <u>failed</u> to explain our dilemma, since OOM value near 3 is a strong indication for the expectation of nearly full compliance with Benford's Law.

It should be noted here that Tukey's LEFT fence penetrates well below the 0 origin and deep within negative territory, even though all the values in the data set are positive.

But modified Tukey's inner-most fences of the alternative (0.75)*IQR slack rule fully and successfully explains away our dilemma:

$X_i > 924.0 + 0.75*(493.2)$

$X_i < 430.7 - 0.75*(493.2)$

Hence here modified Tukey's inner-most fences are:

$X_i > 1293.9$ and $X_i < 60.8$

The new minimum is 61.0 for St. Louis City in Missouri just above the 60.8 fence. The new maximum is 1290.1 for Santa Clara in California just below the 1293.9 fence. Therefore, robust variability according to modified Tukey's inner-most fences is between 61.0 and 1290.1, and:

$OOM = LOG_{10}(1290.1/61.0) = LOG_{10}(21.1) = 1.3$

Hence modified Tukey's inner-most fences method <u>successfully</u> explains our dilemma with its low OOM value of 1.3 which strongly indicates the expectation of none compliance with Benford's Law.

Conclusion - US County Area Data:

(1) COM measure applying $P_{10\%}$ and $P_{90\%}$ with its value of 0.8, proved successful in explaining the non-compliance of the data set with Benford's Law, although it unnecessarily eliminated many essential data points.

(2) Tukey's inner fences were too liberal by allowing too many points outside the core data portion to be included, leading to an exaggerated OOM value of 2.9, and thus it had failed, proving unsuccessful in explaining the non-compliance of the data set with Benford's Law.

(3) Modified Tukey's inner-most fences of the (0.75)*IQR slack rule leading to the realistic OOM value of 1.3, proved successful in explaining the non-compliance of the data set with Benford's Law, while not eliminating much from the essential core data.

Figure 9.5 depicts the histogram of the US county area data from 0 up to 4000 square miles, with bins of 125-width, not showing the rest of the data over 4000 containing only 117 data points (out of 3143 total) which falls very thinly and sparely on the x-axis. Actually, since the histogram in its entirety is not skewed enough (unless the focus is exclusively from ≈ 500 and above), the other reason it's not Benford may be the lack of sufficient quantitative skewness.

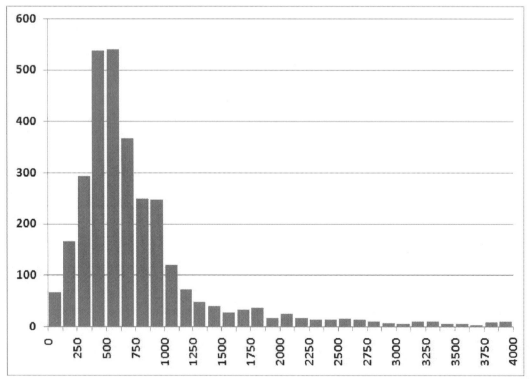

Figure 9.5: Histogram of US County Area Data up to 4000 with Bins of 125-Width

[6] The Frequent Penetration of Tukey's Left Fence into Negative Territory

In both case studies I and II, Tukey's LEFT inner fence penetrated well below the 0 origin and deep within negative territory, even though these two data sets are exclusively of positive values without a single occurrence of negative number.

Numerous real-life data sets of positive values tend to start near 1 or some other low value. This start at low values is quite natural for scientific and physical real-life data sets, as well as for economics, financial, and governmental census data, and which so often do not include any negative values whatsoever.

Hence, a unique very small value such as 0.02 say which some might classify as an apparent outlier in the context of its data set, would typically not be classified as an outlier by Tukey's left inner fence since it so often penetrates well into negative territory whenever upper range is of sufficiently high value. Admittedly, when upper range is of sufficiently low value, Tukey's left inner fence might stand above the 0 origin and may indeed classify 0.02 as an outlier. This aspect of Tukey's inner fences should not be viewed as a weakness, but rather as something statisticians and data analysts should be aware of.

A case in point is the Lognormal Distribution where Tukey's left inner fence always penetrates deep into negative territory whenever shape parameter is high enough (say over 0.75) as it should be for all 'appropriate' and 'authentic' Lognormal cases. Cases where shape parameter is very low (say less than 0.50 or 0.40) are those where the Lognormal is actually almost non-skewed and appears roughly symmetric resembling the Normal Distribution a great deal. It should be noted that the Lognormal is always positive, and any potential outliers on its left side are those just above the zero origin. Hence, having Tukey's left inner fence standing deep within negative territory implies that such a fence can never be effective in eliminating any potential outliers of small values.

Since the Lognormal can be thought of as Euler's number e to the power of some seminal or generating Normal Distribution with certain mean and s.d. values, therefore it can be written as:

$$\text{Lognormal} = e^{[\text{ The Generating Normal }]}$$

Hence the natural log base e of the Lognormal is the Generating Normal itself:

$$\text{LOG}_e(\text{Lognormal}) = \text{LOG}_e(e^{[\text{ The Generating Normal }]}) = \text{The Generating Normal.}$$

Also of note is that:

z-value of +0.675 of the Standard Normal yields 75% to the left of it, hence it is Q_3 there.
z-value of –0.675 of the Standard Normal yields 25% to the left of it, hence it is Q_1 there.
We define Q3 as the 75% percentile of the Lognormal, namely its 3rd quartile.

This implies that $\text{LOG}_e(Q3)$ is the 3rd quartile for the Generating Normal itself. Converting this value to its z-standard equivalent (z-score) yields:

$$Z = (\text{LOG}_e(Q3) - \text{Mean}) \,/\, \text{s.d.}$$

And since $\text{LOG}_e(Q3)$ is the 3rd quartile for the Generating Normal, z must be +0.675. Putting all this together we get:

$$+0.675 = (\text{LOG}_e(Q3) - \text{Mean}) \,/\, \text{s.d.}$$

$$(+0.675)(\text{s.d.}) = (\text{LOG}_e(Q3) - \text{Mean})$$

$$(+0.675)(\text{s.d.}) + \text{Mean} = \text{LOG}_e(Q3)$$

Taking Euler's number e to both sides of the above equation we get:

$$e^{(+0.675)(\text{s.d.}) + \text{Mean}} = Q3$$

In the same vein,

$$e^{(-0.675)(\text{s.d.}) + \text{Mean}} = Q1$$

Since the 'mean' of the Generating Normal is the 'location' parameter of the Lognormal, and since the 's.d.' of the Generating Normal is the 'shape' parameter of the Lognormal, all this can be written as:

The 1st quartile of the Lognormal $= e^{(-0.675)(\text{shape}) + \text{location}}$

The 3rd quartile of the Lognormal $= e^{(+0.675)(\text{shape}) + \text{location}}$

$$\text{IQR} = Q3 - Q1 = e^{(+0.675)(\text{shape}) + \text{location}} - e^{(-0.675)(\text{shape}) + \text{location}}$$

Tukey's left inner fence of the Lognormal $= e^{(-0.675)(\text{shape}) + \text{location}} - (1.5)*\text{IQR}$

And more explicitly, and with 'LEFT' signifying 'Tukey's left inner fence':

$$\text{LEFT} = e^{(-0.675)(\text{shp}) + \text{loc}} - (1.5)*e^{(+0.675)(\text{shp}) + \text{loc}} + (1.5)*e^{(-0.675)(\text{shp}) + \text{loc}}$$

And the brute force method in computer calculations demonstrates that this LEFT value is always negative whenever shape parameter is bigger than ≈ 0.38.

Let us prove the above assertion regarding LEFT for shape values over ≈ 0.38.

$$\text{LEFT} = e^{(-0.675)(\text{shp}) + \text{loc}} - (1.5)*e^{(+0.675)(\text{shp}) + \text{loc}} + (1.5)*e^{(-0.675)(\text{shp}) + \text{loc}}$$

$$\text{LEFT} = (2.5)*e^{(-0.675)(\text{shp}) + \text{loc}} - (1.5)*e^{(+0.675)(\text{shp}) + \text{loc}}$$

$$\text{LEFT} = (2.5)*e^{(-0.675)(\text{shp})}*e^{\text{loc}} - (1.5)*e^{(+0.675)(\text{shp})}*e^{\text{loc}}$$

$$\text{LEFT} = e^{\text{loc}} [(2.5)*e^{(-0.675)(\text{shp})} - (1.5)*e^{(+0.675)(\text{shp})}]$$

No matter what value is assigned to the location parameter, e^{location} is never negative. When the shape parameter is zero, the difference inside the bracket reduces to positive 1:

$$[(2.5)*e^{(-0.675)(0)} - (1.5)*e^{(+0.675)(0)}]$$
$$[(2.5)*1 - (1.5)*1] = +1$$

As the shape parameter increases from zero, the difference inside the bracket reduces monotonically from its initial value of +1, then it reaches zero, and finally it becomes permanently negative. Let us find out the critical value of the shape parameter, namely that particular shape value that reduces the difference inside the bracket exactly to zero:

$$0 = \text{LEFT} = e^{\text{loc}} [(2.5)*e^{(-0.675)(\text{shp})} - (1.5)*e^{(+0.675)(\text{shp})}]$$
$$0 = [(2.5)*e^{(-0.675)(\text{shp})} - (1.5)*e^{(+0.675)(\text{shp})}]$$
$$(2.5)*e^{(-0.675)(\text{shape})} = (1.5)*e^{(+0.675)(\text{shape})}$$

$$(2.5) / (1.5) = e^{(+0.675)(\text{shape})} / e^{(-0.675)(\text{shape})}$$
$$(5/2) / (3/2) = e^{[(+0.675)(\text{shape}) - (-0.675)(\text{shape})]}$$
$$(5/3) = e^{[(0.675)(\text{shape}) + (0.675)(\text{shape})]}$$
$$(5/3) = e^{[(1.35)(\text{shape})]}$$

Taking the natural log base e to both sides of the equation, we obtain:

$$\text{LOG}_e(5/3) = (1.35)(\text{shape})$$
$$(0.511) = (1.35)(\text{shape})$$
$$\text{shape} = (0.511) / (1.35)$$
$$\text{shape} = 0.38$$

Hence, as shape parameter increases from 0 to around 0.38, Tukey's left inner fence is moving leftwards from 1 to 0, and thereafter it moves decisively and monotonically deeper into negative territory as the shape parameter continues to increase.

REFERENCES

Andrews, George (1976). "The Theory of Partitions." Cambridge University Press.

Benford, Frank (1938). "The Law of Anomalous Numbers". Proceedings of the American Philosophical Society, 78, 1938, p. 551.

Buck Brian, Merchant A., Perez S. (1992). "An Illustration of Benford's First Digit Law Using Alpha Decay Half Lives". European Journal of Physics, 1993, 14, 59-63.

Carslaw, Charles (1988). "Anomalies in Income Numbers: Evidence of Goal Oriented Behavior". The Accounting Review, Apr. 1988, 321–327.

Deckert Joseph, Myagkov Mikhail, Ordeshook Peter (2011). "Benford's Law and the Detection of Election Fraud". Political Analysis 19(3), 245–268.

Durtschi Cindy, Hillison William, Pacini Carl (2004). "The Effective Use of Benford's Law to Assist in Detecting Fraud in Accounting Data". Auditing: A Journal of Forensic Accounting, 1524-5586/Vol. V(2004), 17–34.

Freund, John (1999). "Mathematical Statistics". Sixth Edition. Pearson Education & Prentice Hall International, Inc.

Gaines J. Brian, Cho K. Wendy (2007). "Breaking the (Benford) Law: Statistical Fraud Detection in Campaign Finance". The American Statistician, Vol. 61, No. 3, pages 218-223.

Hamming, Richard (1970). "On the Distribution of Numbers". Bell System Technical Journal 49(8): 1609-25.

Kafri, Oded (2009). "Entropy Principle in Direct Derivation of Benford's Law". March 2009, http://arxiv.org/abs/0901.3047.

Kafri, Oded & Hava (2013). "Entropy, God's Dice Game".

Kossovsky, Alex Ely (2012). "Towards A Better Understanding of the Leading Digits Phenomena". City University of New York. http://arxiv.org/abs/math/0612627

Kossovsky, Alex Ely (Dec 2012). "Statistician's New Role as a Detective - Testing Data for Fraud". http://revistas.ucr.ac.cr/index.php/economicas/article/view/8015

Kossovsky, Alex Ely (2013). "On the Relative Quantities Occurring within Physical Data Sets". http://arxiv.org/ftp/arxiv/papers/1305/1305.1893.pdf

Kossovsky, Alex Ely (2014). "Benford's Law: Theory, the General Law of Relative Quantities, and Forensic Fraud Detection Applications". World Scientific Publishing Company. August 2014. ISBN: 978-981-4583-68-8.

Kossovsky, Alex Ely (2015). "Random Consolidations and Fragmentations Cycles Lead to Benford's Law". https://arxiv.org/abs/1505.05235

Kossovsky, Alex Ely (March 2016). "Prime Numbers, Dirichlet Density, and Benford's Law". https://arxiv.org/abs/1603.08501

Kossovsky, Alex Ely (May 2016). "Arithmetical Tugs of War and Benford's Law". http://arxiv.org/abs/1410.2174

Kossovsky, Alex Ely (June 2016). "Exponential Growth Series and Benford's Law". http://arxiv.org/abs/1606.04425

Kossovsky, Alex Ely (Feb 2019). "Quantitative Partition Models and Benford's Law" . https://arxiv.org/abs/1606.02145

Leemis Lawrence, Schmeiser Bruce, Evans Diane (2000). "Survival Distributions Satisfying Benford's Law". The American Statistician, Volume 54, Number 4, November 2000, 236-241.

Lemons, S. Don (1986) "On the Number of Things and the Distribution of First Digits". The American Association of Physics Teachers. Vol 54, No. 9, September 1986.

Leuenberger Christoph, Engel Hans-Andreas (2003). "Benford's Law for the Exponential Random Variables". Statistics and Probability Letters, 2003, 63(4), 361-365.

Miller, Steven (2008). "Chains of Distributions, Hierarchical Bayesian Models and Benford's Law". Jun 2008, http://arxiv.org/abs/0805.4226.

Miller, Steven (Sep 2013). "Benford's Law and Continuous Dependent Random Variables". https://arxiv.org/abs/1309.5603.

Miller Steven, Joseph Iafrate, Frederick Strauch (2013). "When Life Gives You Lemons - A Statistical Model for Benford's Law". Williams College, August 2013. http://web.williams.edu/Mathematics/sjmiller/public_html/math/talks/small2013/williams/Iafrate_SummerPoster2013.pdf

Miller Steven, Joseph Iafrate, Frederick Strauch (2015). "Equipartitions and a Distribution for Numbers: A Statistical Model for Benford's Law". Williams College.

Newcomb, Simon (1881). "Note on the Frequency of Use of the Different Digits in Natural Numbers". American Journal of Mathematics, 4, 1881, 39-40.

Pinkham, Roger (1961). "On the Distribution of First Significant Digits". The Annals of Mathematical Statistics, 1961, Vol.32, No. 4 , 1223-1230.

Raimi, A. Ralph (1969). "The Peculiar Distribution of First Digit". Scientific America, Sep 1969: 109-115.

Raimi, A. Ralph (1976). "The First Digit Problem". American Mathematical Monthly, Aug-Sep 1976.

Raimi, A. Ralph (1985). "The First Digit Phenomena Again". Proceedings of the American Philosophical Society, Vol. 129, No 2, June, 1985, 211-219.

Ross, A. Kenneth (2011). "Benford's Law, A Growth Industry". The American Mathematical Monthly, Vol. 118, No. 7, Pg. 571–583.

Sambridge Malcolm, Tkalcic Hrvoje, Arroucau Pierre (2011). "Benford's Law of First Digits: From Mathematical Curiosity to Change Detector". Asia Pacific Mathematics Newsletter October 2011.

Sambridge Malcolm, Tkalcic Hrvoje, Jackson Andrew (2010). "Benford's Law in the Natural Sciences". Geophysical Research Letters, Volume 37, 2011, Issue 22, L22301.

Saville, Adrian (2006). "Using Benford's Law to detect data error and fraud: An examination of companies listed on the Johannesburg Stock Exchange". Gordon Institute of Business Science, University of Pretoria, South African Journal of Economics and Management Sciences, 9(3), 341-354. http://repository.up.ac.za/handle/2263/3283.

Shao Lijing, Ma Bo-Qiang (2010a). "The Significant Digit Law in Statistical Physics". http://arxiv.org/abs/1005.0660, 6 May 2010.

Shao Lijing, Ma Bo-Qiang (2010b). "Empirical Mantissa Distributions of Pulsars". http://arxiv.org/abs/1005.1702, 12 May 2010. Astroparticle Physics, 33 (2010) 255–262.

Varian, Hal (1972). "Benford's Law". The American Statistician, Vol. 26, No. 3.

Patent: The U.S. Patent Office # 9,058,285. Inventor: Alex Ely Kossovsky.
Date Granted: June 16, 2015. http://www.google.com/patents/US20140006468
Titled: "Method and system for Forensic Data Analysis in fraud detection employing a digital pattern more prevalent than Benford s Law".

GLOSSARY OF FREQUENTLY USED ABBREVIATIONS

CLT   Central Limit Theorem

MCLT   Multiplicative Central Limit Theorem

OOM   Order of Magnitude - LOG(Maximum/Minimum)

POM   Physical Order of Magnitude - Maximum/Minimum

COM   Core Order of Magnitude - LOG($P_{90\%}$/$P_{10\%}$)

CPOM   Core Physical Order of Magnitude - $P_{90\%}$/$P_{10\%}$

GLORQ   The General Law of Relative Quantities

SSD   Sum of Squares Deviation Measure

LDIP   Leading Digits Inflection Point

ES12   Excess Sum digits 1 & 2 of the 1st order

ES04   Excess Sum digits 0 to 4 of the 2nd order

MDD   Modified Dirichlet Density

Printed in Great Britain
by Amazon